NAPOLEON: ON WAR

Napoleon ON WAR

EDITED BY BRUNO COLSON

Translated by Gregory Elliott

OXFORD
UNIVERSITY PRESS

OXFORD

UNIVERSITY PRESS

Great Clarendon Street, Oxford, OX2 6DP,
United Kingdom

Oxford University Press is a department of the University of Oxford.
It furthers the University's objective of excellence in research, scholarship,
and education by publishing worldwide. Oxford is a registered trade mark of
Oxford University Press in the UK and in certain other countries

Published in the United States of America by Oxford University Press
198 Madison Avenue, New York, NY 10016, United States of America

British Library Cataloguing in Publication Data
Data available

Library of Congress Control Number: 2014950237

ISBN 978-0-19-968556-1

Printed in Italy by
L.E.G.O. S.p.A.

Contents

BOOK IV: THE ENGAGEMENT

BOOK V: MILITARY FORCES

BOOK VIII: WAR PLANS

Introduction

You do not need to be French to regard Napoleon as the greatest warrior of all time. He fought as many battles as Alexander the Great, Caesar, and Frederick II of Prussia combined, on terrains, in climates, and against enemies that were very different from one another. His mastery of mass warfare and his ability to raise, organize, and equip numerous armies dramatically changed the art of war and marked the beginning of the modern era. Although his career ended in defeat and exile, this did not affect the esteem in which even his enemies held his military skills.[1] Whereas the reputation of most great military leaders is based on a spectacular success or a few victories, Napoleon won almost all the fifty pitched battles he fought.[2] In a way, officers the world over recognize themselves in him because he imparted to the military profession an intellectual basis and professionalism still affirmed today. At the same time, both academic and popular military history as currently conceived in the West was really born with the study of the Napoleonic Wars and the endeavour to draw lessons for military instruction from them.[3] Yet unlike other great commanders before him, such as Montecucoli and Maurice de Saxe,[4] Napoleon did not write a sustained work on the subject. On several occasions, however, he entertained such a project. Reflections on war pepper his correspondence, *Mémoires*, proclamations, and his writings on Saint Helena, where they have to be pinned down amid so much else.

This work has already been partially done and given rise to numerous collections of 'maxims' in several languages. One of the first was by the Count of La Roche-Aymon, who had emigrated at the start of the Revolution to serve in Condé's army and then in the Prussian army, before re-joining the French army under the Restoration with the rank of brigadier-general.[5] Other editions followed—in particular, those of Generals Burnod and Husson under the Second Empire.[6] That regime made it its duty to systematically publish the great man's words. In addition to the well-known publication of the *Correspondance*, it sponsored an edition of Napoleon's opinions and judgements (not only in the military domain) in alphabetical order, under the editorship of Damas Hinard.[7] In 1898, Lieutenant-Colonel Grouard, author of numerous works on strategy and critical studies of military campaigns, took up some familiar maxims and glossed them with historical examples taken, in particular, from the war of 1870.[8] The origin of these collections was obscure and none of them supplied precise references to the sources of the maxims. The latter were transmitted from one edition to the next. However, it should be pointed out that, in order to feed himself, in 1838 Honoré de Balzac published a *Maximes et pensées de Napoléon* to which he added sentences of his own invention.[9] Frédéric Masson condemned this mystification, but the damage was done and it is likely that certain maxims attributable to Balzac have wormed their way into Napoleonic collections for good.[10] Thus, we probably owe the following sentences to the author of the *Comédie humaine*: 'In war, genius is thought in action'; 'The best soldier is not so much the one that fights as the one who marches.'

These maxims, and various others, are not to be found in the collection by Lieutenant-Colonel Ernest Picard, head of the historical section of the army general staff, who in 1913 published what remains the most serious, properly referenced collection of the Emperor's key military texts.[11] In the intellectual climate favourable to Napoleon that preceded the First World War, Picard returned to the original texts. Of varying lengths, these were grouped into three categories: precepts, men, campaigns. In the first category, the quotations feature by titles, organized alphabetically. This straightforward list is interesting for its initial selection, but does not impose any coherence on Napoleon's scattered ideas and is not accompanied by any commentary. It is also marked by the concerns of its time when it comes to the choice of quotations. Nor does it take account of the diary kept on Saint Helena by General Bertrand, in which the Emperor's reflections on war are especially numerous and interesting. This diary had not been published at the time.

In 1965, a *Napoléon par Napoléon* appeared in three volumes.[12] The first two were a reprise of the dictionary by Damas Hinard, who went unmentioned. A preface by André Maurois was added to them. General Pierre-Marie Gallois provided the preface to the third volume, devoted to the art of war. Once again, what was involved was a reprise of maxims previously published by General Grisot and lacking any references.[13] André Palluel performed more valuable work for the bicentenary of Napoleon's birth.[14] The provenance of the quotations, arranged alphabetically starting with 'abdication', was summarily indicated. No commentary was provided. In 1970, Generals Delmas and Lesouef produced a study on the art of war and Napoleon's campaigns, largely constructed from his letters and referring to the published collections of the correspondence. The work was therefore precise but, besides its limited distribution, it only took account of the correspondence.[15] The appearance of these fairly voluminous works did not prevent the proliferation of small collections of maxims wanting in any references, but very convenient for hard-pressed readers and publishers concerned about profitability.[16]

In England, a portion of the Count de Las Cases's notes containing maxims by Napoleon was published as early as 1820.[17] The presence of this text in the library of the Royal Military College at Sandhurst is attested three years later.[18] In the early 1830s, the British lieutenant-general Sir George C. D'Aguilar translated seventy-eight maxims that form the basis of the most widely diffused edition in the English language, reproduced by David Chandler as late as the start of the twenty-first century. Several editions of military rules and thoughts appeared in Russia in the nineteenth century and were very popular among officers. German, Spanish, Swedish, Venezuelan, and Canadian editions appeared.[19] Torn apart by the Civil War, the USA saw an edition in the North, reprinting D'Aguilar and prefaced by General Winfield Scott, and one in the Southern states.[20] 'Stonewall' Jackson, one of the best Confederate generals, had a copy in his field baggage.[21] The Second World War was the occasion for a new edition by a colonel in the US army, glossed and illustrated by campaigns familiar to Americans like the Civil War or the occupation of the Philippines in 1899.[22] Reinforcing this American tradition, Jay Luvaas, a professor at the US Army War College, was primarily interested in what Napoleon had to say about the art of command.[23] Familiar with Colonel Picard's edition, he took care to provide references and correctly stressed the interest of the letters to Joseph Bonaparte and Eugène de Beauharnais, in which Napoleon sought to convey the essentials of what made for a good general. Luvaas spent several years assembling and translating what Napoleon had said about the art of war. He put together a book where sources are cited, but sometimes second-hand.[24] Equipped with a precious index and a commentary reduced to a minimum, *Napoleon on the Art of War* groups the quotations in ten chapters, ranging from 'Creating the Fighting Force'

to 'The Operational Art'. That the book should end on this note is highly
indicative of the context in which Luvaas completed his work. The 1980s
were years of renewed interest in the 'operational level of war' in the US
army.

Aside from the fact that Luvaas's work likewise takes no account of
Bertrand's Saint Helena diary, my aim is broader and seeks to be more in
tune with contemporary concerns. It will not be restricted to the art of
war—i.e. the conduct of war—but will encompass the phenomenon of war
in its entirety. Collections of texts by Napoleon on war have hitherto
sought to unravel the mystery of his success by striving to identify his
precepts and judgements and his way of seeing things. We shall take
Napoleon not only as a master of the art of war, but also as a privileged
witness of war in all its aspects. In other words, we shall be closer to
Clausewitz than Jomini. We shall be so close to him that we shall organize
the quotations from Napoleon in accordance with the titles of the books
and, so far as is possible, the chapters of Clausewitz's *Vom Kriege* (*On War*).[25]
A comparison between the two on certain points has already been sketched
by Luvaas, who identifies several similarities.[26] Within chapters, we shall
add specific section headings. This will make it possible not only to confer a
certain logic, if only one of presentation, on the Emperor's ideas, but also to
compare them with those of the Prussian general, who is an undisputed
master of reflection on the Napoleonic Wars and war in general. The
exercise might seem to resemble acrobatics, but it will be appreciated that
if a commonality of ideas between these two figures emerges we shall have
some indispensable reflections and analyses.

I am conscious of the difficulties of this undertaking. Napoleon's written
statements, most of them dictated in the heat of the action or in reaction to
things he read on Saint Helena, bring together a number of ideas and are
often bound up with specific events. Various Napoleonic texts could figure
in at least two chapters of *Vom Kriege*. But that work likewise contains
repetitions. We lay no claim to summarize the whole of Clausewitz's
treatise or to conduct a systematic comparison between the ideas of these
two figures. We take from Clausewitz only what helps us to structure
Napoleon's statements. In this regard, we shall also take into account the
draft treatise on tactics that was to be added to *Vom Kriege*, which is devoted
exclusively to strategy. By contrast, Clausewitz's historical works will be
neglected. Prioritizing general reflections, we shall not compare analyses of
any particular campaign. Such an exercise might be attempted in

connection with Frederick II's wars, for example, or the 1812 campaign in Russia. But it would assume such a scale that it would have to be the subject of specific works. Our classification of Napoleon's ideas in accordance with those of Clausewitz is open to challenge. But we have sought each time to grasp what seemed to us to be the dominant idea in the text cited. Instances of repetition are, meanwhile, unavoidable. Sometimes addressed to a general in the field, Napoleon's texts are inevitably much more succinct than the extended elaborations of Clausewitz, who allowed himself several years' reflection to compose a *magnum opus* which, as is well known, remained unfinished.

Readers will arrive at their own judgements, but we have been surprised by the comparative ease of situating quotations from Napoleon in the structure of *Vom Kriege*. That book, it should not be forgotten, was the result of a lifetime devoted to studying war as practised and illuminated by Napoleon.[27] Clausewitz, who read French, was aware of the *Mémoires pour servir à l'histoire de France* dictated by the Emperor to his companions on Saint Helena. In *Vom Kriege*, as we shall see, he makes several references to a specific passage. In his history of the first Italian campaign, he makes ample use of Napoleon's *Mémoires* as a source, comparing them with Austrian accounts and Jomini's narrative. He cites the pages of the edition he read and deplores a 'complete lack of sincerity', above all as regards troop numbers.[28] Less well known is the fact that in his exile Napoleon probably read a text by Clausewitz. It is attested that his library on Saint Helena contained a nine-volume collection of many pieces relating to the events of 1813 and 1814.[29] *Le Mémorial de Sainte-Hélène* refers to it.[30] Attributable to a publicist close to the Prussian military circles occupying Paris, this collection contained a 'Précis of the 1813 campaign up to the armistice of 6 June 1813 by M. de Kleisewitz' (*sic*).[31] This was the first translation of a work by Clausewitz and also the first non-anonymous publication during his lifetime, even if his name got somewhat mangled.[32]

Vom Kriege employs a veritable pedagogy to help its readers to understand war. Clausewitz does not so much teach as stimulate a desire to understand. We find in him not solutions, but incitements to curiosity. 'He is more of an animator than a teacher.'[33] He invites us to ponder the complexity of war, starting out from a few reference points. He is like one of the 'high-performance lenses of the military microscope'.[34] This 'lens' can help us to structure Napoleon's scattered ideas, especially given that the early years of the twenty-first century have witnessed a strong resurgence in Clausewitz studies and *Vom Kriege* is better understood today than ever. The book is regarded as much more open to multiple, changing military situations. It was precisely written to permit of different appropriations. Its main interest consists in its enquiring spirit, its capacity of stimulation for understanding events, but also thoughts on war.[35] Clausewitz wished to lay the foundations of objective knowledge on the basis of which war as such could be analysed. This accounts for his enduring relevance. He provides us with guide lines for conducting our own critical analysis.[36] It is especially apt to use him to reach a better understanding of someone who was not only his contemporary, but also the veritable reference point of all his thinking and his career as a Prussian officer.

If he deals with concepts in less depth, Napoleon broaches subjects barely touched on, or completely absent from, Clausewitz: civil war, naval warfare, occupation, war in Muslim lands, siege warfare, the health of soldiers, the press. These categories will be added, but incorporated into those of *On War*, whose structure we shall follow faithfully, except that we shall omit a few chapters where we cannot relate statements by Napoleon to them. His experience was much vaster than Clausewitz's, not only because he was commander-in-chief, but because he waged war in many more places. For a twenty-first-century reader, the variety of situations encountered remains astounding. The Napoleonic Wars were not merely a series of major battles in which the artillery pounded, the infantry advanced in column or line under a hail of bullets, and the cavalry, swords drawn, charged to the sound of trumpets. In Egypt, Palestine, and Syria, General Bonaparte became acquainted with Muslim combatants with different traditions. In Spain, but also Italy, more so than in Germany or Austria, his troops had to confront the perennial problems of occupation and pacification. They had to develop counter-insurgency techniques. Even if he was not universally present, Napoleon dealt with all this in his correspondence and studied every aspect, geographical and historical, of his armies' theatres of operation. In other words, not only crucial aspects of recent conflicts, but sometimes even the sites of them, were studied by Napoleon. We shall not establish any parallel. Readers can make such comparisons as they wish. They will appreciate that the human aspects of war and command, dominated by the danger and stress of combat, are rediscovered from one age to the next and that Napoleon's words still supply food for thought.

The sources of these words have all been published, but not always with the requisite rigour. External criticism is called for and a return to certain manuscripts dictated. Napoleon's written output was enormous given that his life was so short.[37] It can be divided into three periods: the early writings (reading notes, novels, philosophical, historical, military and political writings, a diary, and the start of the correspondence); the writings of the era of glory or power (the bulk of the correspondence, proclamations, speeches, reports, notes); and the Saint Helena writings.[38] From 1804 onwards, most of his writings are no longer in his own hand, but were dictated to secretaries or members of his entourage. It has to be admitted that nothing irrefutably proves the authenticity of writings by Napoleon. Many were shaped and more or less corrected by others. The Saint Helena writings are indeed by him, but his dictation was too rapid for the result to be literally exact.[39] As regards the letters proper, for its scholarly character and quasi-exhaustiveness the best edition is the *Correspondance générale* which is in the process of being published by Fayard and which now replaces the Second Empire edition of the *Correspondance* and subsequent supplements. Nevertheless, the latter remains useful because it includes texts other than letters: mémoires and notes on a military situation, proclamations to troops, army bulletins, and, in the last volumes, the Saint Helena writings. These had formed the object of a separate publication in 1867, under the title of *Commentaires*. But the version in the *Correspondance* is to be preferred, because it referred back to the manuscripts.[40]

Among the most interesting texts attributed to Napoleon is the 'Note on the political and military position of our armies in Piedmont and Spain' (1794). General Colin cites it at length in *L'Éducation militaire de Napoléon* and makes it a basic element of his work. General Camon, another major interpreter of Napoleonic warfare, cites it in practically all his books. Yet the text is unsigned. Camon specifies that it was sent on 1 Thermidor Year II to the Committee of Public Safety by the young Robespierre (Augustin), and that 'the author is certainly General Bonaparte'.[41] Artillery commander of the army of Italy, Bonaparte is said to have sent the report on 19 July 1794 to Augustin Robespierre, the representative of the people with this army.[42] The note figures in the collection of Napoleon's correspondence in the archives of the Service historique de la Défense (SHD) in Vincennes.[43] The analytical report accompanying the piece does indeed attribute it to General 'Buonaparte' and identifies the writing of Junot, his aide de camp, who stayed with him and who wrote his various memoirs at his dictation. Comparison with other pieces confirms it.[44] The note escaped the editors of the Second Empire *Correspondance*, because it was under the name of the younger Robespierre. Lazare Carnot, member of the Committee of Public Safety, wrote his date of reception on Junot's manuscript: '1 Thermidor Year II' (19 July 1794). The note was published for the first time by Edmond Bonnal de Ganges, who was curator at the Dépôt de la guerre, the predecessor of the SHD.[45] General Colin recognized not only Junot's handwriting, but also the special paper used on the general staff of the artillery of the army of Italy, 'paper of which no sample is to be found in the archives aside from pieces signed by Bonaparte or of which he is the attested author. More than these material details, the style and organization of the ideas reveal the Napoleonic origin of this mémoire. It is impossible to believe that one of the Robespierres, used to a rambling rhetoric, one day discovered the secret of this potent concision.'[46] As a result of having assiduously frequented Napoleon's style in preparing this collection, we are in a position to share General Colin's opinion. Jean Tulard has likewise reprinted the note in his anthology of Napoleon's writings.[47]

Napoleon's dictations on Saint Helena, entitled *Mémoires pour servir à l'histoire de France*, represent an important source for our subject.[48] In them we find a critical account of several campaigns. Of those he conducted, Napoleon had time to recount only the campaigns in Italy, Egypt, and Belgium. But he also described those involving different theatres, like Germany in 1796 and Switzerland in 1799. He also addressed the wars of Julius Caesar, Turenne, and Frederick II. All these accounts are food for thought on war. Finally, the Emperor dictated commentaries on authors whom he read, among them historians and theorists like Generals Jomini and Rogniat. The latter's *Considérations sur l'art de la guerre* were sent to him on 4 December 1818 in a batch of twenty-eight volumes bought by a British businessman, among which were Mathieu Dumas's *Précis des événements militaires*, the first volume of *Victoires, conquêtes... des Français, de 1792 à 1815*, and the history of the Portuguese campaign by General Paul Thiébault.[49] At the start of the year, Napoleon had drawn up a list of books of whose publication he had learned, and which he wished to receive. He read Rogniat's *Considérations* between February and April 1819 and dictated eighteen notes about them to the mamluk Ali.[50] These notes represent one of Napoleon's most interesting texts on war.[51] Obviously, we must take account of his desire to refute anything that blamed his intentions and his actions. The dictations went through various subsequent stages. Philippe Gonnard and Nada Tomiche have clearly explained the working methods of Napoleon and his collaborators.[52] For these texts, the

Second Empire edition of the *Correspondance* is still essential where it indicates that it was based on the original manuscript.[53] When this edition indicates that it is based on the *Mémoires* dictated to Gourgaud and Montholon, we have referred to them. The first edition of the *Mémoires* (1823–5) was read by Clausewitz.[54] When they are not drawn from the correspondence, most of these 'military maxims' attributed to Napoleon derive from these dictated accounts and commentaries.

Then there are the accounts by the companions at Saint Helena, who recount numerous statements by Napoleon in direct or indirect style. As is well known, Count de Las Cases's *Le Mémorial de Sainte-Hélène* enjoyed enormous success on its appearance in 1823–4. Like the *Mémoires* dictated to the generals, Gonnard is insistent that this work—the product of two hands—does indeed represent what Napoleon meant.[55] He shows himself in a favourable light, as a precursor of liberalism and a Romantic hero.[56] This work is in fact Napoleon's real political testament, the last piece in the construction of his legend. We shall cite it in the 1830 edition, where Las Cases was able to restore passages suppressed under the Bourbons.[57] On each occasion we have compared this text with that of the first edition, reprinted with annotations and critical comments by Marcel Dunan, to note that the text has not been subject to alteration apart from accents or punctuation marks.[58] The narrative of Barry O'Meara, the Emperor's Irish doctor, is also a reliable source when it recounts Napoleon's words.[59] On principle, we must be more distrustful of the Corsican doctor Antommarchi.[60] However, he conveys some interesting remarks by Napoleon, which he manifestly could not have invented.

But among the 'Saint Helenian' accounts, it is the journals of Gourgaud and Bertrand, written on a daily basis, which have emerged as the richest sources for our purposes. Naturally, these generals hailing from the 'educated armed forces'—artillery and engineering, respectively—were particularly attentive to the Emperor's comments on military affairs. They were also better placed to understand and elicit them. For Philippe Gonnard, Gourgaud's natural candour renders his text even more reliable than Las Cases's. The conversations undergo no retouching in it. Not only do they take place between service companions but, unlike Las Cases's text, they were not intended for publication.[61] Belatedly, however, they were published—first of all by Viscount Grouchy and Antoine Guillois, and then by Octave Aubry.[62] These editions contain some errors in people's names. We have corrected them by reference to the manuscript. The latter, conserved in the Archives nationales, contains some lacunae, appears in relative chronological disorder, and does not contain any indication of folios or pages.[63] Gourgaud employed minuscule handwriting to hide his text. This does not facilitate its decipherment.[64] The editions involve a degree of interpretation because the notes do not always form complete sentences. We shall cite Gourgaud's journal in its original manuscript version, with reference to the pages of Aubry's edition in brackets.

As for Bertrand's notebooks, they are ignored by Philippe Gonnard and by all the Napoleonic military collections, even the most complete among them. They were only published from 1949 onwards by Paul Fleuriot de Langle.[65] Conserved in the Archives nationales, they are practically short-hand.[66] The order of the notebooks has been altered and does not always correspond to the published text. Proper nouns only have their initial letter and many words are shortened. We have been able to certify that Paul Fleuriot de Langle took some liberties with the original text. Some passages are omitted; others are inverted. Some abbreviations are poorly conveyed. In the round, however, the Emperor's ideas are correctly re-transcribed. With a few exceptions, there are no significant errors of sense. In several places, the manuscript is so difficult to read that we have not been able to decipher it clearly. Furthermore, Fleuriot de Langle legitimately re-transcribes in direct speech words by Napoleon that Bertrand always gives in indirect speech. We shall do likewise. We shall also supply punctuation in a text written in haste that is practically bereft of any. We shall refer to the manuscript in the fairly disordered state in which we found it in 2010–11, indicating the handwritten pagination added in pencil. Reference to the pages of Fleuriot de Langle's edition will always be given in brackets. The praise bestowed on Gourgaud's manuscript by Philippe Gonnard also applies to Bertrand's text, which is actually the most realistic of all when it comes to Napoleon's last moments. Bertrand remained on Saint Helena to the end, whereas Gourgaud left the island in March 1818. Not only do Bertrand's notebooks cover the whole period of exile but, on a military level, they are even more interesting because their author was older than Gourgaud and had enjoyed more significant commands.

Finally, while they were published before the diaries of Gourgaud and Bertrand, the *Récits de la captivité* of the Count of Montholon were written only around 1846, on the basis of personal notes and a reading of Las Cases's *Mémorial*.[67] They also seem to contain borrowings from Gourgaud's text, but clumsy ones where the meaning is sometimes distorted. Montholon wrote too late, in a context where the Napoleonic legend was already flourishing.[68] A 'court general', he did not have as much experience of war as Gourgaud and Bertrand. We have therefore made very little use of his *Récits*.

The final category of sources is the *Mémoires* of all those who rubbed shoulders with Napoleon prior to his exile to Saint Helena. Readers will appreciate that here we were obliged to include only those characters most likely to have spoken to Napoleon and to have heard him. When it comes to marshals and generals, the circle of chroniclers to whom Napoleon might have confided some profound reflection on war is very small: only Caulaincourt, Gouvion Saint-Cyr, and Marmont have proved interesting. Curiously, the *Mémoires* of civilians are more numerous in recounting remarks of a high calibre on war: Chaptal, Gohier, Roederer, Mme de Rémusat, Thibaudeau. Obviously, we must distrust certain *Mémoires* and have recourse to Jean Tulard's indispensable critical bibliography, with additions by Jacques Garnier.[69] This leads us to the important issue of internal criticism of Napoleon's statements.

Prefacing an edition of the 'maxims', David Chandler acknowledges that their 'practical' value is open to dispute.[70] When he wrote to his subordinates, Napoleon gave them orders or advice in connection with a specific situation; he rarely conveyed considerations of a general kind. It must be admitted at once that Napoleon's thought is difficult to pin down because he himself clouded the issue from the start. He always displayed a great deal of opportunism and, above all, advocated adapting to circumstances in war. It is therefore possible to detect contradictions in him, especially in his correspondence, depending on the context. He always sought to achieve his goals by any possible means; and that precludes accepting at face value any statement of his intentions. Deceiving opponents as to his own intentions was a fundamental aspect of his way of waging war. While, as we shall see in his texts, he extolled honour as an essential value of the military profession, we are entitled to ask if his predominant characteristic was not pragmatism.[71] We know to what extent Napoleon cultivated his legend from his first campaign in Italy, and also how far he created his image not only as a liberal hero, but also as an infallible general, on Saint Helena. In all his statements, we must suspect self-dramatization, self-representation, and self-glorification. On Saint Helena, in particular, he sought to fashion his image for new generations. He also had to respond to the campaigns of denigration organized against him by the 'ultra' political forces of the Restoration.

As Antoine Casanova has aptly put it, 'Napoleon elaborated a multiplicity of reflections in which were intermingled a conscious effort at apologetics, imaginary and illusory solutions to the contradictions he had experienced and that haunted him, and pertinent lucidity.'[72] A dispassionate study of his texts requires making a certain selection from them, offering a critical commentary on them, and comparing them with serious works of scholarship.

Unlike David Chandler, we shall not seek to establish a body of sound advice for military men. It exists, but that is not our objective. We want to identify as accurately as possible what Napoleon understood of war, and how he saw it. We also hope to avoid a pitfall by prioritizing ideas—general and theoretical considerations—over events. The latter will figure solely as background, above all to thoughts derived from the correspondence and shown by specific examples. Like people, they will be elaborated on, or provided with bibliographical references, only if that proves indispensable to understanding Napoleon's thinking or to refuting an allegation whose falsity has been demonstrated by rigorous studies. As a matter of principle, unless indispensable for appreciating an important idea, we shall not enter into a critical analysis of Napoleon's campaigns. The extent to which the Emperor's version has marked the military history of this period, particularly in France, cannot be overstated. A number of myths are hard to dispel. However, various recent works make it possible to correct traditional versions. In this sense too, a critical collection such as the one we are proposing is timely. By focusing on theoretical or general considerations on war, we shall be on firmer ground. Similarly, when Napoleon commented on the campaigns of Turenne, Prince Eugène of Savoy, or Frederick II of Prussia, he was more likely to assess them objectively than his own campaigns. Furthermore, Napoleon constructed his legend more as regards the political aims and principles he attributed to himself than in his account of the facts.[73] We must also distinguish between texts. The journals of his companions on Saint Helena contain self-critical reflections that sometimes contradict the positions adopted in the dictations. Napoleon was more sincere in private conversation than in the texts he wished to bequeath to posterity.

His style offers glimpses of his charisma. Napoleon's sentences are always gripping. They have lost nothing of their magnetic power and can still inspire energy and strength of character not only in an officer, but in any man of action. Jules Bertaut has formulated the powers of attraction of the Napoleonic sentence very well: 'Short or extended, cutting or subtle, spontaneous or undulating, it nearly always surprises us by the experience with which it is charged, as well as by the profound views that it lets slip. It can be discussed, contested, execrated. But what cannot be denied is its robustness or the extent of its impact.'[74] Haste pervades all Napoleon's writings, especially his correspondence. He got into the habit of going straight to the point 'by simple arguments with terse formulas'.[75] He had his favourite expressions and often employed *reductio ad absurdum* and contrasting words or formulas. His way of arguing was that of the patriots of the Revolution and nineteenth-century nationalists. It consisted in systematically employing a crushing rhetoric that makes any incident, even a minor one, an affair of state and an issue of principle.[76] He created a simple, vibrant, highly vigorous style for himself: the 'lion's claw' referred to by Sainte-Beuve.[77] General Lewal, founder of the École supérieure de guerre, was less taken in by the form and more concerned with the substance: 'An admirable stage director, without always bothering about accuracy, he [Napoleon] loved to strike the imagination, to fix in people's minds a concise image so as to lodge it there profoundly. His terse, incisive speech, full of antitheses in the manner of oracles, had an air of mysterious profundity. Thoughts were often cloaked in it and could be variously interpreted.'[78] On Saint Helena, he had more time to spare and access to a lot of books. His dictations were sometimes less vivacious but more elevated, less spontaneous but more suggestive, smacking of self-justification and a desire to write for posterity. Napoleon had undeniable qualities as a writer.[79]

Our goal being not merely to publish fine pages or morsels of eloquence, readers will be invited, in the wake of General Lewal, to prioritize the analytical significance of the Emperor's remarks.[80] Occasional comparisons with Clausewitz will aid this. We shall not include plans of military organization intended to prompt discussion, or whose content touches on unduly specific material details, such as the infantryman's equipment, the size of an entrenchment, or the structure of bridges on stilts. Such considerations do not figure in *On War*. To facilitate reading and sketch an analysis of Napoleon's ideas, we shall link his texts by means of a commentary in which comparisons with Clausewitz will be a regular feature. Not all of Napoleon's texts will be subject to commentary or close criticism.[81] Far too many aspects are mentioned that would require a great deal of cross-checking—for example, as regards the law of nations. Only links with the ideas of Clausewitz, much more clearly defined today, will be subject to sustained treatment, as implied by the title of our book. We have introduced punctuation into manuscript texts where it was lacking; respected the punctuation of edited texts; adapted and standardized the spelling of certain proper nouns (Annibal = Hannibal); and corrected the most disconcerting archaisms. We have placed in square brackets what might be required to render a text intelligible. In total, these minor corrections are very few in number.

I am grateful to Hervé Coutau-Bégarie, Guy Stavridès, and Benoît Yvert, who encouraged and made possible the publication of this collection. The identification of the 'Note on the political and military position of our armies in Piedmont and Spain' (1794) was possible thanks to the help of Vice-Admiral Louis de Contenson, then head of the Service historique de la Défense, Colonel F. Guelton, head of the Département de l'armée de terre, and the heritage protection director Bertrand Fonck, section head of the historical archives at Vincennes. Mme Ségolène Barbiche, in charge of the private archives section in the Archives nationales, granted me permission to consult General Bertrand's original manuscript. I would like to express all my thanks to her, as well as to the section staff. General Lucien Poirier, and my colleagues and friends Martin Motte, Thierry Widemann, and Jacques Garnier, kindly agreed to read this work and made some very useful observations. The English edition would not have been possible without the advice and warm support of Sir Hew Strachan. I thank also my editor at Oxford University Press, Matthew Cotton, who has been the soul of faith, patience, and encouragement alike. Gregory Elliott has done a wonderful and difficult job of translating Napoleon's subtle and sometimes abrupt sentences. I am grateful to Alan Forrest for his careful reading of this fine translation. Only such an exceptional expert on Napoleon and his army would have been able to supervise the rendering of the whole.

BOOK I

The Nature of War

Napoleon already possessed a certain idea of war from his Corsican childhood. Ceded to France by Genoa in 1768, the Island of Beauty was torn apart by opposed parties and remained under French military command during the young Napoleon's early years. From the outset, society must have seemed to him to be naturally stamped by violence. In September 1786, when he had his first semester's leave as a young officer, he returned to his native isle with a trunk of books including the works of Plutarch, Plato, Cicero, Cornelius Nepos,[1] Livy, and Tacitus, translated into French, as well as those of Montaigne, Montesquieu, and Abbé Raynal.[2] Noting thoughts by which he was struck, he expanded his intellectual horizons and conferred a historical and philosophical dimension on his ideas on war, which he regarded as one of the main motors of human activity.[3] The outbreak of the French Revolution only served to confirm him in this impression. But before it broke out, he had undergone training as an officer and resolutely embraced the ethics of this profession, dominated by the sense of honour. Unlike Clausewitz, he was much concerned with the relations between war and law. For Clausewitz war was '*an act of force to compel our enemy to do our will*. Force, to counter opposing force, equips itself with the inventions of art and science. Attached to force are certain self-imposed, imperceptible limitations hardly worth mentioning, known as international law and custom, but they scarcely weaken it.'[4]

I

What is War?

At the age of 22, I had the same ideas on war as I did afterwards.[1]

This sentence establishes a continuity that might be a cause for surprise. In 1791, Napoleon had completed his military training and had his first appointments in an artillery regiment. General Colin has demonstrated the truth of the claim. Prior to undertaking his first Italian campaign, Napoleon already possessed a set of ideas about war that were not to change fundamentally. His early writings contain the main ideas of his system of war.[2] A similar observation has been made of Clausewitz by several of the best specialists on his work.[3]

The officer's ethics

Napoleon was trained in military schools where he learned the code of honour of officers of the *ancien régime*. His correspondence would always preserve traces of this. Thus, he often wrote to his opponents in the most courteous of terms. Before the start of operations in his first Italian campaign, he wrote to the Austrian general Colli[4] that he held a high opinion of his soldiers:

> I have too high an opinion of you to think that you would go to any extreme that would be disavowed by a man of honour and which would make rivers of blood flow. You would be culpable in the eyes of all Europe and of your army in particular.[5]

He wrote for an exchange of prisoners or in the hope that the relevant governments would succeed in agreeing peace.[6] His desire to limit the war also emerges from his orders not to ravage the conquered country.[7] A letter to the commander of the Austrian army trapped in Mantua in October 1796 indicates his humane conception of war, even if it is also an incitement to surrender:

Monsieur, the siege of Mantua is more disastrous for humanity than two
campaigns. The brave man must confront danger, but not the plague of a
swamp. Your cavalry, which is so precious, is without fodder; your garrison,
which is so numerous, is poorly fed; thousands of sick men need fresh air,
medicines in abundance, and healthy nourishment: here are plenty of causes of
destruction. It is, I believe, in the spirit of war and the interest of both armies
to come to an arrangement. Render to the Emperor your person, your cavalry,
and your infantry; surrender to us Mantua: we shall all gain thereby—and
humanity more so than us.[8]

The 'spirit of war' is to come to an agreement. In other words, it does not
involve a blind unleashing of violence, a fight to the death. The notion of
war involves the adversaries adhering to a certain number of rules, which are
advantageous to both of them. Napoleon's conception of war remained
bound up with that of the officers of the *ancien régime*, of the nobility and its
sense of honour. For Napoleon, military men, whatever their camp, belonged
to the same family inasmuch as they shared the same values, the same ethic. On
Saint Helena, he talked with British officers on several occasions. When the
regiment responsible for guarding Napoleon—the 53rd infantry—left on
another mission, its officers came to take their leave of him. The Emperor
questioned them on their years of service, their wounds. He stated that he was
very satisfied with the regiment and would always be pleased to hear good
news of it.[9] A few days later, on the departure of the 53rd, he thought of
mounting his horse in uniform to salute them. But on reflection he told
himself that it would look like he was chasing after the English—something
that would pain his supporters in France.[10] Political considerations prompted
him not to follow his first inclination as a military man.

'I know of no namby-pamby war'

The development of a military ethic based on a code of honour characterized wars in Europe from the late seventeenth century. This ideal came into
contradiction with the natural violence of war:

Turenne was an honest man, but his troops looted. That's the reality of the
history of war, not the romance.[11]

In 1814, the French could have fought harder, challenged the allies more for
victory:

So France conducted itself very badly for me. During Cannae,[12] the Romans redoubled their efforts, but that was because everyone was afraid of being raped, having their throat cut, being pillaged. That is waging war, whereas in modern wars everything happens in namby-pamby fashion.[13]

Gourgaud continues:

Wednesday, 7 [January 1818]—His Majesty is in a bad mood, rises and says that in our day peoples make war in namby-pamby fashion. 'When appropriate, formerly the defeated were massacred or violated or enslaved. If I had done that in Vienna, the Russians would not have arrived in such good shape in Paris. War is a serious business.' I say to His Majesty that if everyone was killed, victory would be more difficult, people would defend themselves more, that the musket has created equality between men, and I cite Spain. There we conducted ourselves as in the past and the whole population rose up and drove us out. His Majesty gets annoyed, says that if He had stayed in Spain, it would have been subjugated [. . .].
    Sunday, 25 [January 1818]—[. . .] The Emperor talks about artillery and would like a cannon firing two feet above the parapet; then he chats about Masséna: 'He could have held out another ten days in Genoa.[14] It is said that people were dying of hunger. Bah! You will never get me to believe that he couldn't have held out another ten days. He had 16,000 men in the garrison and the population was 160,000. He could have found supplies by seizing them from the inhabitants. A few old women, some old men, etc., would have died, but after all he would have held Genoa. If one has humanity, always humanity, one should not wage war. I know of no namby-pamby war.'[15]

These sentences are of capital interest when it comes to understanding the nature of war. Napoleon distinguishes between war as he was obliged to wage it in the context of a civilized modern Europe and the true character of war, where no holds are barred. In a message to the Senate in November 1806, he had already evoked a more violent form of war, as in antiquity, but this time it was a question of justifying the harsh occupation of Prussia until a general peace, as well as the imposition of a blockade on the British Isles:

It has cost us to have the interests of private persons depend on quarrels between kings and to revert, after so many years of civilization, to the principles that characterized the barbarism of the early ages of nations. But we have been forced for the good of our peoples and our allies to employ the same weapons against the common enemy as he employed against us.[16]

In signing the Berlin decree establishing the blockade of Great Britain, Napoleon endorsed the idea that the war should be fought to the bitter end: one of the two adversaries must cave in; no compromise was possible.[17]

Clausewitz likewise reckoned that war as such could not be made sentimentally: 'Kind-hearted people might of course think there was some ingenious way to disarm or defeat any enemy without too much bloodshed, and might imagine this is the true goal of the art of war. Pleasant as it sounds, it is a fallacy that must be exposed: war is such a dangerous business that the mistakes which come from kindness are the very worst.'[18] He too characterized war as 'a serious means to a serious end'.[19] It was right for the European nations to adopt among themselves certain rules limiting violence, but these rules pre-existed war as such. They formed part of a superior socio-political context. Ferocious hatred could always resurface in civilized nations because any war could generate such sentiments. In fact, war was a mutual act in which violence could theoretically be unlimited. As a concept, war naturally went to extremes and could entail unlimited use of force. Clausewitz expressed in minute detail what Napoleon merely evoked. But each of them perceived the same thing. The Napoleonic Wars were not 'total wars' in the sense given the term in the twentieth century, but they began to portend them. We shall return to this key question a little later and again at the end of Book VIII.

Civil wars

Napoleon ventured a few general considerations on civil wars in connection with the Vendée, to which Clausewitz devoted a short study:[20]

> It is because all the parties resemble one another: once the civilian torches are lit, military leaders are merely means to victory; but it is the crowd that governs. [...]
>
> This is the peculiarity of revolts: equality of interests starts them, the union of passions continues them, and they invariably end up in civil war, which establishes itself in the revolts themselves. [...]
>
> In civil wars, it is not given to every man to know how to behave; something more than military prudence is required; what is needed is wisdom, knowledge of men... [...]
>
> In party wars, he who is conquered one day is discouraged for a long time. It is above all in civil wars that good fortune is needed.[21]

At the start of the Consulate, the general charged with pacifying the départements of the west was urged to trust those who submitted, to conciliate priests, and to facilitate the travel of the leaders who wished to go to Paris. He also received the following instruction:

> If you wage war, do so with rapidity and severity; that is the only way of rendering it less prolonged and, consequently, less deplorable for humanity.[22]

On Saint Helena, Napoleon had this to say about the war between Caesar and Pompey:

> It is Rome that needed to be taken care of; that is where Pompey should have concentrated all his forces. At the start of civil wars, all the troops must be concentrated because they electrify one another and take confidence in the strength of the party; they attach themselves to it and remain loyal to it.[23]

Some notes on the restoration of French authority on Saint-Domingue deliver the paradoxical observation that

> [. . .] civil wars, rather than enfeebling, re-temper and toughen peoples.[24]

By contrast, the troops who serve in this kind of war gradually forget how to fight a regular army. From Italy, in October 1796, General Bonaparte wrote as follows of the new generals sent to him:

> Anyone who comes to us from the Vendée is not accustomed to proper warfare; we make the same criticism of the troops, but they get hardened.[25]

Napoleon occasionally characterized the conflicts between European countries as civil wars. During a reception for the diplomatic corps and members of the British Parliament on 15 fructidor (2 September 1802) in a Europe pacified by the Treaty of Amiens, he confided to Charles James Fox, the leader of the Whig party and supporter of a rapprochement with France:

> There are only two nations, the East and the West. France, England, and Spain have the same customs, the same religion, and the same ideas, pretty much. It is simply a family. Those who wish to create war between them want civil war.[26]

Was this reflection, prophesying the 'clash of civilizations' and the unity of Europe, merely an ad hoc quip? Or does it reflect Napoleon's true thinking? The Egyptian campaign had given him experience of war with Orientals. We must turn to other allusions to war and peace between Europeans to define his position more clearly.

War and peace

A long conversation with a state counsellor, at a time when Europe was at
peace following the Treaty of Amiens, but before the proclamation of the
Consulate for life, or between late March and early August 1802, is highly
illuminating on Napoleon's relationship to war and peace:

> FIRST CONSUL: [. . .] England fears us and the continental powers do not like us.
> Given this, how can we hope for a strong peace! Besides, do you think that a
> peace lasting five years or more suits the form and circumstances of our
> government?
>
> STATE COUNSELLOR: I think this break would suit France very well after ten
> years of war.
>
> FIRST CONSUL: You misunderstand me: I am not questioning whether a
> genuine, robust peace is good for a well-established state; I am asking
> whether ours is sufficiently such as not to still have need of victories.
>
> STATE COUNSELLOR: I haven't thought enough about such a serious question
> to answer categorically: all I can say, or rather what I feel, is that a state
> which can only be consolidated through war is in a very unfortunate
> situation.
>
> FIRST CONSUL: The greatest misfortune would be to misjudge its situation,
> because one can cater for it when one knows it. Now, tell me: do you
> believe in the continuing hostility of these governments that have never-
> theless just signed the peace?
>
> STATE COUNSELLOR: It would be very difficult for me not to believe in it.
>
> FIRST CONSUL: So then, draw the conclusion! If these governments always
> have war at heart, if they are bound to renew it one day, it would be better
> for it to be sooner than later. For every day weakens the impression in them
> of their last defeats and tends to diminish among us the prestige of our last
> victories. All the advantage therefore lies with them.
>
> STATE COUNSELLOR: But, citizen consul, do you discount the advantage you
> can take of peace to organize the interior?
>
> FIRST CONSUL: I was going to come to that. Certainly, that major consider-
> ation has not escaped my mind and, even in the midst of war, I have
> demonstrated that I do not neglect what concerns institutions and good
> order at home. I will not stop there; there is still a lot to be done. But aren't
> military successes more necessary to dazzle and contain the interior?
> Remember that a First Consul does not resemble those kings by the grace
> of God who regard their states as an heirloom. Their power has old habits
> for auxiliaries. Among us, by contrast, such old habits are obstacles. Today's
> French government resembles nothing surrounding it. Hated by its

neighbours, compelled to contain in the interior several classes of malcontents, to impress so many enemies it requires dramatic gestures and, consequently, war.

STATE COUNSELLOR: Citizen consul, I admit that you have much more to do to consolidate your government than do the neighbouring kings to maintain theirs. On the one hand, however, Europe is not unaware that you are capable of conquering and, in order to remember it, does not need you to supply new proof every year; and on the other, peacetime activities are not always low key and you can command admiration by great works.

FIRST CONSUL: Regarded from a distance, old victories have scarcely any impact and great works of art do not make a great impression except on those who see them; and that is a small number of people. My intention is indeed to multiply such works; the future will perhaps be more attached to me for them than for my victories. But for the present there is nothing that can resonate as much as military success: that is what I think; it is an unhappy situation. To consolidate itself, a new government like ours must, I repeat, dazzle and astound.

STATE COUNSELLOR: Your government, citizen consul, is not, it seems to me, completely new-born. It has assumed a manly garb since Marengo: ruled by someone strong-minded, and supported by the arms of thirty million inhabitants, it occupies a rather distinguished place among the governments of Europe.

FIRST CONSUL: So, my dear friend, you think that suffices? One must *come first of all or succumb*.

STATE COUNSELLOR: And to obtain that result, you see nothing but war?

FIRST CONSUL: Yes, citizen . . . I shall tolerate peace if our neighbours can keep it. But if they force me to take up arms again before they are blunted by softness or prolonged inaction, I shall view that as an advantage.

STATE COUNSELLOR: Citizen consul, what duration do you assign to this state of anxiety which, in the very midst of peace, causes war to be missed?

FIRST CONSUL: My dear friend, I am not sufficiently informed about the future to answer that question. But I feel that to hope for more robustness and good faith in peace treaties, either the form of the governments around us must approximate to ours, or our political institutions must be rather more in harmony with theirs. There is always a spirit of war between old monarchies and a completely new republic. That is the root of European discord.

STATE COUNSELLOR: But can't this hostile spirit be contained by recent memories and stopped by the attitude you adopt?

FIRST CONSULT: Palliatives are not remedies. In our position, I regard any peace as a brief truce and my ten-year appointment[27] as destined to wage war virtually uninterrupted. My successors will do what they can. (This was before the lifetime Consulate.)[28] Besides, don't think that I want to break the peace; no, I will certainly not play the role of aggressor. I have too much

interest in leaving the initiative to the foreigners. I know them well: they
will be the first to take up arms again or to furnish me with just grounds for
taking them up again. I will be ready for any eventuality.

STATE COUNSELLOR: So, citizen consul, what I feared a few moments ago is
precisely what you are hoping for.

FIRST CONSULT: I am waiting; and my principle is that war is preferable to an
ephemeral peace. We shall see what the latter will turn out to be. At this
moment, it is a great prize. It seals recognition of my government by the
one that resisted it longest: that is the most important thing. As for the
rest—that's to say, the future—depending on circumstances.[29]

In this remarkable dialogue, Napoleon makes a systemic analysis of international relations *avant la lettre*. As Raymond Aron was to write much later,
he regarded the heterogeneous character of the state system as the primary
cause of war.[30] To stay with the theoretical analysis of international relations, we shall not be astonished also to find in Napoleon's mouth acutely
'realist' reflections:

> [. . .] to this moment, I had no idea that Austria wanted to go to war. But the
> military system is to meet force with force, and sound politics has it that one
> squares off as soon as a force appears to threaten you.[31]

In 1807, Napoleon recalled the very Roman idea

> [. . .] that the moment one speaks of peace is the time when it is necessary to
> increase preparations and multiply resources.[32]

We shall not be surprised either that, shortly after Austerlitz, he reprimanded his brother Joseph for having proclaimed rather too soon that peace
was going to be signed, as if the French, who had just won such a stunning
victory, desired it most:

> My brother, it was completely unnecessary to announce the dispatch of the
> plenipotentiaries so emphatically and to fire the gun. That is a good way of
> lulling the national spirit and giving foreigners a false idea of our internal
> situation. It is not by crying 'Peace!' that one obtains it. I had not wanted to
> put that in a bulletin; *a fortiori* it should not have been announced at the
> display. Peace is a meaningless word; what we need is a glorious peace.
> I therefore find nothing more impolitic and more false than what has been
> done in Paris on this occasion.[33]

The concept of 'glorious peace' is in part derived from the *ancien régime*.
In fact, any transaction restricting his hegemony over the continent was
unacceptable to Napoleon.[34] Two days later, he added:

You will see that peace, however advantageous I am able to make it, will be deemed disadvantageous by the same persons who demand it so insistently, because they are idiots and ignoramuses who are incapable of knowing anything about it. It is quite ridiculous of them never to stop repeating that people want peace, as if peace meant something: what matters are the conditions.[35]

In 1807, he offered this definition:

Peace is a marriage that depends on a union of wills.[36]

The same idea was noted by Bertrand on Saint Helena. The notion of peace must meet certain criteria:

The way to have peace is not to say that one can no longer wage war. Peace is the diagonal between two forces; it is a capitulation between two struggling forces. If one is destroyed, there is no peace.[37]

French historians generally acknowledge the role played by Napoleon's impatient, domineering personality in the succession of wars between 1803—year of the rupture of the Amiens peace—and 1815. However, they stress the legacy of revolutionary politics—Napoleon making it a point of honour to preserve the conquests of the *Grande Nation*—and also the continuity with the *ancien régime*'s great power rivalries.[38] British and American historians place more emphasis on Napoleon's thirst for glory and his inability to envisage a peace necessarily based on certain concessions to opponents.[39] Napoleon was not a monster, writes Charles Esdaile. Under his regime, political executions were very rare and the number of political prisoners was very small. Napoleon could be charming and his generosity was well known. However, his behaviour suggests that a policy of peace, with its corollaries of trust and self-limitation, was incompatible with his personality. The febrile nature of his mind, which led him to pass very rapidly from one plan to the next, also intervened. Without going so far as to claim that the prospect of a battle gave him physical satisfaction, Esdaile does not regard it as unreasonable to think that military triumph filled a gap in his personal life. The dialogue with Thibaudeau cited above clearly indicates that Napoleon needed military glory for political reasons. Esdaile believes that he also needed it for personal reasons.[40] For Steven Englund, Napoleon was more of a disciple of Hobbes than of Rousseau. He had a pessimistic view of human relations: the state of nature was a permanent struggle for domination. He was incapable of making peace. It has to be said that, in the value-system of the age, triumph on the battlefield represented the *nec plus ultra* of glory and greatness. Englund rightly stresses the need to take this factor into account. Although the world wars have fundamentally altered attitudes, the 'shiver' still prompted by Napoleon's name is certainly marked by fear and disapproval. But even today it still contains an element of admiration and doubtless also of fear for the attraction it exercises over us.[41]

Napoleon himself recognized that, while it did not 'need' war, his regime could not pass up on opportunities to achieve military success. In June 1813, during the armistice, Napoleon wrote to Archchancellor Cambacérès, who was responsible for carrying out his orders in Paris, a letter in which all the nuances of his position are clearly evident:

> The Police Minister, in his policing notes (with which in general I am very satisfied, because of the details they contain and the frequent proof of his zeal I detect in them), seems to seek to render me pacific. This cannot have any result, and is wounding to me, because it assumes that I am not pacific. I want peace, but not a peace that has me taking up arms again three months later and which is dishonourable. I know the situation of my finances and the Empire better than him; there is therefore nothing to tell me on that score. Get him to understand what is discourteous about his manner. I am not a braggart; I do not make a profession of war and no one is more pacific than me. But the solemnity of peace, the desire that it should be an enduring one, and the circumstances in which my empire finds itself will be decisive in my deliberations on the subject.[42]

Having received Cambacérès' reply, the Emperor continued to elaborate on his thinking twelve days later:

> I have received your letter of 23 June. All the ministers' idle talk about peace does the greatest harm to my affairs. For everyone is aware of it and I have seen more than twenty letters from foreign ministers who write at home that peace is desired at any cost in Paris; that my ministers request it of me every day. This is how one ends up making peace impossible; and the fault is above all the Police Minister's. Instead of this pacific tone, a rather more bellicose tone should be adopted. People in Paris have some very wrong ideas if they think that peace depends on me. The enemies' claims are excessive, and I know full well that a peace which does not conform to people's opinion in France of the power of the Empire would be very unfavourably regarded by everyone.[43]

We can clearly see the core of Napoleon's Machiavellian realism, his highly Corsican sense of honour, his idea that French opinion would support him only if he secured an advantageous peace—all this forming an intellectual system in which he was trapped, preventing him from understanding French war weariness. Napoleon required too much of peace—he made it too lofty an ideal—whereas it is always based on a degree of compromise. Ultimately, his temperament, his culture, his personal history, and the history of France since 1789 had accustomed him to war and had in a sense led him to be afraid of peace, out of a fear of diminution. This was his drama, but also that of millions of European men and women.

2

War and Law

While they increased the involvement of nations, the wars of the Revolution and Empire did not completely call into question the principles of international law and the laws of war, tacitly established in the late seventeenth century. Contrary to the idealized image we are sometimes given, they were not that much better respected under the *ancien régime*.[1] As is well known, on several occasions Napoleon ignored the law of nations, as when he had the Duke of Enghien abducted in the Grand Duchy of Baden, or when he transformed the Italian republic into a kingdom in defiance of the Treaty of Lunéville. Napoleon's armies sometimes crossed through neutral territories: Piacenza's manoeuvre in May 1796 was carried out via the Duchy of Parma and in 1805 Bernadotte's corps passed through the Prussian territory of Anspach.[2] Confronted with operational necessities, Napoleon had few scruples. It is true that territorial divisions were still so complex and principalities so numerous that the armies of the great powers often took no account of them. The allies themselves violated Swiss neutrality in late 1813 and early 1814 to invade France.[3] Alongside this, Napoleon's correspondence indicates that he attributed a certain importance to the law of nations and, even more so, to *jus in bello*—that is, to the legal framework governing certain situations in war.

The rules for surrendering fortified towns

For the Swiss jurist Emer de Vattel, 'wise and humane' generals should persuade the commander of a fortified town not to pointlessly await the bitter end and offer him an honourable, advantageous surrender. 'If he stubbornly resists, and is finally forced to place himself at the mercy of the conqueror', he adds, 'one may employ against him and his people the full severity of the laws of war. But that law never extends to taking the life of an enemy who lays down his arms, unless he has been guilty of some crime against the victor.'[4] If the commander decides to resist an attack, he knows that he is risking the lives of all its inhabitants, military and civilian, as well as their property. For Georges-Frédéric de Martens, if a place is taken by storm, 'the garrison must place itself at the mercy of the victor; the only thing that can be requested is its life and it is not against the laws of war to give the place over to looting'.[5] These are the laws to which General Bonaparte sometimes alluded. In connection with a post wrested with considerable force from the Austrians, he writes:

> General Koebloes himself defended La Chiusa with 500 grenadiers.[6] By the laws of war, those 500 men should have been put to the sword. But this barbaric law has always been ignored and never practised by the French army.[7]

In Egypt, Bonaparte had it explained to the Arab commander of El-Arich fort that 'the laws of war, among all peoples, are that the garrison of a town taken by storm is to be put to the sword'. He enjoined him to send two men to determine the details of a surrender 'in conformity with what is practised in such circumstances among all the civilized peoples of the earth'.[8]

At Jaffa, by way of response, the governor had General Bonaparte's envoy beheaded. When the assault that was then ordered succeeded, Bonaparte wrote:

> At 5 o'clock, we were masters of the town which, for twenty-four hours, was given over to pillaging and all the horrors of war, which have never seemed so hideous to me. 4,000 of the Djezzar's[9] troops were put to the sword; there were 800 gunners. Part of the civilian population was massacred.[10]

War had become crueller. However, even if he denies it, we can ponder Bonaparte's insensitivity to the horrors of war once he deemed it necessary to resort to them for political reasons.[11] For Steven Englund, 'Bonaparte . . . had Caesar's capacity to make morally or spiritually perilous decisions without blinking.'[12] Among the campaigns conducted by him, those in Egypt and Syria were the cruellest. In Italy and Spain too, he prescribed severe measures while invoking the law of war. This applied to the Calabrian rebels in 1806:

> Severe examples are necessary. I imagine that we have had this village pillaged by the soldiers. That is how we must treat villages that rebel. This is the law of war, but it is also a duty laid down by politics.[13]

The town of Cuenca in Spain was taken by storm in 1808:

> The town has been pillaged: it is the law of war, since it was taken arms in hand.[14]

No capitulation in open country

The surrender of a place of war in accordance with the modalities and conditions evoked above has no equivalent when it comes to operations in open country. Napoleon strove to clarify this point in an extended reflection on Saint Helena:

> Do the laws of war, the principles of war, authorize a general to order his soldiers to lay down their arms, to surrender them to their enemies, and to make a whole corps prisoners of war? The issue is not in doubt when it comes to the garrison of a place of war. But the governor of a place is in a separate category. The laws of all nations authorize him to lay down arms when he lacks supplies, his defences are in ruins, and he has resisted several attacks. In fact, a place is a war machine that forms a whole, which has a role, a prescribed, determined, and known destination. A small number of men, protected by this fortification, defend themselves, stop the enemy, and preserve the depot entrusted to them against the attacks of a large number of men. But when these fortifications are destroyed, and no longer offer any protection to the garrison, it is right and reasonable to authorize the commander to do what he deems most appropriate in the interests of his troops. Contrary conduct would be pointless and would also have the disadvantage of exposing the population of a whole town—the elderly, women, and children. When a place is invested, the prince and supreme commander charged with the defence of this boundary know that it can protect the garrison and hold the enemy only for a period of time and that, once this time has passed and the defences have been destroyed, the garrison will lay down its arms. All civilized peoples have been in agreement on this subject, and there has only ever been discussion of the extent of the defence to be mounted by a governor prior to surrendering. It is true that there are some generals—Villars is among them[15]—who think that a governor must never surrender, but *in extremis* blow up the fortifications and force a passage at night through the besieging army; or, when the first of these is not feasible, to at least escape with his garrison and save his men. Governors who have adopted this course of action have re-joined their army with three-quarters of their garrison.
>
> From the fact that the laws and practices of all nations have expressly authorized the commanders of strongholds to surrender their arms in providing for their interests, and have never authorized any general to compel his soldiers to lay down their arms in any other instance, we can venture that no prince, no republic, no military law has authorized them to. The sovereign or country enjoins obedience to their general and superiors on the part of junior officers and soldiers, on everything consistent with the good or the honour of

the service. Weapons are assigned to the soldier with the military pledge to defend them to the death. A general has received orders and instructions to employ his troops for the defence of the homeland: how can he have the authority to order his soldiers to hand over their weapons and receive chains?

There are virtually no battles where a few companies of light infantrymen or grenadiers, often a few battalions, are not momentarily encircled in houses, cemeteries, woods. The captain or battalion leader who, once he has registered the fact that he is encircled, effects his surrender would betray his prince or his honour. There are virtually no battles where the conduct followed in similar circumstances has not determined victory. Now, a lieutenant-general is to an army what a battalion leader is to a division. The surrender of an encircled corps, during either a battle or an active campaign, is a contract all of whose beneficial clauses are in favour of the contracting parties and all of whose onerous clauses are for the prince and the army's other soldiers. To escape danger only to render the position of one's comrades more dangerous is obviously an act of cowardice. A soldier who said to a cavalry commander: 'Here is my musket, let me go home to my village', would be a deserter in the face of the enemy; the law would condemn him to death. What else does a division general, battalion leader, or captain do when he says: 'Let me go home, or receive me among you, and I will give you my weapons'? There is but one honourable way of being made a prisoner of war and that is to be captured alone, arms in hand, when one can no longer make use of them. That is how François I, King Jean,[16] and so many brave men of all nations, were captured. In this way of surrendering arms no conditions are attached and there cannot be any way of doing it with honour. One is granted one's life because one is powerless to take that of the enemy, who leaves you yours as a form of retaliation because the law of nations would have it thus.

The dangers of authorizing officers and generals to lay down arms in personal surrender, in a different context from that where they form the garrison of a stronghold, are indisputable. It destroys the martial spirit of a nation, undermines honour, opens the door to cowards, timid men, even to misled brave men. If military law pronounced severe penalties against generals, officers, and soldiers who laid down their arms by surrendering, such an expedient would never occur to military men to escape from difficult straits. Their only remaining resource would be valour or obstinacy—and how much they have been seen to do! [...]

What, then, should a general who is encircled by superior forces do? We cannot give any other response than that of old Horace.[17] In an extraordinary situation, what is required is extraordinary resolution: the more stubborn the resistance, the greater the chances of receiving aid or breaking through. How many impossible things have been done by resolute men, with no recourse but death! The more resistance you put up, the more of the enemy you will kill, and the fewer troops he will have that day or the next to direct against the army's other corps. This issue does not seem to us to be amenable to any other solution, without sacrificing a nation's military spirit and exposing oneself to the greatest misfortunes.

Should legislation authorize a general who is encircled far from his army by greatly superior forces, and who has put up a stubborn fight, to disperse his army at night, entrusting to each individual his own salvation and indicating a more or less remote rallying point? This is open to debate. But there is no doubt that a general who adopted such a course of action in a desperate situation would save three-quarters of his people and—which is more precious than men—he would save himself from the dishonour of handing over his arms and flags as the result of a contract that stipulates advantages for individuals at the expense of the army and homeland.[18]

Very exercised about this point, Napoleon returned to it:

> One must not surrender in the open. One must accept that as one's most basic principle. In this instance, it is necessary to decimate like the Romans. Dupont and Sérurier surrendered in the heart of the countryside in Italy.[19] They would not have done so with good military laws. [. . .] One can throw down arms, but not capitulate to save one's baggage. One must know how to die. War is against nature. [. . .] A vessel must not surrender in a battle; it causes the battle to be lost, because the guns that were targeted on it are now trained on another. If the vessel is alone, that is different: it can surrender; it is in the position of a fortified place.[20]

'Perfidious Albion'

Issuing from a line of Tuscan and Corsican lawyers, Napoleon had an elevated idea of law—what it authorized and what it proscribed. For a disciple of Machiavelli, he attached a sometimes naive importance to it.[21] In the face of the British, who had a much more pragmatic conception of it in accordance with their interests ('*Dieu et mon droit*'), he waxed indignant, particularly when they disavowed the agreement made by one of their generals with the French army of Egypt—something that overturned 'all the ideas of the law of nations':

> [. . .] it is inconceivable that so much bad faith, so much impudence and ferocity could direct the cabinet of a nation which is so enlightened and worthy in so many respects that it is cited as an example.[22]

Evidence had been furnished to the First Consul of the involvement of the London government in an attempt to assassinate him. He made known his feelings through the intermediary of his Foreign Relations Minister, Talleyrand:

> The state of war that exists between the two peoples has doubtless shattered some of the links that naturally unite neighbouring peoples. But although the English and the French are at war, are they any the less, either of them, a civilized, European nation? And does the law of nations, which softens the evils of war, not proscribe according protection to monsters who dishonour human nature?[23]

Prisoners were normally clothed and fed by the country holding them. But a British commissioner now demanded clothing for French prisoners in England. He found himself reminded of the customs of the law of nations:

> The Minister of Foreign Relations will make it known to this commissioner that the French government will not depart from what is established between the civilized powers of Europe for the sake of its prisoners in England. It maintains and clothes Russian, German, and other prisoners and these governments create no difficulty about maintaining French prisoners. It is therefore for the British government to decide whether it wishes to depart from established customs and conventions.[24]

Himself a prisoner of the British on Saint Helena, Napoleon returned on numerous occasions to their conduct, which he saw as harsh and contrary to continental customs:

> Throughout the war, I never stopped offering an exchange of prisoners. But the British government, reckoning that it would have been advantageous to me, constantly refused on one pretext or another. I have nothing to say to that: in war politics comes before sentiment. But why be barbaric unnecessarily? And that is what they did when they found the number of their prisoners expanding. For our unfortunate compatriots, there then began the dreadful ordeal of the pontoons, with which the ancients would have enriched their hell if their imagination could have conceived them.[25]

Criticism of the British went back to the Convention and even to Bossuet, who was one of the first to accuse them of perfidy.[26] Napoleon merely continued the Anglophobia of the seventeenth century and the arguments of the Jacobins.[27] Nevertheless, as has been observed, he distinguished between the bad faith of the rulers and the qualities of the nation. In substance, he did not accept that British conduct was largely due to the distinction that he himself made between continental prisoners, generally well treated, and those of the United Kingdom. There was an attitude of suspicion on both sides, which happily did not extend to the battlefield, where the troops of the two belligerents, especially in the Iberian peninsula, always fought in chivalrous fashion, as if to demarcate themselves from godless and lawless guerrillas.[28] The British pontoons reminded Napoleon of an inglorious episode in the career of Julius Caesar, who had exhibited gratuitous cruelty towards the Gauls of the Vannes region, massacring the senators and selling the population at auction:

> One cannot but detest Caesar's behaviour towards the senate of Vannes. These people had not rebelled; they had provided hostages and promised to live in peace. But they were fully in possession of their freedom and all their rights. No doubt they had given Caesar grounds to wage war on them, but not to violate the law of nations in their regard and to abuse his victory in such an atrocious fashion. This conduct was not just; it was even less politic. Such means never accomplish their end; they exasperate and revolt nations. The punishment of a few leaders is all that justice and politics permit; to treat prisoners well is an important rule. The British have broken this rule of politics and morality by putting French prisoners on pontoons—something that has rendered them odious throughout the continent.[29]

Respect for the law of nations and military laws

The laws of war were invoked to justify the intervention in Venetian affairs in 1797:

> The republic of Venice was a neighbour of the army of Italy. The laws of war assign the preservation of order in the country which is the theatre of war to the general. As the great Frederick said, 'Where there is war, there is no free country.'[30]

Before commencing the Austerlitz campaign, the Emperor relied on the law of nations to launch the Grande Armée against Austria, which was assembling troops around Bavaria. He told Talleyrand, Minister of Foreign Affairs, to make it known to Vienna:

> [. . .] that in every country of the world unjustified arming on the borders of one's neighbour is tantamount to a declaration of war, and that there is no shadow of a doubt that Austria is arming today.[31]

At the end of September 1806, a war was in the offing with Prussia. But as long as it was not declared, Napoleon stressed to his marshals that language should be pacific and no hostilities should occur.[32]

Having returned from the island of Elba in April 1815, Napoleon was outlawed by Europe. Communications with France were broken, but war was not yet declared. The Emperor fretted about this conduct to Caulaincourt, his Minister of Foreign Affairs:

> Monsieur le Duc de Vicence, you would do well to give orders at Strasbourg, to the prefect and the general, to ask the other side's general and civil authorities why they do not let the cabinet's couriers pass. The object of war being to bring peace, to interrupt communications is to act contrary to the law of nations. Have someone sent to Baden and write to the minister how surprising this conduct is. Ask him if we are at war or peace.[33]

In his army, he was keen that military laws should be scrupulously respected, especially as regards soldiers—in other words, the weakest. In Cairo, he wrote this letter to Berthier:

> Citizen general, you will find attached a report of the French military personnel held in the citadel. See, I beg you, to sending an officer of the general staff there to make a more detailed one for me—above all, whether military law, which grants senior officers the right to imprison soldiers for a certain number of days, has not been broken; whether several soldiers are not being held even though the duration of their detention pronounced by military councils has expired; finally, whether the sentences of the military councils are in conformity with the laws and whether the review councils requested by the condemned have been granted them.
>
> You will appreciate to what extent this officer's mission concerns order and humanity.[34]

Partisans were not entitled to be treated as soldiers. If they were captured, they became prisoners of state, not of war, as Berthier was informed in June 1813:

> The Prussian officers and the officer of Lützow's general staff are to be regarded as prisoners of state, and sent incommunicado to Mainz, where they will be put in a state prison, without being granted permission to write. The same will be done to Captain Colomb and all the partisan commanders. Formerly, the custom of war was to have them hanged.[35]

The law of nations on land and at sea

Napoleon developed his thinking on war and law predominantly in connection with the conflict with Britain. The long passage that follows compares the law of nations on land and at sea. The issue of neutrality is treated in depth in it. These questions were of particular concern to Napoleon. The pages he devoted to them on Saint Helena attest to the salience of this fundamental disagreement between the French and British powers. Such was the backdrop to the wars of the Revolution and Empire:

In the centuries of barbarism, the law of nations was the same on land as at sea. The individuals of enemy nations were taken prisoner, either because they had been captured arms in hand or because they were mere inhabitants; and they escaped slavery only by paying a ransom. Movable property, and even landed property, was wholly or partially confiscated. Civilization soon asserted itself and completely changed the law of nations in land war, without having the same impact on war at sea. So that, as if they were two forms of reason and justice, things are regulated by two different systems of law. The law of nations in land war does not lead to the despoliation of private individuals or a change in the status of persons. War only impacts on the government. Thus, properties do not change hands; shops and their merchandise remain intact; people remain free. The only ones treated as prisoners of war are individuals seized with arms in hand and forming part of the military corps. This change has greatly reduced the evils of war. It makes conquering a nation easier and war less bloody and less disastrous. A conquered province makes a pledge and, should the victor require it, provides hostages and hands over weapons. Taxes are collected for the benefit of the victor, who, if he deems it necessary, establishes an extraordinary tax either to take care of the maintenance of his army or to indemnify himself for the expenses occasioned by the war. But this tax has no relationship with the value of the goods in shops; it is simply a greater or lesser proportional increase in ordinary taxes. This is rarely equivalent to a year of those raised by the prince and it is imposed on the whole state, so that it never entails the ruin of any individual.

The law of nations governing maritime war has remained in its state of utter barbarism: the property of private individuals is confiscated; non-combatants are taken prisoner. When two nations are at war, all of their vessels, whether at sea or in port, are liable to be confiscated; and the individuals on board them are made prisoners of war. Thus, in a patent contradiction, a British vessel (in the event of a war between France and Britain) that is found in the port of Nantes—e.g. when war is declared—will be confiscated. The men on board will be prisoners of war, even if they are non-combatants and simple citizens. By contrast, a shop containing British goods belonging to Englishmen living in the same town will be neither sequestrated nor confiscated; and British merchants travelling in France will not be prisoners of war and will be afforded right of passage and the passports required to leave French territory. A British vessel that is sailing and seized by a French ships will be confiscated, even though its cargo belongs to private individuals; the individuals on board it will be prisoners of war, even though they are non-combatants. Yet a convoy of 100 carts of goods belonging to Englishmen, and crossing French territory at the moment of the breach between the two powers, will not be seized.

In land war, even the territorial property owned by foreign subjects is not subject to confiscation; at the very most it will be sequestered. The laws governing land war are therefore more in conformity with civilization and the well-being of private individuals. And it is to be hoped that a time will come when the same liberal ideas are extended to war at sea and that the naval forces of the two powers can fight without occasioning the confiscation of merchant ships and without taking simple merchant seamen or non-military passengers prisoner. Trade would then be conducted between belligerent nations on sea as it is on land, in the midst of the battles fought by armies.[36]

In following the prescriptions of the law of nations outlined by Gentili, Grotius, Vattel, and others, the European powers had succeeded in confining war to combatant forces.[37] This development had begun in the late seventeenth century and become fully established in the eighteenth. Despite the introduction of a stronger national dimension with the French Revolution, war in the Napoleonic era was limited to armies for most of the time. There is a body of evidence proving that the troops of the Grande Armée maintained rather cordial relations with the German populations in 1805 and 1806. The attitude of the Austrians was less welcoming in 1809 and in 1813 the recalcitrance of the inhabitants manifested itself in Silesia and, to a lesser extent, in Saxony. But these populations were never targeted by punitive measures that formed part of a plan. While they suffered from the presence of troops, requisition was subject to compensation. Armies alone were the object of the adversaries' strategic designs. Bombardment of towns was rare. Vienna was briefly subject to one in 1809. Battles did not cause any civilian casualties.[38] The war in Spain involved part of the population, but that mainly concerned organized groups. The relationship between the French armies and the Spanish was not invariably stamped with as much hatred as is generally believed. People often make the mistake of projecting onto the Napoleonic Wars nationalist drives that only flourished in the twentieth century. Obviously, a step was taken in the direction of 'total war', but it was gradual.[39] Like all his contemporaries, Napoleon had been formed in the spirit of the Enlightenment and regarded war as a controlled clash between armies. The extent to which he adhered to the limitations fixed by the law of nations has not been sufficiently stressed. The following little-known text clearly indicates it. Here we also find that

Napoleon stresses an imbalance in the Franco-British confrontation. On land, where France was dominant, the law of nations did not allow people to do as they saw fit. At sea, where Britain ruled, law was practically non-existent. Britain did as it wished. Napoleon pursued his thoughts:

> The sea is the property of all nations; it extends over three-quarters of the globe and creates a bond between the different peoples. A vessel at sea full of merchandise is subject to the civil and criminal law of its sovereign, just as if it were within his estates. A vessel under sail may be regarded as a floating colony, in the sense that all nations have equal sovereignty over the seas. Were the merchant ships of warring powers able to sail unhindered, *a fortiori* no inspection could be made of neutral ones. However, since it is agreed in principle that the merchant vessels of belligerent powers can be confiscated, the right for all belligerent warships to verify the flag of the neutral vessels they encounter was bound to follow. For, were it an enemy, they would have the right to confiscate it. Hence visitation rights, which all powers have recognized by various treaties; hence the right of belligerent vessels to send their longboats alongside neutral merchant vessels, to ask to see their papers and thus verify their flag. All treaties have intended that this right should be exercised with all possible consideration; that the armed vessel should be kept out of gun range; and that only two or three men should embark on the ship being visited, so that there will be no atmosphere of force and violence. It has been recognized that a vessel belongs to the power whose flag it flies when it is furnished with valid passports and consignments, and when the captain and half the crew are nationals. All powers have committed themselves by various treaties to prohibiting their neutral subjects from trading in contraband with powers at war; and by that title they have designated war munitions, such as powder, bullets, bombs, muskets, saddles, bridles, cuirasses, etc. Any vessel with such objects on board is considered to have disobeyed the commands of its sovereign, since the latter is committed to prohibiting his subjects from engaging in such trade; and such contraband objects are confiscated.
>
> Searches made by cruising vessels were therefore no longer a straightforward visit to verify the flag; and, in the name of the sovereign whose flag graced the vessel visited, the cruiser exercised a new right of search to check whether the vessel contained contraband items. Men from the enemy nation—but only combatants—were assimilated to contraband objects. Thus, this inspection was not a derogation from the principle that the flag covers cargoes.
>
> Soon a third case arose. Neutral vessels presented themselves wishing to enter places under siege that were blockaded by enemy squadrons. These neutral vessels were not carrying war munitions, but food, wood, wine, and other merchandise which could be of service to the besieged place and prolong its defence. After lengthy discussions between the powers, they

agreed by various treaties that, in instances where a place was genuinely under blockade, so that there was clear danger in a vessel trying to enter it, the commander of the blockade could forbid the neutral vessel entry and confiscate it if, notwithstanding this interdiction, it used force or cunning to gain admittance.

Thus, maritime laws are based on these principles: 1. The flag covers cargoes. 2. A neutral vessel can be inspected by a belligerent vessel to confirm its flag and its cargo—that it is not contraband. 3. Contraband is restricted to war munitions.[40] 4. Neutral vessels can be prevented from entering a place if it is under siege, provided that the blockade is real and that there is a clear danger in entering. These principles form the maritime law of neutrals, because the various governments have freely and by treaties committed themselves to respecting them and having them respected by their subjects. The different maritime powers—Holland, Portugal, Spain, France, England, Sweden, Denmark, and Russia—have at several times and subsequently contracted their commitments with one another, which have been proclaimed in general peace treaties, such as those of Westphalia in 1646 [sic] and Utrecht in 1712.[41]

It was through his reading that Napoleon acquired his knowledge of international law and here he offers a veritable lecture, prior to his analysis of Britain's conduct:

In the American war in 1778, England claimed: 1. That commodities for building vessels, such as wood, hemp, tar, etc., were contraband; 2. That a neutral vessel did indeed have the right to sail from a friendly port to an enemy port, but that it could not traffic from one enemy port to another; 3. That neutral vessels could not sail from the colony to an enemy country; 4. That neutral powers did not have the right to have their merchant vessels escorted by war ships or, in cases where they did, they were not exempt from inspection.[42]

These claims prompted the indignation of what was not yet called the 'international community', but which corresponded to it. Great Britain did not then dare to apply such measures, but the context of the wars against revolutionary France from 1793 onwards permitted it to do so.

3
Military Genius

Having a genius for war

> Achilles was the son of a goddess and a mortal: such is the image of the genius for war. The divine part is everything that derives from moral considerations of character, skill, the interests of your opponent, opinion, the spirit of the soldier who is strong and victorious, weak and beaten, depending on whether he believes himself to be. The terrestrial part is the weapons, the entrenchments, the positions, the battle orders—everything that pertains to the combination of material things.[1]

We note in this definition that the 'divine' part contains everything that is non-material but important for war. Alongside character and skill, the opponent's interests cover the whole element of interaction in war—what Clausewitz called 'reciprocal action'.[2] The soldier's opinion and spirit also refer to the 'moral forces' to which we shall return in Book III. What Napoleon provides here is a veritable definition of war via its main components. Clausewitz also employs the notion of genius in the sense of a happy combination. By it he understands 'a very highly developed mental aptitude for a particular occupation'.[3] The notion involves a harmonious combination of forces, where one can be predominant, but none must conflict with the others. It assumes a high intellectual level and hence an epoch of high civilization. This was the case in Rome and France, says Clausewitz.

One has a genius for war from birth or one does not. Napoleon lamented to state councillor Roederer that his brother Joseph, placed on the Spanish throne, was not a military man:

> For my part, I am, because it is the special gift I received at birth; it is my life, my habit. Everywhere I have been, I have commanded. At the age of 23, I commanded the long siege of Toulon. I commanded in Paris in Vendémi-aire; I swept off the soldiers in Italy as soon as I presented myself there. I was born for that.[4]

For Napoleon, success in war was not the result of chance, even if the latter was present in events and always had to be reckoned with. The genius of great generals was indisputable:

> No sustained great acts are the work of chance and fortune; they always derive from calculation and genius. One rarely sees great men fail in their most perilous endeavours. Take Alexander, Caesar, Hannibal, the great Gustavus,[5] and others: they always succeed. Is it because they had good fortune that they became great men? No, but because being great men, they proved capable of mastering good fortune. When one wishes to study the springs of their success, one is completely astonished to see that they had done everything to achieve it.[6]

Such 'mastery of good fortune' by the great war leaders is even more clearly explained in a conversation of 1804 reported by Mme de Rémusat:

> Military science, he said, consists in accurately calculating all the possibilities first of all, and then exactly, well-nigh mathematically, calculating the share of chance. This is the point where one must not make a mistake and where a decimal more or a decimal less can alter everything. This division between science and chance only suits the head of a genius, for wherever there is creation it is required, and certainly the greatest improvisation of the human mind is conferring existence on what does not possess it. Chance therefore always remains a mystery for mediocre minds and becomes a reality for superior men. Turenne scarcely thought of it and only had a method. I think, he added smiling, that I would have beaten him. Condé[7] suspected it more than him, but it was out of impetuousness that he gave himself to it. Prince Eugène [of Savoy][8] is one of those who appreciated it best.[9]

Napoleon also confided to Mme de Rémusat how he had subsequently developed his spirit of geometry, and then his spirit of finesse, during his training:

> I was raised, he said, at the École militaire and only showed an aptitude for the exact sciences there. Everyone said of me: 'That's a child who will be suited only to geometry.' [...] When I went into service, I was bored in my garrisons; I began to read novels and this reading interested me greatly. I tried to write some and that added some vagary to my imagination; it intermingled with the positive knowledge I had acquired and I often amused myself by dreaming and then assessing my dreams by the compass of my reason. I threw myself mentally into an ideal world, and I tried to find out precisely how it differed from the world I lived in.[10]

We must return to what Napoleon said about the respective shares of science and chance. He knew of Pierre Simon Laplace's research into probability. He had had him as a teacher at the École royale militaire in 1785. For Laplace, 'the theory of chance consists in reducing all events of the same kind to a certain number of equally possible cases—that is, such that we are equally undecided about their existence—and determining the number of cases conducive to the event whose probability we are investigating. The relationship between this number and that of all the possible cases is the measure of this probability, which is thus only a fraction, whose numerator is the number of favourable cases and whose denominator is the number of all possible cases.'[11] In 1812, Laplace dedicated his *Théorie analytique des probabilités* to Napoleon, who received a copy of it at the start of the Russian campaign and replied to Laplace thus:

> Monsieur Comte Laplace, I have received your treatise on the calculation of probabilities with pleasure. There was a time when I would have read it with interest; today, I must confine myself to expressing the satisfaction I experience every time I see you producing new books that perfect and extend this foremost science. They help to exemplify the nation. The progress and perfection of mathematics are intimately bound up with the prosperity of the state.[12]

We shall encounter chance again in Chapter 7, in connection with Clausewitz's notion of 'friction'. Meanwhile, let us note that the integration of chance into calculations is an essential mark of military genius and, in particular, of Napoleon's genius. Marshal Marmont explained this very well: 'The more elements you include in your calculations, the more you dominate events. Planning must embrace the possible and probable alike; you can even ensure yourself against fortuitous risks. That is how you avert major catastrophes on a day of setbacks. Such planning was one of Napoleon's greatest faculties at the height of his glory. His opponents having been virtually always wanting in it, the results he obtained astonished the world.'[13] Planning is based on hard work. Genius is not sheer improvisation:

> In war nothing is obtained without calculation. Anything that is not profoundly thought through in detail yields no result.[14]
>
> If I always seemed ready to react to anything, to face anything, it is because before undertaking anything I thought about it for a long time; I foresaw what might occur. It is not some genie that reveals to me all at once, in secret, what I must do or say in a situation unanticipated by others: it is my reflection, it is meditation.[15]

Here we find ourselves at the heart of the secret of Napoleonic strategy, if we are to believe a professor at Columbia Business School whose works have been of particular interest to the US army. Napoleon is the most successful general in history and Clausewitz attributed this to his '*coup d'œil*'—he uses the French expression in his text. It signifies a sudden intuition that indicates what line of action to choose; the *coup d'œil* is based on what made it possible to succeed in past situations and therefore assumes knowledge of history. It instantaneously makes it possible to find the appropriate ploy for the current situation. Napoleon had reflected on the campaigns of the great commanders.[16] For Clausewitz, *coup d'œil* 'merely refers to the quick recognition of a truth that the mind would ordinarily miss or would perceive only after long study and reflection'.[17]

Recent scientific advances confirm the existence of the 'intuition of the expert'—a kind of sixth sense based not on visionary dreams, but on study of the past and solid knowledge. These discoveries confirm Napoleon's words, his division between the share of calculation and that of chance. The knowledge of what others have done in similar situations is fundamental: Napoleon's intuition is not based on the excessive self-confidence of a 'great man', but on a form of humility whereby he takes account of the experience of others.[18] Napoleon's primary source of inspiration was books.[19] He read an enormous amount in all literary genres: this has been firmly established.[20] Science no longer contrasts analysis and intuition as two separate functions, located in two different parts of the brain. They are so interlinked that it is often difficult to distinguish between them. There is no good analysis without intuition and no good intuition without analysis. They function together in every situation. Scientists now believe that the mind assimilates data through study, that it places them in the short- or long-term memory, and that it selects and combines them in 'flashes of insight'. Some refer to 'intelligent memory',[21] others to 'emotional intelligence'.[22]

Military genius is innate. Napoleon writes thus to Eugène de Beauharnais, who commanded the army of Italy in 1809 and who was in the process of retreating in the face of the Austrians:

> War is a serious game, in which one can jeopardize one's reputation and one's country. When one is reasonable, one must sense, and know, whether one is made for this profession or not. I know that in Italy you affect to scorn Masséna; if I had sent him, what has happened would not have occurred. Masséna possesses military talents that people should bow down before; his faults should be forgotten, because all men have them. In giving you

command of the army, I made a mistake; I should have sent you Masséna and given you command of the cavalry under his orders.[23]

Masséna had the rare talents of taking advantage of the enemy's faults and rapidly making good a failure—something that assumes even more sang-froid, as Chaptal reports:

> Any man, he said, can form a plan of campaign, but few are capable of waging war, because it only pertains to the truly military genius to conduct himself in accordance with events and circumstances. That is why the best tacticians have often enough been poor generals.[24]

Masséna was of modest origins, without much culture, and loved money. But on a battlefield he had the spirit of decisiveness. He was 'used to great events':[25]

> Joubert had a genius for war, Masséna an audacity, an insight that I have seen only in him. But he craved glory and could not bear being deprived of the eulogies he thought he merited.[26]

The whistle of bullets transformed Masséna, who had no equal when it came to leading an engagement or deciding a move. At such times, the leader is everything. Napoleon wrote multiplied sentences on this subject and was not thinking only of him:

> In war, men are nothing; it is one man who is everything.[27]
> The general's presence is indispensable: he is the head, the whole of an army. It was not the Roman army that subjugated Gaul, but Caesar; it was not the Carthaginian army that made the republic tremble at the gates of Rome, but Hannibal; it was not the Macedonian army that was on the Indus, but Alexander; it was not the French army that carried the war onto the Weser and the Inn, but Turenne; it was not the Prussian army that defended Prussia for seven years against the greatest powers in Europe, but Frederick the Great.[28]
> [...] in war, only the leader understands the importance of certain things and, by his volition and superior understanding, can conquer and overcome all difficulties alone.[29]
> For ultimately, whatever one does, whatever the energy displayed by a government, however vigorous the legislation, an army of lions commanded by a stag will never be an army of lions.[30]

Clausewitz developed this idea very clearly: 'As each man's strength gives out, as it no longer responds to his will, the inertia of the whole gradually comes to rest on the commander's will alone. The ardor of his spirit must rekindle the flame of purpose in all others; his inward fire must revive their hope.'[31]

Napoleon's highly centralized methods of command would display their limitations from 1812 onwards, when he had to entrust autonomous armies to certain marshals and they, used to having virtually everything laid down for them, were not capable of assuming such responsibility. The following words, addressed to Marshal Berthier, chief of staff of the Grande Armée and hence his right arm, have been brandished by numerous commentators as evidence of excessive centralization:

> Stick strictly to the orders that I give you; carry out your instructions promptly; let everyone stay on guard and remain at their post. I alone know what I must do.[32]

We find an echo of the final sentence in Marshal Marmont's *Mémoires*. During the night of 12–13 September 1813, he discussed the allocation of commands over the vast theatre of the German campaign, where the allies mobilized several armies. Leaving him, Napoleon said to Marmont:

> The chessboard is very muddled; I alone can make things out.[33]

Berthier was merely a subordinate. The Emperor was his own chief of staff, in the sense that he was alone in preparing and forming his decisions while in the field.[34] If we are to believe General Bonnal,[35] Napoleon had realized this defect in his methods of command, when applied to a large army, when he let slip in front of Count Roederer in 1809:

> It is possibly a bad thing that I command in person, but that is my essence [privilege].[36]

In defence of Napoleon, it might be said that in the revolutionary armies and the Austrian army he had seen the flaws in group decision-making. It encouraged a dilution of responsibility and the hierarchical nature of military structures also resulted in groups lacking in audacity, taking nothing but conformist, routine decisions. It must also be said that the structure in terms of army corps and armies was new; that it was only really understood by Napoleon and, to a lesser extent, Berthier. Even in 1813, writes General Riley, there had not been enough experience or, above all, time to train officers to produce a general staff working as it subsequently developed.[37] But did the Emperor really want to develop military talents other than his own, which were the foundation of his political power?[38]

The qualities of the leader:
more character than spirit

Napoleon proliferated considerations on the qualities required of a general,
above all on Saint Helena. Montholon, Gourgaud, O'Meara, and Las Cases
were the recipients of statements that overlap:

> The first quality of a supreme commander is having a cool head, which
> receives accurate impressions of objects, which never gets heated, does not
> let itself be dazzled, or carried away by good or bad news; the successive or
> simultaneous sensations it receives in the course of a day are ordered in it and
> only occupy the place they warrant. For good sense and reason result from a
> comparison between several sensations taken into equal consideration. There
> are men who, on account of their physical and moral constitution, make a
> picture of everything. Moreover, whatever knowledge, spirit, courage, and
> good qualities they have, nature has not summoned them to the command of
> armies and the direction of major war operations.[39]

Once more, Clausewitz offers some very similar observations. The leader
must have 'an intellect that, even in the darkest hour, retains some glim-
merings of the inner light which leads to truth; and second, the courage to
follow this faint light wherever it may lead'.[40] Such courage in the face of
responsibilities, of moral danger, is '*courage de l'esprit*'—in French in Clausewitz's text—although it is not, strictly speaking, an initiative of the spirit,
but of temperament. Clausewitz also refers to 'determination': its function is
to remove the torment of doubt and the dangers of hesitation. Napoleon
continues in the same register without excessive modesty:

> To be a good general, one must possess knowledge of mathematics. It serves to
> rectify one's ideas in countless circumstances. Perhaps I owe my success to my
> mathematical ideas. A general must never create mental pictures; that is the
> worst thing of all. Just because a supporter has abandoned a post, it should not
> be thought that the whole army is involved. My great talent, what most
> distinguishes me, is seeing everything clearly. It is even my kind of eloquence;
> it is seeing the substance of the issue straight away, in all its aspects. It is the
> perpendicular that is shorter than the oblique.[41]

Napoleon's passion for mathematics dates back at least to the age of 8. He
shone in the subject at school in Brienne. His tastes did not incline him
towards literature, Latin, languages, and ornamental arts. His eminently

practical turn of mind pushed him towards the sciences, which seemed to him necessary for the profession of war he had chosen. At the École militaire in Paris, it was once again in mathematics that he was the best.[42] His library on Saint Helena contained several works in the discipline, including Monge's *Géométrie descriptive* and Laplace's *Exposition du système du monde.*[43]

A good general, he said, must not 'create pictures': he must not allow himself to be easily impressed; he must keep a cool head, for war is made up of dramatic, unexpected events. In other words, the general must have a firm character. Spirit is a different quality, which makes it possible to see clearly in a confused situation—something that approximates to the *coup d'œil*:

> The mind of a general ought to resemble and be as clear as the field-glass of a telescope, *et ne jamais se faire de tableaux*. Of all the generals who preceded him, and perhaps all those who have followed, Turenne was the greatest. Maréchal Saxe, a mere general, *pas d'esprit*; Luxembourg *beaucoup*; le grand Frédéric, *beaucoup*, and a quick and ready perception of everything. Your Marlborough, besides being a great general, *avait aussi beaucoup d'esprit*. Judging from Wellington's actions, from his dispatches, and above all from his conduct towards Ney, I should pronounce him to be *un homme de peu d'esprit, sans générosité et sans grandeur d'âme*. Such I know to be the opinion of Benjamin Constant and Madame de Staël, who said, that except as a general, he had not two ideas. As a general, however, to find his equal amongst your own nation, you must go back to the time of Marlborough, though as anything else I think that history will pronounce him to be *un homme borné.*[44]

Character and spirit must be in balance, as Las Cases reports:

> It was rare and difficult, he said on another occasion, to combine all the qualities required for a great general. What was most desirable, and immediately distinguished someone from the run of the mill, was for spirit or talent to be in balance with character or courage: that is what he called being *broad* at the bottom and top alike. If, he continued, courage was overly dominant, a general would venture recklessly beyond his designs. By contrast, he did not dare to accomplish them if his character or courage was inferior to his spirit. He then cited the *Viceroy*,[45] whose sole merit was this balance, which nevertheless sufficed to make him a very distinguished man.
>
> After this, we spoke much of physical courage and moral courage; and on the subject of physical courage the Emperor said that it was impossible for Murat and Ney not to be brave; but that you could not have less of a brain than them—especially the former.

As for moral courage, he said he had very rarely met with the moral courage of the early hours; that is to say, the courage of the improviser who, despite the most sudden events, nevertheless allows the same freedom of spirit, judgement, and decision-making. He did not hesitate to pronounce that he had found himself to possess the most of this kind of courage, and that he had seen very few people who had not remained far behind him.

After that, he said that people had a very inaccurate idea of the fortitude required to fight, in full awareness of the consequences, one of those great battles on which the fate of an army, a country, the possession of a throne, are going to depend. Also he observed that one rarely found generals in a hurry to give battle: 'They took up position, established themselves, meditated their stratagems; but then began their indecisiveness; and there was nothing more difficult—and yet more precious—than being capable of making your mind up.'[46]

Military genius is a gift from heaven, but the most essential quality of a supreme commander is firmness of character and a determination to win at any cost.[47]

As we saw above, after *coup d'œil*, Clausewitz cites determination as the second quality essential to a general.[48] He also refers to firmness, strength of character, self-control—all qualities evoked by Napoleon:

[. . .] it is will-power, character, application, and audacity that made me what I am.[49]

It is through vigour and energy that one saves one's troops, that one wins their esteem [and] commands respect from the malicious.[50]

The essential quality of a general is strength of character and that is a gift from heaven. I prefer Lefebvre to Mathieu Dumas.[51] Lefebvre had fire in his belly. You see that in the last instance he wanted to defend Paris and he was certainly right; it could have been defended. Turenne did not shine for his spirit, but he had the genius of the general.[52]

Strength of character, vigour, spirit of decisiveness are therefore the predominant qualities—even more necessary than talent and spirit, notions bound up with a capacity for intuition, imagination, and also intellectual training, education, and knowledge:

In war, one must not have so much spirit. The simplest is the best.[53]

Prince Jérôme received the following reply in 1807:

Moreover, your letter contains too much spirit. It is not required in war. What is needed is accuracy, character, and simplicity.[54]

The crux of the matter must be identified. Everything impels the human mind to create too many pictures, to break the rule of what Jean Guitton calls 'the sole essential thing'—all the more so, he says, in that throughout the ages armies divide, segment space and time, teach the allocation of duties, and always tend to sub-divide.[55] During the Seven Years War (1756–63), the leaders of the French army had a great deal of spirit, but were completely lacking in character. Never was this more evident than in the shameful defeat of the Prince of Soubise[56] by Frederick II of Prussia at Rossbach in 1757:

> People did not know how to wage war in those days. In the French armies, there were too many men of spirit, reasoners, discoursers. What was required was a leader of pronounced character to stick to his decisions and mock the clever-witted gentlemen and make himself obeyed. Marshal de Saxe was not a great eagle, but he had character and made himself obeyed. Today, there is not a general, colonel, or battalion leader who would not conduct himself better than Soubise.[57]
>
> Those court folk were the true bane of the French armies. They were only lacking generals; they could not get them. When, by birth and without effort, you have every rank, every favour, you do not need to make much effort to earn them. Doubtless they were brave and more! They wanted to conduct a little campaign and return to Versailles in October. War is not waged like that. It is a tough profession, which demands continuity, constancy, character. There were good second colonels, good majors; only generals were wanting.[58]

This was a major development sealed by the wars of the Revolution and the Empire: winter no longer necessarily interrupted operations.

The supreme commander must possess a moral fibre highly superior to that of the generals subordinate to him. He must be more demanding with himself. The principle applied at all levels:

> It is when generals set an example that subalterns do their duty.[59]
>
> In an army corps, the eye of the chief must find a solution for everything. Commanders, officers, whatever their merits, are in a constant state of insouciance if the leader's presence does not continually make itself felt.[60]

Courage is required—but not just any courage. Napoleon also says this of his brother Joseph, who was not a military man:

> He has courage, but it is the courage of resignation, not activity. He has more courage than is required to die rather than dishonour himself. But this precisely does not involve dying! It is necessary to save oneself and others.[61]

Finally, a leader must be loved by his men:

> The soul of all armies, especially naval forces, is the manifest attachment of all parties to the leader.[62]
>
> Monsieur Mouton, my aide de camp, I want you to speak to Rear-Admiral Allemand: he is too severe; his commanders and officers do not love him and are abandoning him. Try to get him to understand that it is advantageous for the good of the service to be loved.[63]

In fact:

> It is not enough to give orders; it is necessary to make oneself obeyed.[64]

A leader needs friends:

> Friends in war, my dear Bertrand—that is the only thing. Nothing more is needed. Above all, friends—that does duty for lots of things.[65]

Paradoxically, in order to command, 'civilian qualities' are needed:

> Commanding is a civilian matter today. The soldier wants his general to be the wisest and among the bravest. It is through civilian qualities that one commands. One of a general's qualities is calculation: it is a civilian quality; another is knowledge of men: a civilian quality; another is eloquence, not of the jurists, but the eloquence that electrifies: a civilian quality . . .[66]

Portraits of generals

Napoleon had already evoked the difference between spirit and character at the start of the 1812 campaign in Russia, during an interview reported by Marshal Gouvion Saint-Cyr:

> Napoleon said next that the art of war was the most difficult of all the arts; that this was why military glory was generally regarded as the greatest glory; that the services of soldiers were rewarded by a wise government over and above all others; that a general needed spirit and—what was even rarer—a great character. For a term of comparison, he took a vessel and said: 'Spirit is the sails; character is the draught. If the latter is considerable and the masts are weak, the vessel makes little progress, but it withstands the blows of the sea. By contrast, if the sails are strong and high, and the draught low, the vessel can sail in good weather, but will be submerged in the first storm. To sail well, the draught and the sails must be in exact proportion. I have sent Marmont to Spain. He has a lot of spirit. I do not yet know his draught, but shall soon assess it, because he has now been left to his own devices.'[67]

A few days later, on 22 July 1812 Marmont was defeated by Wellington at Salamanca. Hailing from the artillery, Marmont had education. Officers in the artillery and engineers were those who had most. General Bertrand was a brilliant engineer, but he did not thereby possess the qualities to command an army corps:

> In different circumstances, you have demonstrated eminent talents. But war is only waged with vigour, decisiveness, and a constant will; one must not scrabble around or hesitate.[68]

The image of the ship was taken up again on Saint Helena in connection with General Schérer:[69]

> Schérer was wanting in neither spirit nor courage; he lacked character. He spoke of war boldly but vaguely; he was not right for it. A warrior must have as much character as spirit. Men who have much spirit, but little character, are the least suited for war. It is like a ship that possesses sails disproportionate to its ballast; much character and little spirit would be better. Men who have little spirit and a well-proportioned character will often succeed in this profession. One must have as much base as height. The general who has a lot of spirit and character to the same degree is Caesar, Hannibal, Turenne, Prince Eugène, and Frederick.[70]

Jourdan, the victor of Fleurus in 1794, was not without qualities:

> General Jourdan was very brave in the face of the enemy and in the midst of the firing during a day's engagement. But he did not have mental courage in the still of the night, in the early hours of the morning. He was not wanting in penetration and intellectual faculties, but was without determination and was imbued with utterly false principles of war.[71]

But reckless audacity, like that of the Austrian General Beaulieu in Italy,[72] is not required either:

> We have finally crossed the Po. The second campaign has begun. Beaulieu is disconcerted; he calculates poorly and constantly falls into the traps set for him. Perhaps he would like to give battle, for this man has the audacity of fury, not that of genius.[73]

Valour is needed to win battles, but it can have its converse if not accompanied by prudence:

> Lefebvre is the reason for the victory at Fleurus. He is a brave man who does not concern himself with the major manoeuvres made on his right and left, and

who thinks only of fighting well, who is not afraid to die. That is good. But then such people find themselves in a risky position, surrounded on all sides; they surrender and afterwards become eternal cowards.[74]

Education counts. French generals frequently lacked it:

We have no good generals-in-chief in the French army; none of them has sufficient education. They are all naturals; no doubt fire in the belly counts for a lot, but it is not everything. The Austrian general staff is much more learned than ours.[75]

A certain quantity of knowledge was indispensable. The Russian Suvorov was bereft of it:[76]

Marshal Suvorov had the soul of a great general, but not the head. He was endowed with a strong will, great activity, and fearlessness in the face of any trial. But he had neither genius nor knowledge of the art of war.[77]

Murat needed 'luxury, women, a groaning table' on a daily basis:

It is a great fault in an army commander not to know how to control his passions or his tastes; you can cost thousands of men their lives thus.[78]

Napoleon returned to this fault of Murat's in a passage reviewing several generals. In it he first of all reckons that in the various posts of an army the leaders can possess different qualities:

Men are like musicians in a concert; each has his part. Ney was priceless for his valour, his obstinacy in retreats. He was good for leading 10,000 men. Otherwise he was a true idiot. Lannes was, I believe, like him on the battlefield. You will not find anyone who approximates to these two men for valour under fire except Rapp.[79] Murat too was very brave.

Murat understood better than Ney how to conduct a campaign and yet he was a very poor general. He always waged war without maps. During Marengo, I had sent him to take Stradella. He had indeed sent his corps there and it was already fighting, but he had stayed in Pavia to collect a paltry tax of 40,000 francs. I made him leave straight away, but that cost us 600 men. The enemy had to be expelled from a position that we could have occupied before him. How many mistakes Murat made in order to have his headquarters in a château where there were women. He had to have them every day. So I pretty much tolerate a general having a strumpet with him in order to avoid damage of that kind.

My great reputation in Italy is partially due to the fact that I did not pillage and thought only of my army. The role of supreme commander is a very big one. The smallest mistake costs the life of thousands of men.[80]

Murat, Lannes, and Ney were the three bravest men in the army and the bravest are capable of anything, even cowardice, for want of moral courage.[81]

When his brother Joseph was King of Naples, Napoleon told him how to employ his generals on different tasks in accordance with their aptitudes:

> In the profession of war, as in literature, everyone has their genre. Were there to be sharp, extended attacks where a lot of audacity has to be shown, Masséna would be more appropriate than Reynier.[82] To guarantee the kingdom from any raids during your absence, Jourdan is preferable to Masséna.[83]

On Saint Helena, Napoleon completely repented having entrusted military responsibilities to Joseph, especially in Spain. In very animated fashion, he offered this description of their differences when it came to war:

> When Suchet took General Blake[84] and others prisoner, he said that it was impossible for him to be governed by Joseph; that if I wanted to come or give him an able man, he was ready to serve me, but that it was a waste of his time and a pointless sacrifice to serve King Joseph; that he had no character, that he did not want to work, and that a lot of character was required to govern the Spanish—precisely what I possessed. Joseph had no policy. I had written to him a hundred times to summon the Cortès, to call the Cortès to him. He never wanted to, on various frivolous pretexts, [saying] that he did not want to weaken royal authority or some other [thing]. It was necessary to begin by reigning and guaranteeing the throne.
>
> He complained to me that the troops were pillaging.
>
> —But it is you who command them that I shall lay into, I say. It is remarkable that you make this criticism of me, when it is for me to make it of you.
>
> When I was in Spain, Joseph wanted me to combine my operations with him and for me to call him to my councils. He thought he had military talents:
>
> —You mock me, I said to him, I have no councils and do not consult anyone. I go on the road, I question passers-by, prisoners, soldiers, I adopt a course of action and I march. How do you want me to consult you? You have no idea of war.
>
> He thought himself a military man and talented! [. . .] He offered dinners to officers of the Guard and had them granted favours. They flattered him and he believed himself a military man. He had spirit, letters, some very good qualities; but he was no soldier.
>
> —In the end I'm worth more than Masséna, than Lannes?
>
> —You're not worthy to untie their shoe laces. They are heroes. Even the captain at my door—you're not worth him. If I told you to go with ten men onto this height, you would not know what to do to get there. You have never commanded a battalion! War is not a profession of so much spirit—

except for the sublime of the profession. You say: the enemy is there and
I am going to march there. But knowing if the enemy is there, and in what
strength, is already an art. You think nothing of Masséna or Lannes. You are
very different from Alexander.[85] When I travelled by carriage, he brushed
the flies from the figure of Lannes, who was sleeping. He was a hero in his
eyes. He was questioning him all the time: Is that a position? How would
you defend that? How would you attack that? He had the solicitude and
coquetry of a mistress for him. [. . .]
We are two very different men, I said to Joseph. Everything goes to your head;
you must be impassioned. Nothing goes to my head. Were I at the top of
Milan cathedral and precipitated head first to the ground, I would fall calmly
looking around me.[86]

The 'sublime of the profession' in the penultimate paragraph recalls the
'sublime parts of war' referred to by Marshal de Saxe, who distinguished
them from the 'the detail'.[87] We shall return to this distinction in Book II,
Chapter 2.

The naval general and the land general

Are the same qualities required to command at sea as on land?

The commander-in-chief of a naval army and the commander-in-chief of a
land army are men who need different qualities. People are born with the
qualities required to command a land army, whereas the qualities required to
command a naval force are only acquired with experience.
Alexander[88] and Condé were able to command from their earliest youth;
the art of land war is an art of genius, of inspiration. But neither Alexander nor
Condé would have commanded a naval force at the age of 22. There nothing
is genius or inspiration; everything is positive and experience. The naval
general requires but one science—that of navigation. The land general
needs them all or a talent tantamount to all of them: a talent for taking
advantage of all experience and knowledge. A naval general does not have
to divine anything; he knows where his enemy is and knows his strength.
A land general never knows anything for certain; never sees his enemy clearly;
never positively knows where he is. When the armies are face to face, the
slightest accident of terrain, the smallest wood, can hide part of the army. The
most practised eye cannot say if it sees the whole enemy army or only three-
quarters. It is with the mind's eye, through a whole process of reasoning, by a
kind of inspiration that the land general sees, knows, and judges. The naval
general only needs a practised glance; none of the enemy's forces are hidden

from him. What makes the profession of the land general difficult is the need to feed so many men and animals. If he lets himself be guided by administrators, he will not move and his expeditions will fail. The naval general is never troubled; he takes everything with him. He has no reconnaissance to carry out or terrain to examine or battlefield to study. The Indian Ocean, the American seas, the Channel—it is always plain liquid. The more skilful will have an advantage over the less skilful solely by their knowledge of the prevailing winds in such and such waters, by forecasting those that will prevail, or by atmospheric signs: qualities that are acquired by experience, and solely through experience.

The land general never knows the battlefield where he has to operate. His *coup d'œil* is one of inspiration; he has no positive intelligence. The data for knowledge of the locale is so contingent that one learns virtually nothing by experience. It is a skill first of all to grasp the relationships between terrains depending on the nature of the regions. Ultimately, what is called the military *coup d'œil* is a gift received by the great generals from nature. However, the observations that can be made from topographical maps, the aptitude provided by education, and the habit of reading these maps can be of some help.

A naval supreme commander depends more on the commanders of his vessels than a supreme commander on land does on his generals. The latter can take direct command of his troops, to take himself to all points and rectify false moves by others. The naval general personally has influence only over the men on the vessel he is on. Smoke prevents signals being seen. Winds change or are not the same over the whole area covered by his line. Of all professions, it is therefore the one where subalterns must take on most themselves.[89]

Napoleon identified a key difference between those who see action and others. He profoundly respected all those who took risks and had nothing but contempt for 'desk job types':

I prefer, said His Majesty, a good artillery commander, who knows how to exploit the terrain to position his pieces and is brave on the battlefield, to all the officers in charge of workers, artillery parks, etc. The former has fire in his belly, cannot be bought, whereas all the rest can be bought. I think, continued His Majesty, the same of engineering. A good officer of engineers is one who wages war, conducts sieges and defences of fortified places, and who knows how to adapt to each terrain the fortification suitable to it. Certainly, Haxo or Rogniat would have constructed a fortification better than Fontaine.[90] The former are military men; the latter is merely a mason. War alone furnishes experience. Carnot would not have written his system if he had known the effects of the cannonball.[91] True nobility lies in the one who advances in the direction of the firing. I could have given my daughter to a fighting officer; I would never have given her to an artillery park officer. The latter is an organizer, an administrator. Masséna only found his spirit amid the firing and dangers. Only then did he make good dispositions. I loved Murat because of his shining valour and that is why I forgave him so many idiocies. Bessières was a good cavalry general, but rather cold. He had a lack of the warmth that Murat possessed to excess. Ney was a man of rare valour. During the siege of Danzig, Lefebvre only wrote me idiocies at first. When the Russians disembarked, he was in his element again and his reports were those of a man who sees clearly. In France, we shall never want for men of spirit, makers of plans, etc. But we shall lack men of character and vigour—ultimately, men who have got fire in their belly.

[. . .] What Nelson possesses which the building engineer does not is not something acquired: it is a gift of nature. I accept that a good director of an artillery park is very useful, but it is repugnant to me to reward him as one does those who have shed their blood. For example, I appointed Évain a general despite myself.[92] You are the reason for it. I cannot bear a general who advances in an office. I know one needs generals who have not seen a fuse burn. But it repels me. [. . .] I only esteem those who wage war.[93]

4

On Danger in War

Anxieties and dangers

On the eve of battle, men experience particular feelings. Napoleon found them well expressed in the *Iliad*:

> Homer must have waged war: he is true in every detail of his battles. Everywhere we have the very image of war. In the night before the engagement at . . . ,[1] I think I am on the eve of Jena and Austerlitz.
>
> It is the same anxieties about the great event that is about to happen, the feelings that disturbed him, and which are experienced by all military men. The terrain is always to be found there. It is painting the truth.[2]

Immediate danger is what soldiers must face. The leader sometimes has to look further ahead:

> [. . .] this is because men think only of avoiding a present danger, without bothering about the impact their behaviour might have on subsequent events; this is because the impression of a defeat is erased from the mind of most people only gradually, over time.[3]

Danger in war is a factor of equality between men:

> [. . .] nothing is more conducive to equality than war, where everyone shares the same fate and runs the same risks.[4]

The supreme commander must demonstrate that he is exposing himself to danger. Napoleon wrote as follows to his brother Joseph, who was setting off to command the army of Naples:

> Do not listen to those who would like to keep you far away from the gunfire; you need to prove yourself. If opportunities arise, expose yourself conspicuously. As to real danger, it is everywhere in war.[5]

Napoleon genuinely adopted this line of conduct. No monarch of his time exposed himself to gunfire as much as he did. On the vessel that took him to the island of Elba, he offered this explanation to his entourage:

> Few men, he said, have exercised greater influence over the masses than me. But only stupidity would make me say that everything is written on high and that, were a precipice to be placed in front of me, I should not turn away to avoid falling down it. My belief is that of any reasonable being and in war, where the danger is pretty much equal everywhere, one should not quit a place that is known to be dangerous to go and position yourself where death can also reach you and resign oneself to the fate of one's situation. Confirmed in this thought, you must become master of your courage and sang-froid, which is communicated to the men under your command; the most gutless of them will do themselves the honour of courage.[6]

The dangers encountered at sea are not the same as on land:

> In general, land war swallows up more men than maritime war; it is more dangerous. The naval soldier fights but once in a campaign; the land soldier fights every day. Whatever the fatigues and dangers attached to the sea, the naval soldier experiences many fewer than the land soldier: he never suffers hunger or thirst; he always has his lodging, his kitchen, his hospital, and his pharmacy with him. In the armed services of France and England, where discipline maintains cleanliness and experience has revealed the measures to be taken to preserve good health, naval armies have fewer sick than land armies. Independently of the danger of the engagements, the naval soldier faces the peril of storms. But art has so reduced it that it cannot be compared with those of land, such as popular riots, piecemeal killings, and surprise attacks by enemy light troops.[7]

During his long military career, Napoleon personally ran the risks of war. He suffered no serious injuries, but had several brushes with death. At the siege of Toulon, he was wounded in the forehead (15 November 1793), thrown to the ground by the force of a cannonball (16 December), and pierced by a bayonet or polearm thrust into his left thigh (17 December).[8] On 23 April 1809, in front of Ratisbon, he was struck in the heel by a bullet. Although not deep, the wound was very painful.[9] The following 6 July, at Wagram, a shell exploded in front of his horse. Another grazed General Oudinot beside him.[10] At the Battle of Arcis-sur-Aube, on 20 March 1814, the Emperor indicated various emplacements to be captured and a shell fell just in front of a company. To show them he was afraid of nothing, Napoleon pushed his horse forward onto the shell and kept it stationary near the smoking projectile. It exploded and the horse was ripped apart, and collapsed amid the smoke with its rider, who emerged unscathed and got on to a new mount.[11] This list of episodes is not exhaustive. In total, Napoleon had eighteen horses killed or wounded under him.[12] On Saint Helena he discoursed on the risks run by generals:

> There are men who are lucky in war. From the Empire onwards, Murat was never wounded,[13] and every day made jabs at the advance guard. Ney was never wounded. Masséna was never wounded. I wasn't dangerously wounded when I was hit in front of Ratisbon. That light contusion gave me a fever. Had it been necessary to give battle that day, I would probably not have been what I was in other circumstances. Such is man: an indisposition affects his head.
>
> I had my leg stretched out in the palace of Archduke Maurice in Ratisbon when I received a delegation from Switzerland. I was in a bad mood. I was suffering.
>
> In our battles, the generals are much more exposed than they used to be. When people fought hand to hand, they could only be wounded close up. Alexander does not seem to have been. Today, a girl of 15, dressed as a hussar, will kill a hero, the most vigorous of men. Artillery spares no one. There are scarcely any battles where I haven't had some people killed in my group. When I occasionally approached an entrenchment in a moment of crisis, it was not pleasant for the person in command: 'Does the Emperor think we require his presence to do our duty?'
>
> The value of a soldier is not that of a captain; the value of a captain is not that of a divisional commander; the value of the division general is not that of a supreme commander. If the latter is killed or wounded, the day's outcome will certainly change or, at any rate, probably change. He must not therefore expose himself gratuitously.[14]

The last sentence qualifies or clarifies the advice given above to Joseph.

Death and mourning

One of the worst moral dilemmas faced by Napoleon in war involved the plague victims in Jaffa in March 1799. A situation of this kind can be met with in any era of history and puts one in mind of certain American Westerns and war films. Given the impossibility of transporting the soldiers stricken by the plague, and in order to avoid them falling into Turkish hands and suffering appalling tortures, General Bonaparte probably had poison administered to them, which resulted in a painless death, as Bertrand recounts:

It must be remembered that it was a question of not leaving them prisoners in the hands of the Turks, who in their remaining twelve hours of life would have cut them into pieces, applied molten lead to them . . . etc. Had it been my wife or son, I would have behaved similarly if I could not take them with me, because the first principle of charity is to do to others what we would have done to ourselves [. . .]. On this one must consult not civilians, but soldiers. Ask the 53rd. They would speak with one voice.[15]

The British 53rd infantry regiment was responsible at the time for guarding the prisoner of Saint Helena. The appeal to the opinion of soldiers of another nationality testifies once more to Napoleon's proximity to all members of the profession. His lack of hesitation over taking the decision at Jaffa will always be open to discussion. He did not concern himself with the prevailing morality, or the teaching of the Church, when adopting the course of action that he sincerely believed to be least painful for his men.

Death in battle is an outcome that can be anticipated by any soldier. It is not part of our intention to study Napoleon's views on the subject in depth, but there was unquestionably a certain fatalism about him.[16] As we have already seen in some quotations, and as we shall see later, while his way of waging war did not spare men, he sincerely believed rapid operations invariably avoided greater suffering. Frequently confronted with the death of men close to him, he sometimes gave vent to his compassion. We know the deep impression made on him by the spectacle of the battlefield of Eylau after the terrible clash of 8 February 1807. He had these words to say:

A father who loses his children savours no charm in the victory. When the heart speaks, even glory has no more illusions.[17]

He wrote to Josephine:

My friend, I am still at Eylau. The ground is littered with the dead and wounded. This is not the best part of war; one suffers and the soul is oppressed at the sight of so many victims.[18]

On 26 June 1813, at Dresden, Napoleon showed a different side of himself to Prince Metternich. If we are to believe the Austrian diplomat's memoirs, the Emperor, in a moment of anger it is true, shouted:

I have grown up on the battlefields and a man like me hardly concerns himself about the lives of a million men.[19]

On Saint Helena, he gave it to be understood that he had ended up being accustomed to frequenting death:

> It is quite true that the idea of God is a very natural idea. At all times, among all nations, people have had it. But one dies so quickly; in war I have seen so many people die immediately and pass so rapidly from the state of life to that of death that it has made me familiar with death.[20]

On another occasion, when speaking of a book by the naturalist Buffon, he explained:

> What he says about death is good. It is not to be feared, because five-sixths [of men] die without suffering and those who seem to be in agony suffer little, for those who have recovered have no memory of it. So the machinery is disrupted and pain is not felt as acutely as people believe, because it leaves no traces. Charles XII, it is said, carried his hand to his sword when a cannon ball or bullet struck him dead.[21] So the pain was not such as to deprive him of the desire to defend himself; it was not extreme.[22]

We owe to Napoleon some of the most beautiful letters of condolence that have ever been written:

> Your nephew Elliot has been killed on the battlefield of Arcola. This young man had familiarized himself with weaponry; he marched at the head of the columns several times; he would have made an admirable officer one day. He died with glory in face of the enemy; he did not suffer for one moment. What reasonable man would not envy such a death? Who, amid the vicissitudes of existence, would not want to leave a world that is so often despicable thus? Who among us has not regretted a hundred times not being shielded thus from the impact of calumny, envy, and all the odious passions that seem well-nigh exclusively to govern the conduct of men?[23]
>
> Your husband was killed by cannon fire, while fighting alongside it. Without suffering, he died the gentlest death, the one most envied by soldiers.
>
> I feel your pain acutely. The moment that separates us from the object we love is terrible; it isolates us from the earth; it causes the body to experience the convulsions of agony. The faculties of the soul are destroyed; it retains its links with the universe only through a nightmare that changes everything. In this situation we feel that, if nothing compelled us to go on living, it would be much better to die. But when, after these initial thoughts, we press our children to our heart, tears and tender feelings revive nature and we live for our children. Yes, Madame, you will cry with them; you will raise them in childhood and cultivate their youth; you will speak to them of their father, of your grief, of the loss they have suffered, of that suffered by the Republic. Having attached your soul to the world through filial love and maternal love,

appreciate the friendship and keen interest I shall always take in my friend's wife as something. Be persuaded that he is one of the men, few in number, who warrant being the hope of grief, because they feel the sorrows of the soul acutely.[24]

This piece of eloquence, remarkable for its humanity, evinces a profound sensitivity. Rarely have such appropriate words been found to console a loss. Napoleon understood the pain of others. On several occasions, he is known to have cried after a battle. He also made a major contribution to developing a rhetoric of military heroism and glorious death. If the bloodbaths of the First World War have doubtless destroyed the credibility of this kind of discourse in Western Europe for ever, we must not be anachronistic. In the late eighteenth and early nineteenth centuries, life was so hard and life-expectancy so short that death in battle could seem like an outcome which, if not enviable, was at least acceptable.

5

On Physical Effort in War

Illnesses, fatigue, and privations

At the end of his first Italian campaign, General Bonaparte wrote:

> My health is completely ruined and health is indispensable and cannot be replaced by anything else in war.[1]

Four days after the Battle of Eylau, Marshal Lannes, who was ill, received these words:

> Augereau was so ill that he could not mount a horse; he wanted to be there out of zeal. But in war one needs health, since it is necessary to remain on horseback part of the night in order to know one's business. So turn your thoughts to getting cured, so that you can resume your command in a fortnight.[2]

As we know, in the field Napoleon could sleep when and where he wished and did not need much sleep. It was during the night that he took his major decisions. With the years, however, fatigue won out and he no longer rode on horseback as much at night:

> A general must work mainly at night. If he tires himself needlessly during the day, he is overcome by tiredness. At Vitoria we were beaten because Joseph slept too much.[3] Had I slept the night of Eckmühl, I would not have carried out that fine manoeuvre, which is the best I ever did.[4] With 50,000 men, I beat 120,000. I multiplied myself by my activity. I woke Lannes up by kicking him, he was so fast asleep. A commanding general should not sleep. Ah, my God! Perhaps the rain of 17 June had more influence than is thought on the loss at Waterloo. Had I not been so tired, I would have been on horseback all night. What can seem the most minor events often have the largest outcomes.[5]

From the 1812 campaign onwards, Napoleon felt the fatigue of war much more. He confided to Caulaincourt in the sledge that brought him back from Russia:

> I am becoming heavy and too fat not to love rest, not to need it, not to regard the movement and activity required by war as a great burden. Like other men, my physique necessarily has an influence on my morale.[6]

War was synonymous with fatigue and privations, especially since the stores could not keep up with the pace of the rapid marches inaugurated with the first Italian campaign and it was therefore necessary to 'live off the land', making do. Veterans knew it full well when the First Consul addressed them on the day the constitution of Year VIII was adopted:

> There is not one of you who has not waged several campaigns, who does not know that the most essential quality of a soldier is to be capable of steadfastly putting up with privations.[7]

During a session in Paris one day, Napoleon said to the state counsellors:

> Messieurs, the profession of war is not a bed of roses; you know that here in your seats only after reading the reports or accounts of our triumphs. You do not know our bivouacs, our forced marches, our privations of all kinds, our sufferings of every variety. I know them because I see them and sometimes I share them.[8]

For Clausewitz, '[i]f no one had the right to give his views on military operations except when he is frozen, or faint from heat and thirst, or depressed from privation and fatigue, objective and accurate views would be even rarer than they are. But they would at least be subjectively valid, for the speaker's experience would precisely determine his judgement.'[9]

Physical effort in war is such that one should not engage in it beyond a certain age. In January 1813, having returned from Russia, the Emperor exclaimed in front of Count Molé:[10]

> Ah! Believe me, M. Molé, from the age of 30 one begins to be less suited to waging war. Alexander died before sensing any decline.[11]

Conscripts were called up at the age of 20 to serve for five years:

> It would be appropriate to extend the term of service to ten years—i.e. to the age of 30—with five years in the army and five years in the reserves. It is from 25 to 30[12] that a man is at full strength; it is therefore the most propitious age

for war. Soldiers must be encouraged by all means to remain under the flag; we shall achieve this by holding old soldiers in high esteem [...].[13]

Caring for the men

On several occasions Napoleon recommended to his officers that they should take good care of their men:

> The supreme commander warns the officers in command of detachments that he will bring those whose solicitude is not focused on all the men under their orders before a council of war. We must wait for dawdlers and the job of a leader is to march with his full complement intact.[14]
>
> You must ensure that the corps of your army are properly clothed and well administered. One should not always rely on brigade leaders, but see the soldier and, if he has complaints to make, render him justice.[15]
>
> Recommend to Marshal Jourdan once again that he attend to the health of the troops. That is his most dangerous enemy.[16]
>
> To this end, as I did in the army of Italy—and he must remember it— General Marmont must establish small, well-ventilated, and healthy depots for convalescence, where he will direct anyone leaving the hospitals of Dalmatia, and from there send them in detachments of 100 to Cattaro and Ragusa by sea. [...] It is only by constantly concerning oneself with such minor forms of aid that one prevents the destruction of an army.
>
> [...] Half of the art of war consists in rapidly concentrating one's army, sparing pointless rushing and, consequently, the health of the soldier.[17]
>
> If there are tired men in these various columns, send them to a convalescent hospital which you will set up in Berlin and leave them there for a week. That way, one saves men; one prevents illnesses.[18]
>
> Take care that the recruits who pass are well treated at Mont Cenis and have wine. You will look them over on their arrival in Turin and allow them, if necessary, to rest for one, two, or three days when they are tired. It is with such treatment that one preserves a soldier's health; two or three days' rest for a tired man prevent an illness.[19]

In the Napoleonic era, soldiers more often died from illness than wounds received in battle.[20] The need for troops to camp in healthy places was a constant concern of Napoleon:

> Above all else, I recommend the troops' health to you. If they are located in unhealthy places, the army will dissolve and be reduced to nothing. This is the first of all military considerations.[21]

It is better to fight the bloodiest battle than to locate one's troops in an unhealthy place.[22]

Stick to the spirit of the order I have dictated and dispense with the letter. A day or two's distance is nothing when it is a question of the troops' health; of healthy regions above all. What use are sick men, with whom one can do nothing when the enemy presents himself?[23]

6

Intelligence in War

'Military genius' was based on an enormous quantity of information assembled and stored in the memory to aid decision-making. Napoleon constantly kept himself informed whether in Paris, or in the field, or visiting.[1] He loved doing this and took a particular pleasure in knowing the exact state of his armed forces:

> The good condition of my armies derives from the fact that I concern myself with them on a daily basis for an hour or two, and when each month I am sent inventories of my troops and fleets—which amounts to twenty thick booklets—I abandon every other occupation to read them in detail, in order to identify the difference between one month and another. I take more pleasure in this reading than a young girl takes in reading a novel.[2]

Knowing what the enemy is doing

It is always difficult to form an idea of the enemy's conduct. The theoreticians, who reason too abstractly about what the enemy should do in accordance with their interests, are to be distrusted. At Schönbrunn in July 1809, shortly after the Battle of Wagram, Napoleon let fly at Jomini, who had come to give an account of the operations in Galicia:

> Here we have the mania of Messieurs the tacticians; they assume that the enemy will always do what he should do! But if things were thus, one would not dare go to bed at war, since that is the best time for the enemy to attack an army.[3]

It is always very difficult to assess the numerical strength of the army facing you:

> The hard thing in war is to know the enemy's strength. It is the instinct for war that supplies that. There is no mind in war, especially given the rapidity

with which it is fought today. That was good, at the most, when people stayed
in the same position for two months. Prince Charles [of Austria] took eight
days to do what I did in a quarter of an hour.[4]

At Ligny, there were as many opinions on the opponent's strength as there
were people. Brave little Gourgaud said there were only 10,000 men. Wanting
to know what cavalry force I had in front of me, I asked Bernard:[5] 'How many
of them are there?'—'6,000 men, Sire, you have 6,000 men, six squadrons at
most, the enemy is clearing out.'—'Bernard, said Gourgaud, sees the first line,
but in the wood, remember, there is a second line.' Everyone judges according to his character, his possibilities.

At Bautzen, when I took a look at the battlefield the following day, when
the enemy had been depleted by 30,000 men, we were very uncertain if the
enemy was weaker. Some people saw fewer positions; others saw the same
number of fires. In short, we could hardly make out anything. That is why the
profession of general is so difficult. This is the difference between land and sea.
At sea you know what you are dealing with. On land you never know.[6]

However that may be, one takes one's decisions depending on the enemy:

> In war, you adopt your course of action in the face of the enemy. You have
> always got the night to yourself to prepare. The enemy does not take up position
> without your knowing it. But you should not calculate theoretically what you
> want to do, because that depends on what the enemy is doing and will do.[7]

Knowing one's opponent is essential in war. But sometimes his reputation
does not correspond to reality, as General Bonaparte told the Directory
à propos of Archduke Charles of Austria after he had fought him in Italy:

> Prince Charles has hitherto manoeuvred worse than Beaulieu and Wurmser;[8]
> he has made mistakes at every stage—and extremely crude ones. This has cost
> him dearly, but it would have cost him much more if his reputation had not
> impressed me to a certain extent and prevented me from convincing myself of
> certain faults I spotted, assuming they were dictated by views which did not in
> fact exist.[9]

What emerges from this complex sentence is a crucial aspect of war: the
constant interaction and retroaction between opponents—in this instance,
as regards the reputation of one of the most renowned supreme commanders of his time.

Napoleon indicates how he informed himself of the enemy's position in
the field:

> One cannot know the enemy's position in large-scale manoeuvres—only
> when he is [in] camps, in places where he stays for several days, never when

he is marching. Or at least [one does not know it] through espionage then.
[...]

I learnt about Wurmser's attack on Corona in my carriage, returning from
Brescia to Peschiera with Josephine.[10] Noticing the carriage, a soldier sus-
pected that it was the general and said to me:

—Ah! You are here, so much the better! People are waiting for you; people
are asking for you. You are well, march on!

—Where?

—To Corona—we're fighting there.

—You don't know what you're saying.

—Oh! I was wounded at the crack of dawn; we're fighting hard; it's hot
there.

I enter Peschiera. I go to find General Guillaume:[11]

—So, they're fighting in Corona.

—No.

—There's no doubt about it. Have you not seen this soldier?

—Oh! People say the same thing every day.

Thinking that the general had been informed, the soldier had crossed the town
and continued on his way. He had been wounded and, having got himself
bandaged, he re-joined his depot.

Wounded soldiers and sutlers often travelled four to five leagues and thus
provided news. I have often learnt a lot of news thus.

I set up my headquarters at a junction on a road and questioned all the
passers-by. True espionage is as follows: 1. Interrogating prisoners and desert-
ers is the best way. They know the strength of their company, battalion,
regiment in general, the name of the commanding general, even the division
general, the places where they have slept, the route they have taken. Thus one
learns to know the enemy army; 2. Peasants and travellers. One profits:
travellers always pass; some arrive. 3. The letters that one intercepts, especially
if they are by an officer on the general staff, then they are important. But the
ones that are usually captured are by troop or outpost commanders, who do
not know much. When you can get hold of an inventory of the army, as in the
case of Czernitchev,[12] no doubt that is good, but rare.

Hence my habit of having a bivouac fire lit on a road, at the end of a bridge,
at the entry to a village, at the crossroads, and questioning.[13]

Those in command at sea cannot glean information in the same way:

[...] as on land, it would have to be the general himself who could observe
the enemy first. But at sea an admiral can never leave his army, because he is
never sure of being able to re-join it once he has left.[14]

However, it was in connection with the British fleet, of whose dispersion he
was aware, that Napoleon, at a time when he still thought he could invade

the British Isles, wrote to his Navy Minister a sentence that also applies to land war:

> Nothing vouchsafes more courage and clarifies ideas more than clearly knowing the position of one's enemy.[15]

Interrogating prisoners

The interrogation of prisoners had to be systematic:

> Every day you must have prisoners taken by your outposts and in all the directions that threaten the enemy: that is the way to get news of the enemy. There is no other effective way.[16]

When he had not yet gone to Spain to take matters in hand, the Emperor made the following observations on the way to obtain intelligence in 1808:

> We have no intelligence as to what the enemy is doing. It is always said that we cannot have information, as if this position was extraordinary in an army, as if spies were ordinarily to be had. In Spain, as everywhere else, parties must be sent to abduct now the priest or mayor, now the head of a convent or the postmaster, especially all the letters, now the head of the staging post, or the person who performs those duties. You put them in custody until they talk, by having them interrogated twice a day. You hold them hostage and order them to send foot passengers and supply news. When we know how to take strong and severe measures, we shall have information. All posts, all letters, must be intercepted.
>
> For the sole purpose of having information, one might create a large detachment of 4,000–5,000 men who, presenting themselves in a large town, take the letters from the post, seize the most affluent citizens, their letters, papers, gazettes, etc.
>
> There is no doubt that even on the French line all the inhabitants are informed of what is happening; a fortiori outside the line. So what is to stop us from seizing prominent figures, carrying them off, and subsequently returning them without maltreating them? When one is not in a desert but a populated area, if the general is not informed, it is because he has proved incapable of taking the requisite measures. The services rendered by the inhabitants to an enemy general are never performed out of affection or even for the sake of money. The most genuine services are obtained in return for safeguards and protection, for preserving someone's property, life, town, monastery.[17]

Three days later, a new note on Spanish affairs took up virtually the same arguments. Waging war above all consisted in obtaining intelligence:

We must finally wage war—that is, have news via priests, mayors, heads of convents, the main property-owners, the post. Then we shall be perfectly informed.

The reconnaissance to be carried out every day around Soria, Burgos in Palencia, and Aranda can daily furnish three intercepted posts, three reports by arrested men, who shall be treated well and released when they have provided the desired intelligence. We shall then see the enemy coming; we shall be able to concentrate all our forces, rob him of the marches, and fall on his flanks just as he is meditating a plan of attack.[18]

It will have been noted that in these two texts Napoleon demands that the men under interrogation should not be maltreated. There was a good way of interrogating that succeeded in extracting intelligence. In June 1809, Eugène de Beauharnais still had to improve in this respect:

It is unheard of for a brigadier-general captured ten days ago not yet to have arrived with me. You think that you have profited to the utmost and interrogated him. You are wrong: the art of interrogating prisoners is one of the fruits of experience and tact in war. What he told you seemed of little interest to you. If I had interrogated him, I would have prised out the maximum intelligence about the enemy.[19]

A notion likewise employed by Clausewitz, 'tact' refers to an intellectual capacity acquired by practice. It facilitates the most appropriate reaction in any new or urgent situation. This mental faculty is no mere irrational impulse or product of extraordinary inspiration. Instead, it stems from reflexive experience.[20] Intelligence is obtained with 'tact', never by maltreating prisoners—not even in Egypt:

The barbaric custom of beating men accused of having important secrets to reveal must be abolished. It has been recognized from time immemorial that this way of interrogating men, by subjecting them to torture, yields no positive results. The victims say anything that comes into their heads and anything they realize one wants to know. As a result, the supreme commander prohibits the use of methods condemned by reason and humanity.[21]

As for the enemy, he must not obtain intelligence. We know the extent to which Napoleon controlled the press. Several critics criticize him for not communicating his plans to generals, but he knew how readily men speak. He made secrecy a veritable mode of command. At the start of the Russian campaign, he wrote to his brother Jérôme:

No one must be taken into your confidence, not even your chief of general staff.[22]

7
Friction in War

Accidents, circumstances, delays

In Clausewitz, friction refers to the set of imponderables, chance events, and accidents that always prevent war from corresponding to what was planned. This aspect was particularly stressed by Napoleon:

> From triumph to collapse is but one step. In the most important circumstances, I have seen that a mere nothing has always determined the greatest events.[1]
>
> One must have waged war for a long time to conceive it; one must have undertaken a large number of offensive operations to know how the least event or sign encourages or discourages, determines one operation or another.[2]
>
> [...] in war, uncalculated events can happen every fortnight.[3]
>
> [...] such is the outcome of battles—they often depend on the smallest accident.[4]

Analysing Clausewitz, Alan Beyerchen explains these sentences of Napoleon. Friction refers to the way that causes which are minor to the extent of being imperceptible can be amplified in wartime to the point of producing disproportionately major effects; and the latter are not open to being anticipated. Interactions at all levels within armies, as between adversaries, amplify micro-causes to the extent of generating unanticipated macro-effects. The precise information required to anticipate such effects is impossible to obtain.[5] When waging war, it is necessary to give chance its due:

> All the naval expeditions undertaken since I have been head of the government have always failed, because the admirals have double vision and have found—God knows where—that one can wage war without taking any chances.[6]
>
> In all affairs, something must be left to circumstances.[7]

This does not prevent the role of chance being reduced to a minimum through the foresight of the commanding general, as Murat found himself reminded in March 1808:

> If I take so many precautions, it is because my habit is not to leave anything to chance.[8]

We shall return to the 'military genius' discussed above, which calculates the role of chance exactly. One of the most important manifestations of friction in war is delays. The time factor is essential:

> Loss of time is irremediable in war; the reasons given are always bad, because operations fail solely on account of delays.[9]
> In politics, as in war, the lost moment never comes again.[10]
> In a combined war like this, days are of great importance.[11]

The last sentence was written during the campaign of autumn 1813, when the combined war reached its highest degree of complexity, for this time the numbers were such that the Emperor entrusted several army corps to a marshal, creating the echelon of an 'army'. Meanwhile, although in overall command, he placed himself at the head of a 'strategic reserve' able to come to the aid of one of his marshals' armies. It is superfluous to add that this development only increased the degree of friction in operations. The expression 'combined warfare' had appeared in the second half of the eighteenth century to refer to the 'elevated parts of war'—those that are not reducible to a geometrical approach, but which appeal to the dialectic.[12]

In Clausewitz, friction also refers to the confusion generated by noise and rumour. The 'fog of war' refers not so much to a lack of information as to the way that distortion and an overload of information create uncertainty about the true situation.[13] One should not lend credence to all the intelligence obtained or—above all—rely on the first rumour. Here too it is a question of tact. Napoleon tried to get those of his subordinates who were less used to major responsibilities—his brother Joseph in 1808, Generals Lauriston and Arrighi de Casanova, Duke of Padua,[14] in 1813—to understand this:

> In war, spies and intelligence count for nothing. It would be staking men's lives on very flimsy calculations to rely on them.[15]
> You go too quickly and get alarmed too soon; you lend too much credence to every rumour. More calm is required in the conduct of military affairs, and before lending credence to reports, they should be discussed. Everything spies

and agents say, if they have not seen it with their own eyes, is nothing; and
even when they have seen it, it is often not much.[16]

One should not lightly cry alarm: one should not let oneself be scared by
chimeras and one must have more firmness and discernment.

Write to the Duke of Padua that he gets alarmed too easily, and that he is
too quick to accept all the false rumours circulated by the enemy. That is not
how a man of experience should act: one must have more character than
that.[17]

Clausewitz generalizes what Napoleon says: 'Many intelligence reports are
contradictory; even more are false, and most are uncertain. What one can
reasonably ask of an officer is that he should possess a standard of judgement,
which he can gain only from knowledge of man and affairs and from
common sense.'[18]

Seizing the moment

Friction can be overcome by the sense of timing that is indispensable in the
war leader:

An army always has a head. When the head is absent, someone commands in
his place. He cannot be instructed not to fight if his foragers are attacked. He
must indeed support them. War is an affair of the moment.[19]

It is possible to exploit friction, as Las Cases reports:

He said that war was composed solely of accidents and that, although obliged
to comply with general principles, a leader could never lose sight of anything
that might put him in a position to exploit these accidents. The vulgar would
call this luck, but it is simply the attribute of genius . . .[20]

All great events only ever hang by a thread. The skilful man makes the most
of anything, neglects nothing that might give him extra chances; sometimes by
scorning a single one, the less skilful man causes everything to fail.[21]

We are back with military genius. Napoleon confided that he had long
meditated on the interaction between chance and historical circumstances:

I have always liked weighing up the chances that are mixed up with some
events.[22]

A great general will seek to take advantage of the disproportionate effects
and unforeseen situations created by non-linearity.[23] Jean Guitton refers to a

general who in a sense summons up the aleatory because he possesses the faculty of the counter-aleatory in his genius.[24] It is necessary to seize the moment, because it is possible that it will not come again:

> Make the most of fortune's favours, when her caprices are for you; be afraid lest she change out of spite: she is a woman.[25]
>
> Success in war, he continued, is so bound up with the *coup d'œil* and the instant that the Battle of Austerlitz, won so completely, would have been lost had I attacked six hours earlier.[26]

Once again, it is a question of tact. One must learn to see clearly in the 'fog of war':

> Before making a move, it is necessary to see clearly; and it is because I saw that you were acting too rapidly, before having seen your enemies' plans develop, that I forbade my troops to leave Hanau. Experience will teach you the difference between the rumours spread by the enemy and reality. Never, in the sixteen years I have commanded, have I issued a countermand to a regiment, because I always wait until an affair is mature and I understand it before starting manoeuvres. My troops will only leave Hanau when I know what they have to do.[27]

In April 1809, Prince Eugène abandoned northern Italy to the Austrians too readily. One should not pre-empt the enemy by ceding to him what he is not demanding. The enemy does not know one's exact situation. Here friction represents an advantage and it must not be squandered:

> Piava was a good enough line for you to have tried to hold it. The Austrians are so unaccustomed to waging war thus that they were astonished that you did not hold the line of Livenza, which was a good rallying line for you. So they cannot believe that you have abandoned Piava. In war, one sees one's own woes, but not those of the enemy; you must display confidence.[28]

Adapting orders to circumstances

Taking friction into account, Napoleon, to a much greater degree that has been said, counted on his lieutenants' ability to make decisions in the light of circumstances. Several orders issued while campaigning tend in this direction:

> Moreover, as events require it, I shall give you all the instructions you will need, not doubting that, in all circumstances, you will act in conformity with the spirit of the war we are waging.[29]

General, at the distance at which you find yourself, and with your rank, you must sense that it is not a literal order which will make you move, but the mass of events.[30]

Monsieur le Duc de Feltre, your instructions for the march of the Barcelona convoy are too precise. General Baraguey d'Hilliers[31] must be allowed to alter it in the light of circumstances. Recommend to him that he have the convoy arrive at Barcelona, but that he choose the moment well. Five battalions, which do not amount to 2,000 men, might not be sufficient to go to Barcelona. When they come from the ministry, orders of this kind need to be given vaguely and circumspectly. My troops must be spared and situations where they are put in jeopardy avoided. It would be an irreparable loss if these five battalions of good troops were to experience a setback.[32]

In 1812 in Russia, the distances between the components of the Grande Armée were greater than ever. Thus Berthier had to make Jérôme, King of Westphalia, who commanded a group of three army corps and a cavalry corps, understand that unduly precise orders could not be issued:

Inform him that the letter you wrote him on the 10th makes my intentions sufficiently clear; but that we are so far apart that it is now for him to manœuver, depending on the circumstances, in the general spirit of his instructions.[33]

Berthier wrote the same thing a little later to Marshal Macdonald:

The Emperor simply cannot give you positive orders, but only general instructions, because the distance is already considerable and is going to become even greater.[34]

Conclusion to Book I

Clausewitz took part in four campaigns conducted by Napoleon: those of 1806, 1812, 1813, and 1815. In 1806, Clausewitz was a captain and aide de camp to Prince August of Prussia. In 1812, having resigned from the Prussian army, he served as a lieutenant-colonel on various Russian general staffs. He was a Russian liaison officer with Blücher's general staff in 1813 and then chief of staff of a multi-national corps of 22,600 men. He only donned Prussian uniform again in 1815, as colonel and chief of staff of the 3rd corps.[1] He only found himself facing Napoleon in four major battles: Moskowa (Borodino), Lützen, Bautzen, and Ligny. Yet at the end of this first book one is struck by the similarity between their considerations on the character of war. It is a serious thing that involves shedding blood. It must not mask its true identity. While civilized nations have introduced certain rules, the notion of war as such is in tension with that of humanity. In itself, war is not a 'namby-pamby' business, but in practice, Europe no longer fights with its erstwhile ferocity. However, that can always resurface, as was the case in the Vendée. In the circumstances in which France found itself under the Consulate and Empire, Napoleon recognized that his regime needed military success to maintain itself. These were circumstances bound up with the upheavals of the Revolution, but Napoleon's personality was itself attuned to the restlessness of war.

More than Clausewitz, Napoleon was concerned with the relations between war and law because he was a head of state in search of legitimacy, but also no doubt because he had retained something from his jurist father and his Latin culture. He based himself on law to justify the looting of a town taken by storm, or the ban on surrendering in open country. He was revolted by the liberties taken by the British and by the legal vacuum that prevailed on the high seas. Clausewitz did not broach all these dimensions, but approximated to Napoleon as regards 'military genius' and the qualities of the supreme commander. The latter must calculate a good deal and, at the

same time, give chance its due. Strength of character is more important than
mental subtlety, even if the latter can prove precious. A leader must be able
to decide amid uncertainty and demonstrate superior moral courage—the
courage of 'the early hours'. For the British general Riley, who fought in
Iraq and Afghanistan in the early twenty-first century, the essence of
Napoleon's conception of generalship is still valid: in a military organization, the general is the man capable of grasping the problematic involved in
its entirety; of defining what is liable to determine the decision; of implementing that decision and altering the situation advantageously so as to win.
Riley also reckons that Napoleon's intuitive decision-making is more
appropriate than ever to contemporary wars: a general who blindly follows
procedures, and allows himself to be submerged by the flood of intelligence
generated by contemporary information systems, would be a prisoner of
these rigid processes and schemas supposedly capable of foreseeing everything. Removing instinctive leadership means relying exclusively on technology and pure reason. In a competitive activity like war, where survival is
at stake, that is not only against nature, but extremely dangerous.[2]

War is full of danger. In it death is on the prowl and one has to become
accustomed to this. It involves fatigue and privations. Napoleon certainly
did not do enough to relieve his soldiers' sufferings—what leader can be
sure of having done the maximum?—but his correspondence attests to his
efforts in this direction, particularly in connection with encampment in
salubrious places. Intelligence is indispensable to war, since it can only be
waged in line with what the enemy does. Intelligence is obtained with tact
and assessed with 'a standard of judgement, which [one] can gain only from
knowledge of men and affairs and from common sense'.[3] War is full of the
unforeseen and nothing goes according to plan. The skilful leader must
exploit such accidents and grasp the favours of fortune, because they do not
recur. Experience teaches one to see clearly in the fog. 'In war the experienced soldier reacts rather in the same way as the human eye does in the
dark: the pupil expands to admit what little light there is, discerning objects
by degrees, and finally seeing them distinctly. By contrast, the novice is
plunged into the deepest night.'[4]

BOOK II

The Theory of War

N apoleon invented a new practice of war. He was the first to be capable of leading large masses of men towards a single, precise objective: putting enemy armies out of action in record time. For this man of action, the theory of war was not the main concern, even though he had read a few books on the subject in his youth. Count Jacques de Guibert was the foremost theoretician at the time. In his *Essai général de tactique* (1772), he stressed speed in operations and the possibility of separating the army into several corps during marches and then concentrating them at the point where it was necessary to give battle. Napoleon re-read Guibert on Saint Helena and made some comments on other theoreticians. He had no particular taste for classifications and definitions, but distinguished between different levels in the art of war.

I

Classifications in the Art of War

Tactics, grand tactics, major operations

On Saint Helena, Napoleon offered the following definition of tactics, as reported by Gourgaud:

> Lagrange[1] said to His Majesty one day that he believed that tactics were the art of getting people who did not want to fight to fight. HM replied that tactics were the art of getting a large number of armed individuals to act simultaneously to the greatest advantage, as quickly and obediently as possible.[2]

For Clausewitz, tactics was the theory of 'the use of armed forces in the engagement', while strategy was the theory of 'the use of engagements for the object of the war'.[3] Peter Paret has stressed that Clausewitz's definitions were not descriptive, but functional: they took account of the end and means and were applicable to all periods and all types of armed conflict.[4] Narrower in scope, Napoleon's definition of tactics is likewise functional. End and means figure in it. Tactics imply order. They call on intelligence, knowledge, and organization:[5]

> Two mamluks faced down three Frenchmen, because they were better armed, better mounted, more expert, had two pairs of pistols, a musket, a carbine, a helmet with a visor, a coat of mail, several horses, and several men on foot to serve them. But 100 French cavaliers did not fear 100 mamluks; 300 did not fear 400; 600 did not fear 900; and ten squadrons routed 2,000 mamluks—such is the major influence of tactics, order, and movements! The cavalry generals Murat, Lasalle, and Leclerc presented themselves to the mamluks in three lines and with a reserve. Just as the first was about to be outflanked, the second advanced by squadron to the left and right into battle; the mamluks stopped short to outflank this second line, which, as soon as it was extended by the third, charged them. They could not withstand the shock and scattered.[6]

Tactics were taught in arrangements called 'schools', corresponding to a type of unit:

Cavalry has more need of order, tactics, than infantry. Moreover, it must be able to fight on foot, be practised in the school of squads and battalions.[7]

Like Guibert and Jomini, when he started out, Napoleon used the expression 'grand tactics' to refer to what the latter subsequently called major operations—that is, joint moves by the major units of an army, divisions then corps, in pursuit of a single objective. These moves are carried out before and after battle, which they bring about and whose outcome they make it possible to exploit. A continuum is therefore established between marches or manoeuvres, engagements and pursuit. This sequence was Napoleon's essential contribution to the art of war. In the twentieth century, the level of grand tactics assumed the title of operative or operational art.[8] The Empire's marshals were often very good on the battlefield, but they frequently did not understand grand tactics. Marshal Lannes might perhaps have achieved such an understanding:

> He had had little education; nature had done everything. He was superior to all the generals of the French army on the battlefield when it came to manœuvring 15,000 men. He was still young and would have improved; perhaps he would even have become skilful at grand tactics.[9]

Napoleon denigrated General Moreau because he was his rival for glory and then his political opponent. Moreau had beaten the Austrians at Hohenlinden on 3 December 1800 more decisively than the First Consul at Marengo on 14 June. We should therefore distrust the following judgement:

> Moreau had no system, either in politics or in military affairs. He was an excellent soldier, personally brave, capable of moving a small army on a battlefield well, but absolutely lacking in any knowledge of grand tactics.[10]

We know that at the start of the Consulate Napoleon was not yet strong enough politically to impose his views, even in military matters. He had to leave the command of the principal army—that of Germany—to Moreau. He could not prevent Moreau and his lieutenants from letting slip opportunities that he himself would have seized had he been present. If we are to believe the Saint Helena conversations, the First Consul often said:

> There it is. They knew nothing more; they did not know the secrets of the art or the resources of grand tactics![11]

Questions of grand tactics cannot be reduced to rules. To stop the invasion of one's country, how is the capital to be defended? Should the army cover it directly, manoeuvre, or enclose itself in a defensive camp, subsequently falling on the enemy's flank from there?

> After the business of Smolensk in 1812, with the French army marching straight on Moscow, General Kutuzov covered that city by successive manoeuvres, until, having arrived at the defensive camp of Mojaisk, he held firm and accepted battle.[12] Having lost it, he continued his march and crossed through the capital, which fell into the hands of the victor. If he had withdrawn in the direction of Kiev, drawing the French army towards him, he would have had to protect Moscow with a detachment, he would have been further weakened. Yet nothing would have stopped the French general from having this detachment followed by a superior detachment, which would have forced it too to evacuate this important capital.
>
> Similar questions addressed to Turenne, Villars, or Eugène of Savoy would have greatly embarrassed them. But ignorance entertains no doubts about anything;[13] it seeks to resolve a problem of transcendental geometry with a second-rate formula. All these questions of grand tactics are indeterminate physico-mathematical problems, which cannot be solved with the formulae of elementary geometry.[14]

The first edition of this text, read by Clausewitz, contained the following variant for the penultimate sentence:

> To be dogmatic about what one has not practised is the prerogative of ignorance: it is to believe it possible to resolve by a second-rate formula a problem of transcendental geometry that would make Lagrange or Laplace quail.[15]

In the final reprise of the arguments against General Rogniat's book, this edition added the name of the Swiss mathematician Leonhard Euler:

> Alexander, Hannibal, Caesar, Gustavus Adolphus, Turenne, Prince Eugène, and Frederick the Great would all have been very ill at ease if they had had to decide on this question—a problem of transcendental geometry, which has a large number of solutions. Only a novice could believe it simple and easy: Euler, Lagrange, Laplace would spend many nights before formulating it as an equation and clearing away the unknowns.[16]

Clausewitz slightly confuses the names of the scientists—we cannot exclude the possibility that he consciously refused to cite the French among them— but indicates that he had read this edition: 'Bonaparte rightly said in this connection that many of the decisions faced by the commander-in-chief resemble mathematical problems worthy of the gifts of a Newton or an Euler'.[17]

From 1807 onwards, Jomini called the book he had initially entitled *Traité de grande tactique* (1805–6) *Traité de grandes opérations militaires*.[18] Napoleon rarely made the same substitution, but we do hear it from his lips in Moscow in 1812:

> The Duke of Reggio [Oudinot] is brave on a battlefield, but the most mediocre and incapable general there is. [Gouvion] Saint-Cyr is a superior man, but systematic; he only wants to see the point where he is, whereas everything must be connected in a system of major operations like these.[19]

Strategy

Napoleon only used the term strategy on Saint Helena, mainly when commenting on a book by Archduke Charles of Austria, who employed the word:[20]

> Prince Charles's work on strategy, which I believed I would read with little interest—I hardly bother with scientific words and cannot care less about them—has, however, greatly interested me, for I have read the three volumes in a day. I have skimmed over lots [of things], because I scarcely understand what the Archduke means: the distinction between strategy and tactics, between the science and the art of war. These definitions are bad. Those given by Jomini in a note are better, though still mediocre. Strategy, he says, is the art of moving troops and tactics the art of engaging them. It would be better to say: strategy is the art of plans of campaign and tactics the art of battles.
>
> This can serve to open people's minds and give them a few ideas, but it cannot teach war. There is too much intellect: there shouldn't be so much in war. Perhaps these principles are today's. War is an affair of the moment: what was good at midday is no longer so at two o'clock. What can books teach about that?
>
> Castiglione is an object of admiration.[21] If there is something to praise, it is the course I adopted of lifting the siege [of Mantua], abandoning its execution, and concentrating my troops. Wurmser's conduct was simple. He was ordered to get the siege of Mantua lifted. He believed I wanted to attack Mantua; he marched on Mantua. I spotted the storm and sensed I could only avert it with difficulty. I abandoned the siege, marched on Peschiera's column, beat it, and returned. Instead of that, I could have marched on Montechiaro's column. I would have beaten and repelled it, but by evening I would have been encircled and my victory would simply have led me to my doom. Instead of that, I did a very simple thing: my communications were threatened by

Peschiera's column and I marched to ensure my rear. An imbecile would have done that. There is nothing to particularly admire in it and the 'spirit' that would have led me to attack Montechiaro would have hastened me to my doom.

How to teach these distinctions of the hour, the moment? The fact is this: without doubt I beat the enemy without so much intellect and without using Greek words. I did not do so at Leipzig, but I could have if I had cavalry and known the enemy's movements.[22]

This extract from Bertrand's *Cahiers* is transcribed from the original manuscript. Fleuriot de Langle's edition omits several sentences, in particular where Napoleon offers his definitions: 'strategy is the art of plans of campaign and tactics the art of battles'. This passage, hitherto unpublished, can be compared with Clausewitz's definitions, referred to at the start of the chapter. The division between the two spheres obeys the same criteria. Tactics concerns the use of armed forces in engagements, strategy the use of engagements for the war's objective.[23] Like Napoleon, Clausewitz links strategy with planning: the strategist 'will draft the plan of the war, and the aim will determine the series of actions intended to achieve it: he will, in fact, shape the individual campaigns and, within these, decide on the individual engagements'.[24] Employed since 1771 by the Frenchman Joly de Maizeroy as a synonym for grand tactics, the word strategy only began to become established in 1799 with the Prussian Dietrich von Bülow.[25] Archduke Charles then consecrated it and Jomini introduced it into France. In October 1819, Napoleon added the following point about the same book by Archduke Charles:

> What is historical is valuable and well done. There are some analyses—it is a veritable history. As to strategy, I do not understand much about it. I do not know what a strategic position is, however much I attend to it. I do not even understand the word strategy very well. Jomini provides a clearer definition in his note, when he says that tactics are the art of directing troops and masses on a battlefield and that strategy is the art of directing troops in the moves of a campaign. That is what used to be called grand tactics. Jomini's book is good.[26]

2

On the Theory of War

The desire to theorize war

For Napoleon, some aspects of war can be subject to theoretical considerations, but this is much less true of the 'elevated parts'. Clausewitz was of the
same opinion in *Vom Kriege*, recalling that the first books on the theory of
war concerned attacking and defending fortified places—that is, the sphere
of engineering and artillery, where the material aspects were predominant.

In the first edition of Napoleon's *Mémoires*, he could read this passage:

> Supreme commanders are guided by their own experience or by their genius.
> Tactics, movements, the science of the engineer and the gunner can be learnt
> in treatises, rather like geometry. But knowledge of the elevated parts of war is
> acquired solely through experience and by studying the history of the wars and
> battles of the great commanders. Does one learn to compose a book of the
> *Iliad*, a tragedy of Corneille, in a grammar?[1]

Here we once again encounter the distinction between the material and
immaterial parts referred to in Book I in connection with genius for war.
The first corresponds to the training received by Napoleon at Brienne,
Paris, and Auxonne, where he acquired a profound knowledge of all the
material details of war: manufacture of powder, permanent fortification,
firing of guns, infantry drill, etc.[2] The second refers to reading and experience. His reading included a history of the Seven Years War by the
Welshman Henry Lloyd. A veritable embodiment of eighteenth-century
cosmopolitanism, the latter was at once a spy and an officer in the French,
Prussian, Austrian, and then Russian armies. He too distinguished between
'two parts; one mechanical, [which] may be taught by precept. The other
has no name, nor can it be defined, or taught. It consists in a just application
of the principles and precepts of war, in all the numberless circumstances,
and situations, which occur; no rule, no study, or application, however
assiduous, no experience, however long, can teach this part: it is the effect of

genius alone.'[3] This classical Enlightenment conception is met with in most military thinkers of the time, like the Frenchman Bourcet and the Prussian Scharnhorst.[4] Clausewitz made a similar distinction: 'Instructions exist for internal operations; there are instructions on how to establish camp, how to leave it, how to employ trench instruments, etc. But there are no instructions for how to conduct a campaign, give battle, or assemble a machine.'[5]

The Second Empire edition of the correspondence adopted the original manuscript of Napoleon's text and offered a slightly different version. The introductory sentence no longer figures. The end of the paragraph is also altered:

> Tactics, developments, the science of the engineer and the gunner can be learnt in treatises, rather like geometry. But knowledge of the elevated parts of war is acquired solely by studying the history of the wars and battles of the great commanders and through experience. There are no precise, defined rules. Everything depends on the character with which nature has endowed the general, his qualities, his defects, the nature of his troops, the range of his weapons, the season, and the thousand circumstances that mean that things are never the same.[6]

Knowledge of military history furnishes the '*coup d'œil*' of the military genius mentioned in Book I with content. What Napoleon has to say about the *coup d'œil*, determination, and historical examples is confirmed by modern research on strategic intuition. The latter can be defined as 'the selective projection of past elements into the future in a new combination as a course of action that might or might not fit your previous goals, and the personal commitment to work out the details along the way'.[7]

As early as 1807, Napoleon wanted to see military history being studied at the Collège de France:

> Next would come the history of French military art. The professor would convey the different plans of campaign adopted in the various epochs of our history, either to invade or to defend; the origin of our successes, the cause of our defeats, the authors and memoirs in which the factual details and the evidence of the outcomes are to be found. This part of history, a subject of curiosity for everyone, and so important to military men, would be of the greatest utility to statesmen. At the special school of engineering, the art of attacking and defending fortifications is demonstrated. One cannot demonstrate the art of war on the grand scale, because it has not yet been created, even though it can be. But a chair in history, which would make it known how our borders have been defended in different wars by the great commanders, would be bound to have very great benefits.

[. . .] Without this, military men will, for example, not possess the means to learn to make the most of the errors that have caused reverses and to appreciate the dispositions that would have averted them. The revolutionary wars in their entirety could be rich in lessons, and to garner them it is often necessary to employ unnecessary diligence and prolonged research. This is not because the facts have not been written down in detail, because they have been everywhere, but because no one is concerned to facilitate research and to provide the leadership required to do it with discernment.[8]

Much more so than theory, history and experience are required to know 'the art of war on the grand scale'. Scharnhorst said the same thing.[9] We shall return to the value of historical examples in Chapter 6 of the present book. In November or December 1799, the young First Consul confided one evening at the Luxembourg Palace:

I have fifty pages to write on the art of war which, I believe, would be new and useful.[10]

If the possibility of theorizing tempted him from time to time, this did not prevent him from reminding his brother Jérôme in 1807:

War is learnt only by advancing in the direction of the firing.[11]

If Jean Colin is to be believed, Napoleon read relatively few military authors in his youth—essentially Feuquière, du Teil, Guibert, Lloyd, and Bourcet.[12] But he read other ones throughout his life, especially on Saint Helena, as is attested by several comments.[13]

Machiavelli, Folard, Maurice de Saxe

Before broaching the authors criticized by Napoleon, it should be known that he stated as a matter of principle that he did not owe anyone anything in any respect.[14] Thus, his reading notes are generally negative. In Machiavelli's *Art of War*,[15] he found

Little worthy of note. [. . .] Machiavelli spoke of war without having waged it and was careful to refuse a command offered to him by a prince of his time.[16]

Chevalier Jean-Charles Folard was the most discussed military writer of the first half of the eighteenth century. A veteran of Louis XIV's last wars, he had written *Commentaires sur Polybe* (1727), in which he advocated infantry attacks in columns rather than lines, because for him shock had more impact than fire. He drew inspiration from the tactics of the Greeks and Romans.[17] On Sunday, 12 January 1817, Napoleon had this to say about him:

> It is easy to write and digress on the wars of the Ancients and the Moderns. I would like to have Folard[;] I would pulverise him in four words. My reflections on the war of the Ancients and Moderns cannot be the same as those of a lot of people. There are some decisive things.[18]

As is well known, Folard's *Commentaires sur Polybe* nurtured the tactical reflection of one of the Empire's best marshals, Davout.[19] Napoleon meant here that Folard, who had taken part in the War of Spanish Succession, was manifestly dated and that, were he himself to get down to writing a military work, he would leave those of others far behind. But the comment reported by Bertrand also signifies that Napoleon did not at the time (January 1817) possess Folard's *Commentaires*. Dr Antommarchi's memoirs, although to be distrusted on many points, corroborate those of Bertrand. On 25 September 1819, he said that Napoleon had just received a crate and hoped to find Folard's *Polybius* in it:

> If only I had *Polybius*! But perhaps it will come to me by a different route.[20]

The book arrived only a few months before the death of the Emperor, who read it, notably on 12 March 1821.[21]

Marshal de Saxe's *Rêveries* were still read by officers of the Grande Armée:[22]

> Marshal de Saxe's *Rêveries* occupy military men every day. But Marshal de Saxe was not educated. For a start, he did not know the language; he was neither an engineer nor an artilleryman.[23]

In a letter to Marshal Berthier in early 1812, Napoleon had shown rather more consideration for the man and his work:

> My cousin, among many extremely mediocre things, Marshal de Saxe's *Rêveries* contain some ideas about how to tax enemy countries without tiring the army that seemed good. Read them and put their content in an instruction for my generals in Spain.[24]

Frederick II, Guibert, Lloyd, Bülow

Napoleon was not highly appreciative of Frederick II of Prussia's famous *Instructions* to his generals. Gourgaud recounts:

> I said to him that I have Frederick's *Instructions*. [His Majesty] read them and said that they were charlatanism.[25]

The Emperor added shortly afterwards:

> There are some good things in those instructions. But they were written in haste and not in sufficient depth. [. . .] Frederick did not want to say everything; he left a good deal vague in his instructions. He could have done better, but did not want to.[26]

Which book, precisely, is being discussed?[27] Gourgaud's journal provides no more than inventories of the Saint Helena library to clarify things for us. No doubt it was a genuine text by the king and not the *Secret Instructions* wrested from him, which were in fact attributable to the Prince of Ligne.[28] Frederick II's *Instructions* to his generals were widely diffused in the Grande Armée. Even a rather crude and uncultured general like Boussart had a copy.[29]

Since General Colin's interesting study of Napoleon's military education, which unfortunately is lacking in precise references, Count de Guibert's *Essai général de tactique* is reputed to have prophesied and inspired Napoleonic war. Guibert 'was bound to satisfy Napoleon's youthful enthusiasm. Everything led him to read Guibert, which he found everywhere. There was not a single public or private library that did not possess the *Essai général de tactique*.'[30] Napoleon did indeed read Guibert.[31] But it is going too far to say that he rated him highly.[32] He granted a pension to his widow and was a friend of his nephew, who was his aide de camp in Egypt, where he was killed. On Saint Helena, he said nothing positive about Guibert. When Gourgaud read him one of Guibert's works—probably the *Essai général de tactique*[33]—he prompted this remark from the Emperor:

> It is a spirited work, but utterly Prussian; it wanted to introduce the bastonnade among us. In Paris, one finds men of spirit under the paving stones, but not men of war. Guibert was a braggart.[34]

In his notebook for August 1818, General Bertrand notes a conversation on Guibert's *Tactique*—a book deemed 'empty' by the Emperor, but the author was only 24.[35] Recent re-editions of Guibert's works all mention, as Napoleon's one and only judgement, this laudatory sentence: 'The *Essai général de tactique* is a book for forming great men.'[36] In truth, it is drawn from the preface to the re-edition of Guibert's military works in 1803. The exact text runs as follows: 'Frederick II included the *Essai général de tactique* among the very small number of books that he advised a general to read; this judgement is a eulogy in itself. Washington made this book the companion of his glory; and Bonaparte, who carried it with him in the camps, has said that it was a book for forming great men.'[37] All the historians who see Guibert behind Napoleon rely on this passage, to which they attribute 'evidential value' because it could not have been published without the First Consul's authorization.[38] That is possible. But no one has yet explained why Napoleon, who had access to that same 1803 edition on Saint Helena, explicitly disavowed this passage written by Bertrand Barère, of whom Robespierre said: 'As soon as a piece of work presents itself, Barère is disposed to take it on. He knows everything, he knows everyone, he is right for everything'.[39] As for Napoleon, he said precisely this in November 1818:

Guibert composed his *Tactique* at the age of twenty-four, which explains everything. Barère said in his prosopopoeia of Guibert that Washington and General Bonaparte had Guibert's *Tactique* in their pockets and thus that they owe him some of their success. He is completely wrong! [...]

Guibert writes well; he has vague ideas, spirit, nothing positive, nothing to teach. People carry too much baggage [he writes]. Okay. How much is required? He does not say. He proscribes firms [for supplies] and prefers regimental [suppliers]. There is a case to be made for this, but it has to be argued, not dictated. It is an important issue. I employed regimental [suppliers] and I had an excellent man, M. Maret.[40] But I am not persuaded that it is a good thing. [...]

Guibert does not know the history of Frederick or Caesar. He says that Caesar, going to Africa, did not take food supplies with him and that such is the characteristic of a great man. This serves only to convey false ideas of war. Why not carry food supplies with you? [...]

Guibert says that the Romans were not concerned about food supplies. He is wrong. These details are often missing in the history. [...] Guibert says that the soldier must fast. This is an idiocy. He cannot fast; he must live like country folk: potatoes, meat, etc. He has to eat.

He [Guibert] seems to mean: give me an army to lead and you shall see how I do it. He would probably have led it badly. He had too much spirit. Yet there is so much vagary in nature that it may be he would have been a company commander. I have looked up several things in his book and found nothing positive in it. Appointed to the Council, Guibert sent into the regiments second colonels, majors—something that would have done damage to the artillery and engineering, which had consulted.[41] They were refused in the artillery. A lot of officers would have been killed. The court folk wanted to take our places as in the line[42] [...]

There is too much spirit in Guibert's book. There should not be so much of it in war, but good sense instead. You have to stick to your path. Above all, it is a profession of good sense. He [Guibert] says that formerly people did not know how to march; that war is in the legs. What does that mean? That no positions are required?

[He says] that oblique order was invented by Frederick, when it has always existed and battles have only ever been won by means of oblique order.[43] This is a pile of ill-digested stuff. Good sense indicates that when one wants to attack a point, one should bring all one's strength to bear against it; and the simplest men who have ever attacked have done so.

He [Guibert] wants fortified places in each province. What does a province, which is a political district, have to do with places, which are a physical thing? Why should it be a place per province rather than per diocese or department?[44]

One cannot compare wisdom and folly. Yet Kolin's move has been presented as a masterpiece.[45] With his fanatical cult [of Frederick II] and knowledge of the [German] language, Guibert seeks to make it a miracle. One cannot abuse the people one is addressing in more ridiculous fashion.[46]

The judgement seems final. Nevertheless, Guibert was possibly the most perceptive military writer of his time. He understood that future wars would witness a mixture of the strategic and tactical levels.[47] Nor should it be excluded that Napoleon, to enhance his own stature, diminished Guibert and what he owed him.

Henry Lloyd wrote a history of the Seven Years War that enjoyed some success.[48] Napoleon read it, but was also very critical of it:

> I have never read a more mediocre book or seen so many false ideas brought together. He wants to arm the infantry as a pike party. He shortens the musket by twelve inches, rendering the third rank's fire null and reducing the weapon's range and especially its accuracy. He places the pikes in the fourth rank, so that half the soldiers do not shoot; thus the battalion loses half of its strength. [...]
>
> According to Lloyd, cavalry is useless, except for mounting guard and patrolling. It is useful before, during, and after battles. Without cavalry, infantry is carried off—for example, at Nangis, Champaubert,[49] [?], etc. Cavalry is the only way of making the most of victory. [...] He strips the cavalryman of the musket or carbine. A very dangerous system: how will the cavalry guard itself in its bivouacs [?].[50]

On 13 April 1817, Napoleon characterized Lloyd as mad: he claimed that the musket was no good, when it was the best weapon that had ever existed, especially with its bayonet.[51] As for the Prussian Bülow, Napoleon read his history of the 1800 campaign on Saint Helena. A British officer had lent him the book. The Emperor made some notes in it in pencil, all of them very harsh on the author: bad, very bad, absurd. The British officer published it all and thought it fitting to select this annotation of Napoleon's as an epigraph:

> If you wish to learn how to beat a superior army with an inferior army, study this writer's maxims; you will glean some ideas on the science of war. He prescribes the opposite of what should be taught.[52]

Jomini

Jomini was Marshal Ney's chief of staff in mid-August 1813, when he changed sides and offered his services to the Russians. At the time, Napoleon wrote as follows to Cambacérès, Arch-Chancellor of the Empire:

> Jomini, chief of staff of the Prince of Moskowa, has deserted. He is the one who published some volumes on the campaigns and whom the Russians long ago bought. He succumbed to corruption. He is a military man of little value, but a writer who has grasped some sane ideas on war. He is Swiss.[53]

Once Jomini became the Tsar's aide de camp, Napoleon did not register fewer errors on the part of the allies:

> The enemies did not display talent at Dresden,[54] which proves that Jomini, albeit useful, might well not have grand ideas about war.[55]

However, in their Saint Helena journals Bertrand and Gourgaud report several favourable references to Jomini. Gourgaud speaks first:

> I say that I think that we were wrong to let a book like Jomini's be printed.[56] His Majesty looked at me and said: 'Isn't it true that it is a highly singular work?'—So much so, sire, that I attribute the successes of our enemies to that work.' HM answered me animatedly: 'I assure you that I had not read it when I waged [the] Ulm, Austerlitz and Jena campaigns. But it is truly astonishing: one might believe I had followed his advice.'[57]
> Jomini elaborated on my system well; in that he did me a bad service.[58]
> In Jomini, there are plenty of good things.[59]

Napoleon primarily regarded Jomini as a historian of Frederick II's campaigns and often agreed with his critical analysis:

> There are good things in Jomini's book. I regret not having consulted it.[60]

On another occasion, the Emperor explained himself on that point:

> Looking at Frederick's battles today, I can understand how people could have the idea they did of the Prussian cavalry. Previously, I had but a slight knowledge of Frederick's campaigns. Had I studied them more, it would have been very useful to me in the Prussian campaign.
>
> I assume I had put Jomini in my cabinet for that, so that he could point out the battlefields to me. But he was unable to tell me.[61] He had some difficulty explaining himself, so I did not pay too much attention to him. I had so little time to give! The result was some discontentment. He [Jomini] clearly understood my system as early as Castiglione. He made some good observations on Dumouriez's campaign.[62]

However, Jomini made some errors in his history of the first Italian campaign:

> Jomini wrote a good book on Italy, with good intentions. But it was full of mistakes. Initially, I wanted to point them out to him, but there were too many. In general, he follows the Austrian reports more than ours. He does not have enough faith in our accounts and believes them exaggerated; he is wrong. He knows Rivoli and la Favorita. But he does not know Arcola and Castiglione. [...]
>
> In general, the tone is good. He writes as a gracious man who expresses his doubts when he does not know.[63]

Napoleon wanted to make some notes to correct certain of Jomini's mistakes:

> That will still be good, in case my Italian campaigns do not appear. By identifying errors, I will get people to read Jomini and it is good that he should be known.[64]

Jomini enabled Napoleon to rediscover the name of opponents:

> His book is very enjoyable, because it contains the names of all the enemy generals whom I did not know or had forgotten.[65] [...]
>
> It is good that Jomini wrote his book: it makes my campaigns well known. It is to be hoped that he will finish, because he possesses enemy memoirs and materials that I cannot have. I have also forgotten a lot of things. The events and several dates are confused in my head and I have corrected them several times in line with Jomini's book.[66]

On 26 April 1821, a few days before his death, the exile of Saint Helena asked who could write the history of his age. He thought of his brother Lucien and of Jomini:

Lucien should abandon his poetry and concern himself with composing a history of the Revolution and my reign; being a worker, he could easily construct fifteen or twenty volumes around that. But who will do the military history? Hitherto Jomini is the only one who has shown talent, but he is currently sold to Russia.[67]

In sum, as we can clearly see, Jomini was much more highly regarded as a historian than a theoretician.

Rogniat and Marbot

Napoleon was very hard on General Rogniat's *Considérations sur l'art de la guerre*:[68]

This book contains a large number of incorrect ideas liable to cause a regression in military art.[69]

The 'Dix-huit notes sur l'ouvrage intitulé *Considérations sur l'art de la guerre*' were dictated for publication. In them, Napoleon sharply criticized Rogniat. Bertrand's journal contains some initial, more nuanced reflections by the Emperor, which are followed by equally sharp criticisms:

Rogniat's work is the one that struck me most. It is the work of a man of spirit who possesses knowledge and merit. But it is not the book of a lieutenant-general of engineers. It is good for a commander who simply says what he likes, without any consequences, a general-in-chief who has commanded, but not a lieutenant-general of engineers, from whom precise things about their profession are required. He has dreams and should not. There are too many fortifications [he says]. That is not the issue. Should they be demolished? To be in a position to judge his system, he should have said: the border has places here and here; it should only have them here and here. Then one could compare the two systems. He wants small places that could be guarded by small company, with redoubts in between. He wants four forts around a place. Why four rather than two? Isn't that enough to support an army? He thinks that an army can withdraw and take shelter, not from the cannons, but the fusillade and hail of bullets and the cavalry. On that he has some sane, wise ideas. He is familiar with the ancients' manner of camping, conducting themselves in battle, and fighting, their weapons, their weight, their dimensions. It would be very useful to us for the book we are composing.[70]

He criticizes Austerlitz and Jena. He does not know them.[71] He says that Marengo was a fine manoeuvre, but that I did not invent it and Hannibal was the first to perform it. It is not clear what Hannibal has in common with the

Army of Reserve's campaign, when I cut the communications of Melas's army, positioned myself between Milan, Austria, and him, and forced him to surrender. Did Hannibal cut off Scipio or anyone else? These criticisms are tiresome on the part of a lieutenant-general of engineers who should concern himself with his own compass and level.

In general, this book has affected me painfully. [...]

It is an extravagant work on the part of a general officer who commanded at Dresden and Waterloo. He writes under the influence of bitterness and ambition, or perhaps some fault, or some sense of fault, if it is patent, and wants to escape it. Is it a finished work? An ordered one? To denigrate Jena, Essling, nearly all my battles! At Waterloo, to enhance the glory of Wellington! By a man who commanded the engineers at Waterloo! That a man who has not commanded a company[72] should witter on about war is understandable, [but] to speak of his system on artillery, cavalry, advance guards to people who have been practising the profession for twenty years is revolting. [...]

What idiocies preached as marvels by a man without experience! This work is shameful for him. These are not interesting 'musings', like those of Marshal de Saxe. It is shameful for the national character to see a lieutenant-general who commanded at Waterloo erecting such a critical monument against me and glorifying Wellington.[73]

Marbot's refutation of Rogniat pleased Napoleon enormously:[74]

This is the best book I have read for forty years—the one that has given me most pleasure. He [Marbot] was not able to have sight of my work.[75] I will happily entrust it to him for a second edition. There are things that he says better than me. He knows them better, because, at bottom, he was more of a corps commander than me.

Then there are the major parts, where he does not reach me. He does not know that. The artillery is greatly inferior. But when he speaks of companies formed in one row and a battalion formed in three rows with three companies, then he really fights, then he is on his ground. I am always of his opinion. He has led me to his. In one passage, he says that one cannot be a corps commander of 9,000 men; one can be their general—but that's another thing. Whoever says corps commander, says a man who goes into all the details, knows all the individuals.[76]

Books have their uses. To Marmont, who commanded Alexandria in Egypt, Bonaparte wrote:

Carefully re-read the regulation on the servicing of places under siege: it is the fruit of experience; it is full of good things.[77]

Gourgaud reports a conversation in exile on 16 December 1816:

> The Emperor spoke to me of weapons in antiquity and reckoned that in the infantry the first row should bury themselves a few inches and the third, composed of the tallest men, stand on galoches. Each soldier should carry a steel-tipped stake. In Egypt, the French infantrymen were equipped with them; it was a great advantage against cavalry and surprise attacks. I said to him that Gay Vernon refers to the time required to withdraw. I ask for the book, HM reads it for two hours, [and declares that] no work is so bad that one does not find in it something to learn from.[78]

If we are to believe Bertrand's journal, a few days before his death Napoleon had the notes of his military works burnt. He wanted no trace of them to survive. But Montholon also said to Bertrand that the Emperor wanted to 'leave something good' on his military ideas; that he had himself received them from Marshal de Saxe or Frederick of Prussia; and that one could give Marbot back his notes on Rogniat; that ultimately there were a lot of things that Marbot knew better than he did.[79]

On 12 January 1817, the Emperor described the contents of these military notes that have not come down to us:

> It will be an unassuming work of detailed things that I am promising myself and without doubt I can compose a work in two volumes which will be very interesting, instructive, and will serve to fix ideas on many points on which people ramble daily, which gets in the way. [...] I am very accustomed to war, have made a lot of campaign plans, meditated much.[80]

3

Art of War or Science of War

Napoleon still uses the word 'art' in the sense of a body of technical knowledge peculiar to a profession—here that of war. He designates thus the works of fortification intended to support his tactical position around Dresden in 1813.[1] Such technical knowledge is the peculiarity of civilized peoples, as the Emperor implies when commenting on Caesar's wars:

> Caesar had three wonderful campaigns: the civil war, the African war, which was his masterpiece, the Spanish war against Pompey, and even the one against the Gauls. But if he had only had the latter, it would not have sufficed to justify his great reputation. They were barbarians, armed multitudes, brave but undisciplined, without knowledge of the art of war, whereas in the other three campaigns he was fighting against skilful, disciplined armies like his own.[2]

Techniques can be rapidly mastered by all peoples:

> Voltaire said that Europe could no longer be invaded by the Barbarians because of the art of modern warfare. A false idea. [. . .] The Barbarians will soon know the significance of artillery. During the siege of Cairo, an Armenian moulded some guns in twenty days. Would a European officer have done that?[3]

Napoleon went beyond the initial sense of the art of war, envisaging it as a way of proceeding, an intelligent manner of waging war:

> The art of war is simply the art of increasing one's own chances.[4]
> With an inferior army, the art of war consists in always having more forces than one's enemies at the point one is attacking or at the point being attacked. But this art is not learnt in books or by habit: what really constitutes a genius for war is a form of behavioural tact.[5]

The definition can be negative, as Berthier wrote in 1812 to Marshal Marmont:

His Majesty is not satisfied with the direction you are giving to the war: you possess superiority over the enemy and yet, rather than taking the initiative, you are always on the receiving end. You move your troops around and tire them. This is not the art of war.[6]

Experience therefore teaches no more than books—which seems to contradict the sentence quoted above to the effect that 'war is only learnt by advancing in the direction of the firing'. Napoleon means that experience is insufficient to make a good general. As we saw in Book I, 'military genius' is a gift. The Emperor returned to this point:

War is a singular art. I assure you that I have fought sixty battles. Well, I have learnt nothing that I did not know in the first one. Take Caesar: in his last battle he fought as he did in his first, and he was on the point of losing it.[7]

This supports our first quotation at the head of Book I, Chapter 1. Napoleon had also spoken thus at Dohna, in Saxony, at the start of September 1813. Marshal Gouvion Saint-Cyr recounts it in his *Mémoires*. The Emperor was dining with him and Murat when one of his aides de camp, General Lebrun, informed him of Ney's defeat at Dennewitz:[8]

Napoleon questioned him and went into the smallest details of the manoeuvres performed by the various corps with utterly imperturbable sang-froid. In a way that seemed to us as clear as it was precise and correct, he then explained the causes of the reverse, but without the least fit of temper, without an ill-sounding or ambiguous expression against Ney or any of the generals who were his collaborators. He put everything down to the difficulties of the art, which (he said) were far from being known. He added that, were he to have the time one day, he would write a book in which he would demonstrate the principles so precisely that they would be accessible to all military men and war could be learnt just as one learns any science. I said to him that it was much to be desired that the experience of a man such as he should not be squandered for France; but that I had always doubted whether anyone could accomplish this work; and yet, if it was possible, no one was as entitled as he was to make the claim. I added that it had hitherto seemed to me that the most extensive experience or practice was not the best way of acquiring this science; that of all the generals, whether friends or enemies, who had been seen at the head of Europe's armies during the long wars occasioned by the French Revolution, none seemed to me to have learnt much from experience; and that I did not make an exception of him, still regarding his first Italian campaign as his military masterpiece. He told me that I was right and that, given the scarce means available to him at the time, he too regarded it as his best campaign; that he himself knew but one general who had constantly gained from experience and that was Turenne, whose great talents were the fruit of the most profound studies, and who had come closest to the goal that he intended to demonstrate should he one day have time to compose the work to which he had just referred.[9]

Readers will have noted the use of the word 'science' and, once again, a contradiction in Napoleon, who thinks here that war could be taught in a book, were he himself one day to write it . . . which he did not, justifying our enterprise.

Recounting an anecdote from the end of the Hundred Days, Napoleon repeated that war was a matter of education by habit and intuition, and that it was difficult to make it understood:

> After my abdication, when I was at Malmaison, they came to tell me that the [enemy] advance guard was there, etc. I did not budge.
>   —But that might be the case, said Joseph. Why won't you have it?
>   —It might indeed be the case, but it is not. Only the education of war, if I can express myself thus, tells me this. With the battalion that is here, I can brave anything which might happen. There is nothing, obviously nothing to fear here, according to the enemy's marches and what we know from the initial posts.
> Making things understood: that is not possible. Such is war: the intuition of the profession.[10]

4
Method and Routine

Principles

On several occasions, Napoleon assumed the existence of principles of war:

> The art of war possesses some unchanging principles, whose main objective is to ensure armies against leaders' mistakes about enemy strength—a mistake which more or less always occurs.[1]
>
> All the great commanders of antiquity, continued Napoleon, and those who have worthily followed in their footsteps, accomplished great things only by obeying the natural rules and principles of war—that is, by the correctness of stratagems and a well thought out relationship between means and their consequences, between efforts and obstacles. They succeeded only by obeying them, whatever the audacity of their enterprises and the extent of their success. They made war a veritable science. Solely in this respect are they our great models and it is only by imitating them that we can hope to approximate to them.
>
> People have attributed my greatest deeds to good fortune and they will not fail to impute my reverses to my mistakes. But if I write up my campaigns, people will be astonished to find that in both instances my reason and my faculties were invariably exercised solely in conformity with principles, etc.[2]

Arthur de Ganniers has clearly defined Napoleon's conception of the principles of war. While his initial reading had rather inclined the young Bonaparte towards politics, the siege of Toulon made him glimpse the possibility of a promising military career. He discovered the *Mémoires* of Feuquière, Maillebois, Villars, Catinat, Rohan, and several works by the old staff corps.[3] His political studies had provided him with a solid basis for establishing the elevated parts of war in them and his gifted mind made it possible for him to almost instantaneously connect the main truths of military science scattered throughout these writings. Far from searching for rules to apply, thanks to his acute, profound understanding, he effortlessly distinguished not only the basic principles, but also the whys and

wherefores of the principles and their effects. He would have been embar-
rassed to identify whence he had derived a particular principle. But he
developed a highly personal way of divining them in the sources, extracting
them, conceptualizing them, and then applying them.[4]

What are the principles identified by Napoleon? Some date back to
antiquity:

> Caesar's principles were the same as those of Alexander and Hannibal: keep
> your forces unified, do not get exposed on any point; rapidly deploy to the
> significant points; trust moral methods, the reputation of your arms and the
> fear it inspires, and also political means to keep your allies loyal and conquered
> people obedient; give yourself every possible chance to ensure victory on the
> battlefield; to do that, concentrate all your troops there.[5]
>
> [. . .] concentration of forces, activity, and firm determination to perish
> with glory. These are the three great principles of military art that have always
> rendered fortune favourable to me in all my operations.[6]

The comments on Turenne's campaign make it possible to bring out a more
precise principle, bound up with the armies of the modern age:

> It is one of the most important principles of war, which is rarely violated with
> impunity: *assembling one's billets at the point furthest removed and best sheltered from
> the enemy.*[7]

This principle is restated, along with two others, in 'Observations sur les
opérations militaires des campagnes de 1796 et 1797, en Italie'. The notion
of lines of operations figures in them. It will be developed in Book V, but let
us indicate here that it involves the route linking an army in the field with a
centre grouping together food supplies, munitions, and hospitals:

> Is not the march into Germany via two lines of operations—Tyrol and
> Ponteba—contrary to the principle *that an army must have only one line of
> operations*? Is not the concentration of these two army corps in Carinthia, so
> far from the starting-point, contrary to the principle of *never concentrating one's
> columns in front of, and close to, the enemy*? [. . .]
>
> The march of the Bernadotte division on Groetz, which, had it been
> accomplished without impediments, might have had some advantages, would
> have been against the rules; by contrast, the march it carried out conformed to
> the principles of concentration that are the true principles of war.[8]

Marches that leave the flank open to attack by the enemy are to be avoided.
Charles XII of Sweden did not guard against this during his invasion of
Russia in 1708:

That is the great fault he committed against the principles of war. This convoy exposed the flank to enemy for three hundred leagues. [...] This is what knowing the profession is for. Some things are due solely to the inspiration of genius, but there are some precepts which can be known, such as this one.[9]

Principles and rules are to be construed in a very general sense, meaning that it is necessary to calculate, forecast, meditate, and proportion one's ambitions to one's means:

> Any war must be methodical, because any war must be conducted in conformity with the principles and rules of the art,[10] with reason, and have an objective; it must be waged in line with the forces one possesses. There are two kinds of offensive war: one that is well conceived, conforming to the principles of science; and another that is poorly conceived, violating them. Charles XII was beaten by the Tsar,[11] the most despotic of men, because his war was ill thought out. Tamerlane would have been beaten by Bajazet[12] if his war plan had resembled that of the Swedish monarch.[13]

The same idea emerges from this passage, along with the need for secrecy:

> In war, the supreme commander's first principle is to hide what he is doing, to see if he has the means to overcome the obstacles, and to do everything to overcome them when he has resolved to do so.[14]

What is needed is a 'system'—i.e. calculation—repeated the Emperor to his brother Joseph, King of Naples, in 1806:

> The Sicilian expedition is easy, since there is only one league's journey to make. But it needs to be done by a system, because chance does nothing for success. [...] In war, nothing is obtained except by calculation. Anything that is not profoundly thought through in its details does not yield any result.[15]

On the axes to which a curve is related

The space accorded to principles is relative. They are guides to action, but are far from saying everything:

> It is true that Jomini positively establishes principles above all. Genius acts on inspiration. What is good in one circumstance is bad in another. But principles must be regarded as axes to which a curve is related. It is already something that on a particular occasion one thinks one is deviating from principles.[16]

The allusion to the axes and the curve is taken up again to situate the role of theory in a comment on naval warfare:

> The principle of not making a move until after a signal from the admiral is all the more erroneous in that a vessel's commander is a past master at finding reasons to justify having badly executed the signals he did receive. In all the sciences necessary to war, theory is good for imparting general ideas, which form the mind. But their strict implementation is always dangerous. It is the axes that must serve to trace the curve. Moreover, rules themselves require us to reason, to judge if we must deviate from the rules, etc.[17]

Clausewitz developed a very similar conception. He criticized Jomini and Bülow because their principles were prescriptive and their strict implementation always dangerous, to employ Napoleon's terms. But he was convinced that certain principles emerged from relations of cause and effect, and that they established models discernible in time, and made it possible to construct a scientific theory of war. On the other hand, their application was a matter of judgement and experience; it required a certain sensibility and tact. Unfortunately, in war, in the situations of stress characteristic of engagements, there was a tendency to rely on lists that said what to do, rather than to exercise judgement.[18] Principles became defective only when they became unduly rigid, when they stopped functioning as guides to thinking and became prescriptive rules.[19] As General Lemoine wrote, 'while not attaining an absolute character, principles remain tendencies from which one can sometimes deviate, but which one must strive to render paramount'.[20] It is impossible to get closer to Napoleon's formulation.

Napoleon would have liked to write instructions for his generals in the manner of Frederick II, assembling various principles. But he ultimately desisted from doing so:

> I had wanted to write on this subject, but then some generals were beaten saying that they had followed the principles. There are so many factors. Yet one can write on war to bring out the difference from the writers of antiquity.[21]

While some principles are useful as axes, there must not be too many. They must not be too precise and are valid only when applied:

> What good is a maxim that can never be put into practice and which, when put into practice inappropriately, is often the cause of the army's doom?[22]

When some people elevate specific moves into principles, they are mistaken. One cannot prescribe in detail:

That was Clarke's refrain and this is what is droned on about by Dumas, who understands nothing about war: turn the wings. These are the false principles and words void of ideas that are correctly attacked by Jomini.[23]

Ultimately, Napoleon avows:

I would like not to posit any principle; it is too problematic: circumstances change everything. Every two hours, an army is in a different position. What was wise and skilful at 5 o'clock in the morning is folly at 10 o'clock. One can therefore offer nothing but general precepts: that one should not perform flanking marches in front of the enemy. Nothing adventurous will be recommended: that is to lead a general to his doom, because everything is a *matter of the moment* and, above all else, depends on the way it is carried out.

Thus, in leaving Verona, I planned to wage my campaign in the vicinity of Bassano.[24] But that depended on what happened. I won the Battle of Roveredo and took 9,000 prisoners. That changed the enemy's position. Before leaving, I received news from Trento which informed me that, according to reports from my spies, there was nothing at Vicenza and that Wurmser's headquarters were not yet at Passeriano. If I had known about a move on Verona by the enemy, I would probably have returned from Trento. I do not recall the precise placement of my troops, but am convinced that I had a corps marching on Verona, seven or eight leagues away, which would have served me as an advance guard and that I would have arrived in Verona during the night. The news I received of the enemy's tranquillity determined me to pursue my plan. This plan was thus pursued, meditated, recalculated, every twelve hours. This way of implementing a programme is conducive to wisdom. I was therefore unable to retain my original plan.[25]

A good army would be one where each officer knows what he must do depending on the circumstances. The best is one which approximates to that.[26]

Such is precisely what Clausewitz wanted Prussian officers to do in *Vom Kriege*: 'Theory exists so that one need not start afresh each time sorting out the material and plowing through it, but will find it ready to hand and in good order. It is meant to educate the mind of the future commander, or, more accurately, to guide him in his self-education, not to accompany him to the battlefield; just as a wise teacher guides and stimulates a young man's intellectual development, but is careful not to lead him by the hand for the rest of his life.'[27]

5

Critical Analysis

Napoleon practised critical analysis in Clausewitz's sense of an analysis of campaigns. In 1786, he returned from Paris to Corsica with a trunk crammed with books. Before the officers of the barracks in Bastia, he commented on the wars of antiquity, the successes and the errors of the great commanders, in a critical spirit that surprised them.[1] In the field, Napoleon always sought historical precedents, examples of operations in the same theatre, at the same time as he had generals and officers in whom he had complete trust recognize it. On 5 July 1794, when he commanded the artillery of the army of Italy, he learned that an officer had waged war in this region in 1747. He ordered him to approach and gleaned the maximum intelligence from him.[2] Before beginning his first Italian campaign, he had the Bibliothèque nationale deliver the *Mémoires pour servir à la vie de Catinat*, two books on Prince Eugène of Savoy,[3] several maps and descriptions of Piedmont and Lombardy, the campaigns of Vendôme and those of Maillebois by Pezay—in other words, accounts of all the recent wars in the theatre where the army of Italy was to begin its campaign.[4] At the end of August 1805, he sent Murat to cross through Bavaria as far as Bohemia incognito and ordered him first of all to get hold of a copy of the account of Marshal de Belle-Isle's campaign in these regions during the Austrian War of Succession. He also had to see the battlefield of Mösskirch, where Generals Moreau and Lecourbe had beaten the Austrians on 5 May 1800.[5] In 1808, he ordered his librarian to construct a portable library for him that facilitated a precise, rational, and rigorous approach to, and understanding of, the times and spaces of past events.[6] Commenting on Saint Helena on Frederick II's campaigns, he said of the Battle of Kunersdorf, where the King of Prussia had been trounced by the Austrians and Russians:

> I cannot see that Frederick is to be criticized for any major mistakes. I regret not having visited this battlefield when I was on the scene and not having

known his campaigns. It would have been beneficial to me in the Dresden campaign to know these positions.[7]

According to General Colin, Napoleon had studied the King of Prussia's operations in a four-volume *Vie de Frédéric II*, produced anonymously but attributed to a certain Professor Laveaux and published in 1787 in Strasbourg by Treuttel.[8] It was not in fact the best work on the king's campaigns. But in 1788 the young Bonaparte took what he got hold of. On Saint Helena, he ventured his reflections on these campaigns and once more regretted not having made better use of them:

> Frederick had great moral audacity. I am currently dictating my observations on his campaigns; this will be very interesting. I should have had his campaigns explained at the École Polytechnique and the military academies. Jomini would have been good for that. It would have prompted a lot of ideas in those young heads.[9]

As Napoleon was aware, Jomini had written critical history. It is only possible when one is in possession of each camp's perspective. Naturally, Napoleon was conscious of this:

> I have begun, you know, a book which, if I finish it, will be of great interest. Absolutely great man that he was, Frederick made many mistakes—for example, Kolin. But his historians were Prussians. One would need to read his campaigns as written by an officer of Daun.[10]

Napoleon was very conscious of the need to confront his opponents' perspectives—a point that some historians of battles still do not appreciate. He stressed the importance of a critical history for training generals, reports Gourgaud:

> His Majesty said that were He to continue writing his book on his campaigns, it would be the best work for training generals, but that it would not have to be printed. 'Without speaking of general principles,' said HM, 'I would offer a critique of each campaign, the reasons for and against, and people would educate themselves by reflecting on this.'[11]

Materials for reflecting on the 'for and against' were collected during campaigns. Thus, in November 1805 Napoleon asked an officer in the engineers to reconnoitre a route between Steyer and Vienna that had already been travelled, 'marking the path we followed, the path we could have followed, and the one that would have been best'.[12] Clausewitz was to advocate a critical history in the same sense, wherein 'known facts [are not to be] forcibly stretched to explain effects ... We can see that this may sometimes lead to a broad and complex field of enquiry in which we may easily get lost. A great many assumptions have to be made about things that did not actually happen but seemed possible, and that, therefore, cannot be left out of account.'[13] History can provide materials for the exercise of judgement.[14] Such would have been the essentials of an education in advance of the military schools, which the Emperor contemplated:

> I wanted to give a course on war at Fontainebleau. I would have summoned Gérard, Maison, and those whom I would have liked to advance. I would have produced some excellent generals.[15]

When people wish to criticize certain operations, it is essential to clearly distinguish between epochs and moments. Napoleon said this of his campaign against Archduke Charles in Italy in 1797:

> I have often said that war is like an engagement with sabre thrusts. A thrust in the heart is fatal, but if you wait until your enemy has his hand raised and [is] ready to split your head open before delivering it, you will be out of action before you have been able to make your thrust. A moment ago, you could have made your thrust, but now you are forced to parry and restrict yourself to defending, rather than seeking to kill. This is the nuance, the instant of time that must be seized, which is the spirit of war. [...]
>
> I have often heard many operations criticized by people who, not distinguishing between epochs and moments, reasoned utterly falsely, but with a semblance of reason.[16]

6

On Examples

Learning about war through history

General Colin correctly emphasized that the young Bonaparte did not possess as many works of military history as one might think. In the 1780s, complete scientific works on recent wars did not exist. Resources were wanting.[1] Before 1796, Napoleon probably studied the operations of the War of the Austrian Succession in Italy, the campaigns of the War of the Spanish Succession, Feuquière's *Mémoires*, and Ramsay's *Histoire du vicomte de Turenne* (1735).[2] Accounts could differ so much that for Napoleon, even more so than for our contemporaries, history often revealed contradictory truths rather than a single one, wrote Las Cases:

> It has to be admitted, my friend, the Emperor said to me one day, that *veritable truths* are decidedly difficult to obtain for history. Fortunately, they are an object of curiosity, rather than real importance, most of the time. There are so many truths! . . . that of Fouché, for example, and other schemers of his type. Even that of many honest folk will sometimes differ from mine. The historical truth, which we plead for and to which everyone hastens to appeal, is often nothing but a word. It is impossible at the very instant of events, in the heat of entangled passions. And if, later, people are in agreement, it is because the interested parties, the contradictors, are no longer. But what is this historical truth most of the time? An accepted fable, as someone has very ingeniously put it. In all these affairs, there are two essential, utterly distinct parts: material facts and moral intentions. The material facts would seem to have to be incontrovertible. Yet see if there are two reports that are the same: some remain interminable trials. As for moral intentions, how to find one's way around, even assuming good faith in the narrators? And what if they are motivated by bad faith, self-interest, and passion? I have given an order; but who has been able to read my profound thoughts, my real intention? And yet everyone is going to seize on this order, measure it by his standards, twist it to his plan, to his individual system. Witness the different colours that will be given to it by the schemer whose intrigues it might disrupt or, on the contrary,

serve, the twist he is going to give it. The same will be true of the important person to whom ministers or the sovereign have confidentially let slip something on the subject; the same will be true of the numerous palace idlers, who, having nothing better to do than listen at doors, in the absence of having heard anything, make it up. And everyone will be so sure of what they recount! And the inferior ranks, who will have it from these privileged mouths, will be so sure of it in their turn! And then the memoirs, the agendas, the bons mots and salon anecdotes take off! . . . My dear friend, such is history!

I have seen people challenge me over the conception of my battles, the intention of my orders, and pronounce against me. Is this not the creature's denial of his creator? No matter: my contradictor, my opponent will have his supporters. This is what has diverted me from writing my private *Mémoires*, expressing my individual feelings, in which, naturally, the nuances of my private character originated. I was unable to stoop to confessions à la Jean-Jacques, which would have been attacked by the first comer. Thus, I thought I was obliged to dictate to you here only on the subject of public deeds. I know very well that even these accounts can be contested. For who is the man down here, whatever his rights and the strength and power of these rights, whom the opposing party does not attack and contradict? But in the eyes of the wise man, the reflective man, the reasonable man, my voice will, after all, easily be worth that of another, and I have little fear of the final decision. Today, there is so much enlightenment that when passions have died down and clouds passed, I have faith in the radiance that will remain. But what errors in the meantime! People will often attribute a lot of profundity and subtlety on my part to what was perhaps merely the simplest thing in the world: people will assume I had plans I never had.[3]

However, history is the best way of learning about war:

Let my son read and often meditate on history; it is the only true philosophy. Let him read and meditate on the wars of the great commanders; that is the only way to learn about war.[4]

It is necessary to take from history what might be useful:

Why and *how* are such useful questions that they cannot be posed too often. I studied history less than I made its conquest. That is to say, I wanted and accepted only what could give me another idea, disdaining what was of no use, and seizing on certain results that pleased me.[5]

With Caulaincourt, Napoleon reconsidered the years in which he devoured history:[6]

Reading history, he said, early on gave me the sense that I could do as much as the men to whom it assigned the highest ranks, but without a fixed aim and without going beyond the hopes of a general. All my attention was focused on war and knowledge of the service for which I believed myself destined. It did not take me long to see that the knowledge I wanted to acquire, which I regarded as the objective I wanted to achieve, was still very far from what I could arrive at. I therefore increased my application. What seemed difficult to others seemed easy to me.[7]

The example of the great commanders

The principles of war emerge by themselves from the history of the campaigns of the masters of yesteryear:

> The principles of the art of war are those that steered the great commanders whose exalted deeds have been transmitted to us by history: Alexander, Hannibal, Caesar, Gustavus Adolphus, Turenne, Prince Eugène, Frederick the Great.
>
> Alexander waged eight campaigns, during which he conquered Asia and part of India; Hannibal waged seventeen—one in Spain, fifteen in Italy, one in Africa; Caesar waged thirteen—eight against the Gauls, five against Pompey's legions; Gustavus Adolphus waged three—one in Livonia against the Russians and two in Germany against the House of Austria; Turenne waged eighteen—nine in France and nine in Germany; Prince Eugène of Savoy waged thirteen—two against the Turks, five in Italy against France, six on the Rhine or in Flanders; Frederick waged eleven in Silesia, Bohemia, and on the banks of the Elbe. The history of these 83 campaigns would be a complete treatise on the art of war; the principles that must be followed in defensive and offensive war would flow from them as from their source. [...]
>
> But do you want to know how battles are fought? Read, meditate the reports of the 150 battles of these great commanders. [...]
>
> Wage offensive war like Alexander, Hannibal, Caesar, Gustavus Adolphus, Turenne, Prince Eugène, and Frederick. Read and re-read the history of their 83 campaigns. Model yourselves on them. That is the only way to become a great commander and discover the secrets of the art. Thus enlightened, your genius will prompt you to reject maxims opposed to those of these great men.[8]

The Prussian Scharnhorst had delivered a similar speech, based on the same names.[9] The mistakes made by some generals also warranted reflection. On Saint Helena, Napoleon was crushing about General Championnet, who commanded the army of the Alps in 1799:

This general's manoeuvres and moves are to be studied as a series of errors. He did not make a single move that was not contrary to all the principles of war. [*crossed out*: and thereby learn to avoid them].[10]

Napoleon heralded Clausewitz in preferring examples from modern wars to examples from antiquity. Gunpowder had ushered in too many major changes:

As we have already said, forming up into battalions or columns, camps, marches, everything in war is the result of the invention of gunpowder. If Gustavus Adolphus or Turenne arrived in one of our camps on the eve of battle, they could command the army the following day. But if Alexander, Caesar, or Hannibal came back from the Elysian Fields, they would need at least one or two months to understand what the invention of gunpowder, muskets, cannon, howitzers, and mortars have led to, and were bound to lead to, by way of changes in the art of defence, as in the art of attack. During this time, they would have to be kept following an artillery park.[11]

Feuquière

Clausewitz invokes Feuquière to say that, when called on to justify a theory of war, historical examples are not always convincing.[12] Napoleon took the same step before him with reference to the same writer:

In 1797, when Generals Provera and Hohenzollern[13] turned up to lift the siege of Mantua, where Marshal Wurmser was trapped, they were stopped by the lines of circumvallation[14] of Saint-Georges, which gave Napoleon time to arrive from Rivoli, to scupper their enterprise, and oblige them to surrender with their troops.

Should one await the attack of the relief army in one's lines of circumvallation? Feuquière says: *You must never await the enemy in your lines of circumvallation; you must emerge from them to attack him.* He bases himself on the example of Arras and Turin.[15] But the besieging army at Arras continued its siege of Turenne's army for 38 days; it therefore had 38 days in which to take this city. But Prince Eugène was forced to turn all the lines of circumvallation that covered the siege, in order to attack the right where the Duke of La Feuillade had neglected to have them constructed—which proves the importance attached by this great general to the obstacle of the lines.

But if we had to cite all the attacks on lines that have failed, and all the places that have been captured under the protection of lines or in sight of their relief, or after the relief armies had come to reconnoitre them, had deemed them impregnable, and departed, it would be seen that the role they have played is very significant. It is an additional means of strength and protection that is definitely not to be despised. When a general has come upon the siege of a fortified place, and gained a few days over his opponent, he must use them to cover himself with lines of circumvallation. Henceforth he has improved his position and, in the general mass of affairs, acquired a new degree of strength, a new factor of power.

The option of awaiting an attack in the lines should not be ruled out: nothing can be absolute in war. Can your lines not be covered by ditches filled with water, by floods, forests, a river, in whole or part? Might you not be superior to the relief army in infantry and artillery, but decidedly inferior in cavalry? Might your army not have a greater number of brave men than the relief army's, but be short of exercise and not in a condition to manoeuvre in the open? In all these instances, do you think it necessary either to raise the siege, and abandon an enterprise about to succeed, or to hasten on to your doom by setting out with brave, but non-manœuvring troops to confront numerous good cavalry in the open?[16]

Clausewitz possibly adopted the allusion to Feuquière from Napoleon, since this extract comes from the *Mémoires pour servir à l'histoire de France* read by the Prussian. 'Nothing can be absolute in war': he must have been particularly struck by that sentence.

Feuquière was wrong to seek to generalize from examples, when everything depends on circumstances:

Feuquière is wrong when he criticizes circumvallations. They are always necessary; they are indispensable. If the Duke of York had had some before Dunkerque, he would never have lost the Battle of Hondschoote.[17] Feuquière says that at Valenciennes, Arras, and Turin, the lines were broken open. But for these three examples, how many contrary ones are there? Should one wait in one's lines? That is another question. It cannot be answered positively; it depends on a host of circumstances, the strength of the lines, the troops, etc.[18]

The Emperor criticized Feuquière. But he also discovered qualities in him on Monday 13 October 1817, according to Gourgaud:

At 6.30, the Emperor undressed, read the account of the battles of Hochstädt and Ramillies. 'At Hochstädt, Prince Eugene sought to turn the French by the left and then throw them into the Danube. This is poorly recounted in Feuquière. The maps of it are poor.' His Majesty lay down, made me read [the chapter] entitled 'When a general should give battle', approved, and said that Feuquière [had written a] good book.[19]

The history of war cannot be written by just anyone. On Saint Helena, Napoleon read a work by Lacretelle and judged it harshly:[20]

> He understands nothing of war and everything he says about it is misinterpretation. It seems that, in order to write about it, you need to have waged it. The ancient historians were generally soldiers.
>
> He speaks rather well about administration. It seems that the same style does not suit accounts of administration and of war. Administration is composed of different acts whose dates are unimportant. But in war everything is precise: a particular move is good because it was made the following day. It would have been bad the previous evening or two days later.[21]

Military education by example and its limits

In a long letter to the Minister of War, Clarke, in 1809, Napoleon attached importance to the pupils at military academies learning their profession on the basis of examples illustrating the virtues of arms:

> Monsieur General Clarke, our military is uneducated: we must concern ourselves with two books—one for the school at Metz,[22] the other for Saint-Cyr.
>
> The book for the school at Metz must contain prescriptions on fortifications, the judgements earned by all commanders who have lightly surrendered those entrusted to them, and finally the commands of Louis XIV and of our own times which prohibit surrendering a stronghold before it has been breached and crossing the ditch is feasible. [...]
>
> The aim must be to impress the importance of the defence of fortifications and to excite the enthusiasm of young soldiers with a large number of examples; to make known to what extent the schedules that have been advanced as rules of progression have constantly been subject to delays in their implementation. Finally, a large number of heroic facts must be included in this book, whereby the commanders who long defended the most mediocre places rendered themselves immortal, and at the same time the sentences that have universally dishonoured those who did not do their duty should be recalled [...]
>
> As for the book for the École militaire [of Saint-Cyr], I want it to deal with administration on campaign, rules of encampment so that everyone knows how a camp is marked out, and finally the duties of a colonel or commander of an infantry column. It is above all necessary to stress the duties of the officer commanding a detached column; to clearly express the idea that he must never despair; that, even when encircled, he must not surrender; that in open

countryside there is only one way for brave folk to surrender and that is, like François I and King Jean, in the middle of the mêlée and under the blows of musket butts;[23] that to capitulate is to seek to save everything except honour; but that, when one acts like François I, one can at least say like him: *Everything is lost except honour!* Some examples should be cited here, such as that of Marshal Mortier at Krems,[24] and a large number of others that fill our annals, to prove that armed columns have found a way of breaking out by looking to their courage for their resources; that whoever prefers death to ignominy saves himself and lives with honour, whereas those who prefer life die covered in shame. Thus we can draw from ancient or modern history everything that serves to excite admiration or contempt.[25]

The examples to be given to pupils have an edifying value. They should exalt patriotism, sense of honour, energy in combat. They should not indicate a particular tactical disposition, to be reproduced. Even the rules for the siege and defence of fortified places established by Vauban do not possess an absolute value. The study of past events makes it possible to offer observations, but no book can really form a general for war, which is a matter of the moment:

> The books by Jomini and others are doubtless good. The observations I have made are good. But all that cannot make a great general. I will not recommend a volume. War is a matter of the moment.
>
> Rivoli was just at the right time. An hour later, I would have been doomed if, before attacking, I had waited for all the enemy's columns to move, I would have been encircled, attacked by triple forces, doomed. It would have been wiser to withdraw to Castel-Nuovo. If I had been beaten, everyone would have said that I was an ignoramus; that it was obvious that I was turned, that I was doomed. Thus, there are no arguments to be had on past events. They are pointless. Observations are good, but they will not make a general.[26]

Conclusion to Book II

Napoleon and Clausewitz were especially close as regards the theory of war.[1] The latter took up the former's allusion to the difficulty of taking decisions in war—a difficulty worthy of a Newton or Euler. For Napoleon, Jomini's definitions of strategy and tactics were clear and could convey ideas. But they did not teach war. The 'elevated parts' of war were acquired solely through experience and the study of history. The Emperor disdained the writings of theoreticians, while accepting what he deemed good from particular writers. Clausewitz proceeded in exactly the same way.[2] Napoleon was highly critical of Guibert, whom people have sought to make his prophet. He especially appreciated Jomini as a historian of campaigns. The art of war was a question of tact. It was bound up with the talent of the military genius and not learnt in books. For Clausewitz too, the theory of war was more of an art than a science, because it essentially consisted in the power of judgement, in the ability to identify what is important.[3] Napoleon wanted to write a book on the subject, but he never did. The proximity of Clausewitz's ideas to his is not self-evident. Most theoreticians, Jomini first among them, interpreted war in terms of principles and rules that could ensure success. Obviously, Napoleon spoke of principles and systems, but did not in practice define what he meant by them. For him principles were axes to which a curve was related—nothing more. The many editions of his 'maxims' have thus impoverished his thinking. Clausewitz has likewise been subjected to attempts at 'packaging'. It has to be said that he himself employed the notion of principles in his instruction of the Crown Prince of Prussia.[4] But in him, as in the Emperor, it is simply a question of establishing some basic truths about the very notion of war and thus providing some keys for understanding reality.

No doubt General Colin understood this clearly, but he did not find adequate words to express it. It has to be said that the issue is not straight-forward. Colin refers to 'calculation', to 'solutions fixed by reasoning', to 'inflexible rules' expounded by Napoleon himself, to a 'finite number of principles and rules', to 'general laws, logically deduced from incontestable principles', to laws that the Emperor had demonstrated to himself, and which were 'posited once and for all, with the unshakeable solidity of geometrical theorems'.[5] The multiplicity of phrases betrays the difficulty in hitting on the right words. Some are doubtless maladroit, because they might suggest proximity to the ideas of Jomini and the prescriptive sense of the theory of war. Napoleon, wrote the German general von Caemmerer, was no mere improviser, who resolved each issue solely in accordance with the current situation, solely by creative inspiration. He had a method of action based on some basic principles. At the same time, he was conscious that everything is relative in war 'and that it is sometimes even doubtful whether two times two equals four'.[6] The analysis of Clausewitz by Antulio Echevarria II makes it possible to identify more clearly what Colin and Caemmerer perceived. Clausewitz rejected prescriptive theories and doctrines, but believed in certain truths, in the form of laws and principles relative to the conduct of war. The principles emerged from relations of cause and effect in war, from 'laws' that had to be applied by judgement in each circumstance. Clausewitz conceived the theory of war as establishing the main laws that interacted in any war, because they stemmed from the very nature of war.[7] Napoleon's conception approximated to this. The two men stressed the importance of a critical analysis of past campaigns, allowing the circumstances peculiar to them their due share. Historical examples made it possible to reflect and inspire, but they could not establish a theory.

BOOK III

On Strategy in General

Napoleon had a lot to say about the maintenance of an army's moral qualities. This was a sphere in which he was particularly innovative. From 1796 to 1807, unlike their opponents, the effectiveness of his armies was largely based on the individual motivation of combatants. Napoleon was conscious that soldiers' endurance and combativeness depended on the 'fire in their belly' which he inspired—notably by his constant presence among them. Struck by this during Prussia's rout in 1806, and desirous of distancing himself from some theoreticians' geometrical views on the movements of armies, Clausewitz precisely dwelt on moral factors, which he regarded as vital in strategy.

I

Strategy

What is simplest in war is always best and the only good things are simple.[1]

The art of war is a simple art and completely practical. There is nothing vague about it; everything is good sense and nothing is ideology.[2]

As General Lewal rightly stressed, these famous sentences obviously do not mean that the art of war is within easy reach of simple minds; that it does not require special study.[3] They mean that strategic conceptions must be simple and rational. Napoleon clarified this when criticizing the plans of the outgoing Directory in 1799:

> The armies of the Alps and Italy should have been unified under a single head; the lack of coordination between these two armies has been dire. The plans adopted in Paris were contrary to all the rules of the art of war. War being a practical profession, all complex schemes must be excluded from it.[4]

Clausewitz took up this idea of simplicity in strategy. For him strategy refers to the use of engagements for the war's objectives. Militarily, it fixes the objectives corresponding to the war. It establishes a plan envisaging various actions conducive to those objectives. The forms of strategy, the means it employs, 'are so very simple, so familiar from constant repetition, that it seems ridiculous in the light of common sense when critics discuss them, as they do so often, with ponderous solemnity. Thus, such a commonplace manoeuver as turning an opponent's flank may be hailed by critics as a stroke of genius, of deepest insight, or even of all-inclusive knowledge.'[5] Simple good sense indicates that the enemy's flank should be turned as often as possible. Material relations of this kind are easy to understand. 'Everything in strategy is very simple, but that does not mean that everything is very easy.'[6] Like Napoleon, Clausewitz repeats that the tasks are simple in war, but that interactions, friction, and human realities render their execution extremely complex.[7]

We know from Book II that Napoleon only employed the word 'strategy' on Saint Helena, with reference to a work by Archduke Charles. He construed it in Jomini's sense as referring to the manoeuvres of armies outside an engagement but leading to it; this approximated to Clausewitz's sense of the term. We can see this in the following reflection on the defeats suffered in Spain, where the word is deliberately left in German:

> The Anglo-Portuguese army has become as adept at manoeuvring as the French army. We have been beaten as a result of events of war, manoeuvres and errors of *Strategie* at Talavera, Salamanca, and Vitoria.[8]

In another, little known passage, Napoleon used the adjective 'strategistic' positively to compare his Russian campaign with Charles XII of Sweden's:

> 1 Charles II travelled five hundred leagues in enemy country; 2 he lost his line of operations the day after his departure from Smolensk; 3 he remained without news from Stockholm for a year; 4 he had no reserve army.
>
> 1 Napoleon only went one hundred leagues in enemy country; 2 he always maintained his line of operations; 3 every day he received news and convoys from France; 4 from the Vistula to the camp of Moscow, he posted three-quarters of his army in reserve. Finally, Charles II was operating with 40,000 men, Napoleon with 400,000. These two operations were diametrically different. Whereas one conformed to well thought-out rules, with means proportionate to the end, the other was ill thought out in its objective by a non-strategistic mind.[9]

These claims are not false. But they do not indicate how many of the assembled 400,000 men returned. Rather than criticizing the Russian campaign, let us observe that Napoleon, had he lived longer, might have become used to speaking of strategy. Once again, here he clearly understands the term in Clausewitz's sense.

'Manoeuvers designed to turn a flank,' wrote Clausewitz, 'are easily planned. It is equally easy to conceive a plan for keeping a small force concentrated so that it can meet a scattered enemy on equal terms at any point, and to multiply its strength by rapid movement. There is nothing admirable about the ideas themselves. Faced with such simple concepts, we have to admit that they are simple. But let a general try to imitate Frederick!'[10] The real difficulty for great generals, and their merit, consists in the execution. Conscious of this, Napoleon advocated the simplest movement possible:

[...] what is appropriate to war is simplicity and reliability.[11]

The art of war is like anything beautiful: it is simple. The simplest man-oeuvres are the best.[12]

From his first Italian campaign, Napoleon proved himself to be a master of execution. He inaugurated an original 'manner' that disconcerted the Austrians. He spelt things out to the Directory:

> As to divisional generals, unless they are distinguished officers, I beg you not to send me any. For our way of waging war here is so different from others that I cannot entrust a division without having tested the general who is to command it in two or three actions.[13]

2

Moral Factors

Napoleon stressed this dimension of war:

> Three-quarters of war is about moral factors; the balance of real forces only accounts for one-quarter.[1]

Everything is opinion in war

> An intimidated enemy [. . .] makes all the sacrifices required of him. [. . .] One always negotiates more advantageously [. . .] with a sovereign who has not left his capital and whom one is threatening, than with a sovereign who has been forced to leave it.[2]

Napoleon became a master in the art of exploiting the terror he inspired. He tried to get his intimates, to whom he had entrusted armies, to understand the importance of opinion, even if they did not possess the knowledge or qualities required to command them. He had this written for his brother Joseph, who was tottering on his Spanish throne:

> Everything is opinion in war, opinions about the enemy, opinions about one's own soldiers. After a lost battle, the difference between winner and loser is minor. But it is [enormous in opinion],[3] because two or three squadrons are enough to have a major impact.[4]

He tried to raise the morale of Eugène de Beauharnais, who had just suffered a setback against the Austrians on the Piave in northern Italy:

> In war, one sees one's own woes but not those of the enemy. You need to exhibit confidence [. . .]. If you knew history, you would know that gibes are useless and that the greatest battles recorded by history were lost solely because people paid attention to the armies' words.[5]

Troop morale was to be maintained by keeping men in the corps to which they belonged. For esprit de corps to be created, men should be transferred as infrequently as possible:

> I loathe anything that removes a soldier from the corps he belongs to, for among the men morale is everything; and anyone who does not love his flag is not a genuine soldier.
>   Neglect of this principle has led to desertion and the disorganization of the army.[6]

The French had a tendency to become infatuated with all things foreign. The effects of fashion could have unfortunate consequences militarily:

> The French are so disposed to becoming infatuated with foreign things that perhaps school pupils should not be taught foreign languages. One of the obstacles to our fleet's recovery is our sailors' high opinion of British superiority. It was *Prussomania* that led to the Battle of Rossbach being lost.[7]

The number of the enemies

The issue of an army's size was important. The Emperor had Minister of War Clarke write a veritable lecture on propaganda to his brother Joseph:

> I want you to write to the King of Spain to get him to understand that nothing is more contrary to military rules than disclosing the strength of one's army, whether in orders of the day and proclamations or in gazettes; that, when one is induced to speak of one's strength, it should be exaggerated and presented as formidable by doubling or tripling the numbers; and that, when referring to the enemy, his strength should be reduced by one-half or one-third; that in war everything is about morale; that the King deviated from this principle when he said that he only had 40,000 men and made it public that the rebels had 120,000; that presenting the enemy's number as immense demoralizes the French troops, while representing the French as few in number projects a low estimate of them; that this is to proclaim one's weakness throughout Spain— in a word, to give moral strength to one's enemies and deprive oneself of it; that it is in the spirit of man to believe that the smaller number must ultimately be defeated by the greater.
>   The most experienced military men find it difficult to assess the number of men in the enemy army on the day of battle and, generally speaking, natural instinct leads to reckoning the visible enemy to be more numerous than it actually is. But when one is imprudent enough to allow ideas to circulate, to authorize exaggerated estimates of enemy strength oneself, it has the

disadvantage that every cavalry colonel on reconnaissance sees an army and every captain of light infantry sees battalions.

It therefore pains me to see the poor leadership being given to the spirit of my Spanish army, in repeating that we were 40,000 against 120,000. Such statements have had but one result: a reduction in our credit in Europe by making it thought that it is not worth much and weakening our moral resilience by enhancing the enemy's. Once again, in war morale and opinion are more than half of reality. The art of the great commanders has always consisted in creating the impression that their own troops are numerous and conveying it to the enemy, while getting their own army to believe that the enemy is utterly inferior. This is the first time we have seen a leader reduce his resources below the truth while exalting those of the enemy.

The soldier does not judge. But military men of sense, whose opinion is respectable and who judge on the basis of knowledge, pay little attention to orders of the day and proclamations and know how to assess events.[8]

The number of enemies one is dealing with remains a major source of fear in war. Napoleon recounted the following episode from the 1809 campaign in Germany to Gourgaud:

Marching on Landshut, I found Bessières in retreat and made him march ahead. He objected, citing the strength of the enemy. March even so, I told him. He advanced and the enemy, seeing him engaged in an offensive manoeuvre, believed him stronger than he was and withdrew. In war, every-thing happens like that. Soldiers mustn't count the enemy. In Italy, we were invariably one against three, but the soldiers trusted me. It is moral strength, rather than numbers, that determines victory.[9]

Clausewitz likewise said that in war 'men are always more inclined to pitch their estimate of the enemy's strength too high than too low, such is human nature'.[10] Commanding the Alexandria region in Egypt, General Marmont received orders to organize a mobile column with an infantry battalion and two artillery pieces to raise taxes and impose order in two provinces:

This measure will have the advantage of deriving maximum benefit from these two provinces and maintaining a good reserve at a remove from the epidemic in Alexandria. Depending on events, you will have it return to Alexandria, where its presence would raise the morale of the whole garrison. For it is axiomatic in the mind of the multitude that when the enemy receives reinforcements, it must receive them so as to believe itself equal in strength [. . .].[11]

Maintaining morale and control of the press

Napoleon always kept up the morale of his armies and the French population by means of the press, whose importance in influencing public opinion
he was one of the first to understand. In 1797, in Italy, he wrote to Berthier:

> You will wish, Citizen General, to take steps so that no gazette tending to
> demoralize the army, incite soldiers to desertion, and diminish energy in the
> cause of freedom, is introduced into the army.[12]

In April 1805, he wrote from Italy to Vice-Admiral Decrès:

> I want you to have placed in the newspapers that great news has arrived from
> the Indies; that dispatches have been sent to the Emperor; that their content
> has not leaked out, and that it is known only that the affairs of the British are
> going very badly and that everything promised by the commander general of
> Mauritius has been done by him. These minor means have an incalculable
> impact on men, whose calculations are not the result of cool heads and to
> which everyone brings the alarms and prejudices of his clique.[13]

The war in Spain required greater monitoring of information:

> A *Courrier d'Espagne*, written in French by various schemers, is published in
> Madrid and could have the worst possible impact. Write to Marshal Jourdan
> that there should be no French paper in Spain and that this one is to be
> suppressed. Anywhere my troops are, my intention is not to tolerate any
> French paper unless it is published on my orders.[14]
>
> My intention is that newspapers should print nothing about Spanish affairs
> except in accordance with *Le Moniteur*. Take the necessary measures.[15]

During the French campaign, French morale was no longer attached exclusively to the Emperor's brilliant successes. These had also to be accounted
for by exaggerating the strength of the French forces:

> The newspapers are written without thinking. Is it appropriate just now to go
> so far as to say that I had few men and that I won only because I surprised the
> enemy, and that we were one against three? In truth, you must have lost your
> head in Paris to say such things, when I always say that I have 300,000 men,
> when the enemy believes it, and it needs to be repeated ad nauseam. I created
> an office to direct the newspapers; does it not have sight of these articles? This
> is how, with a few strokes of the pen, you destroy all the good that comes from
> victory! You might well read these things yourselves, know that there is no

question here of misplaced vanity, and that one of the first principles of war is to exaggerate one's strength and not diminish it.[16]

Newspapers are not history, any more than bulletins are history. You must always make the enemy believe you have enormous forces.[17]

Moral implications of strategic moves

In Russia, in 1812, Marshal Oudinot operated independently of the main mass of the Grande Armée. He beat the Russians, but he believed that large forces lay ahead of him and began a retreat. Napoleon asked Berthier to write to him:

> After the fine victory he had obtained, it is astonishing that it is the enemy who has remained master of the battlefield. He has retreated; the enemy has advanced; the enemy knew that two divisions had passed the Dvina, it advanced still further. War is a matter of opinion and the art consisted in preserving the reputation he had earned after the great advantage he had secured.[18]

A few days later, the Emperor conveyed the results of his retreat to Oudinot:

> The Russians are everywhere making known the stunning victory they have won over you, since you unaccountably left them sleeping on the battlefield. The reputation of arms in war is everything and equivalent to actual forces.[19]

At the start of 1813, when commanding the remains of the Grande Armée that had withdrawn to Germany after the disaster in Russia, Prince Eugène of Beauharnais positioned himself behind Berlin. The Emperor gave him the following lecture:

> Nothing is less military than the course of action you have adopted of establishing your headquarters at Schöneberg, behind Berlin; it was very clear that this would attract the enemy. If, on the contrary, you had taken up position before Berlin, communicating by convoys with Spandau, and from Spandau with Magdeburg, having a division of the corps come from half-way from the Elbe or constructing redoubts, the enemy would have been bound to believe that you wanted to give battle. Then he would have crossed the Oder only after concentrating 60,000–80,000 men with the serious intention of capturing Berlin. But he was still very far from being able to do that. You could have gained twenty days and that would have been advantageous, politically and militarily. It is even likely that he would not have risked that manoeuvre, because he is well aware of what he is exposed to and could not

ignore the large numbers of troops that we are assembling on the Main and that the Austrians are assembling in Galicia. But the day your headquarters were positioned behind Berlin, it indicated that you did not want to hold this city; you thereby gave up a posture that the art of war consists in knowing how to retain.[20]

Decision-making in the field cannot be done indiscriminately with respect to the soldiers:

Countermands must be avoided. Unless a soldier can see a very good reason for them, he becomes demoralized and loses confidence.[21]

Soldiers must never be the witnesses of discussions between leaders.[22]

3

The Principal Moral Elements

Napoleon regarded the national sentiments of his armies as a significant moral resource:

> A good general, good officers, good organization, good education, good discipline—these make for good troops, regardless of the cause they are fighting for. But it is true that fanaticism, love of the homeland, national glory can inspire young troops positively.[1]

Reputation likewise affords moral ascendancy:

> It seems that the Prussian cavalry [under Frederick II] was excellent. This Seydlitz is their Murat and a formidable man.[2] The others did not dare advance too far in front of him. When the cavalry decisively outstrips the enemy, it is formidable.[3]

Religion can also represent a great moral force in war. On 10 September 1816, Napoleon said this of the paradise promised the faithful by Muhammad:

> It encourages people to fight, promises the pleasure of the blessed to those who perish in combat. The wings of angels cure wounds. Thus, his religion made an enormous contribution to the success of his arms; and since it is to the success of his arms that the rapid establishment of his religion must be attributed, his paradise had great influence.
>
> The Christian religion does not excite courage and, as a general, he [Napoleon] did not like Christians in his armies. Unforeseen death is so dangerous, it is so difficult to go to heaven, we are so insistently reminded of our last moments, that it is incompatible with the passion of war and unexpected death. [. . .]
>
> The Koran is not only religious; it is political and civil. It contains all the ways of governing. The Christian religion only preaches morality. Yet the Christian religion is a greater revolution than the other, which is but a manifestation of it. The Christian religion is the reaction of the Greeks to the Romans, of spirit against force.[4]

Clausewitz identified the skill of the commander, the military virtues of the army, and the latter's patriotic spirit as the principal moral factors.[5] The former were discussed in Book I, Chapter 3 (military genius); the others are the subject of the next chapter. It is in a situation like this that we appreciate the unfinished character of *Vom Kriege*. The chapters are not balanced. This one is much too short and in practice merely recalls or anticipates others.

4

Military Virtues of the Army

Discipline

> What are needed are men of guaranteed morality, who know that
> subordination is the primary military quality.[1]

The first army Napoleon commanded—that of Italy—in spring 1796 hardly
stood out for its discipline. Numerous orders might be cited prohibiting
pillaging. Let us restrict ourselves to those that assess the possible implications and adduce general considerations:

> Pillaging only enriches a small number of men; it dishonours us, destroys our
> resources, and makes enemies of peoples whom it is in our interest to have as
> friends.[2]
>
> Concentrate on the good discipline of your troops; a month of relaxation
> does damage that can only be cured by six months of treatment.[3]
>
> Pillaging destroys everything, even the army that engages in it. Peasants
> desert; this has the double drawback of making them irreconcilable enemies
> who avenge themselves on isolated soldiers and who will swell the enemy
> ranks as and when we ravage them. That deprives us of any intelligence,
> which is so imperative for waging war, and all means of subsistence.[4]

We must separate out discourse and reality here. The testimony of chroniclers
does not make it possible to define Napoleon's position on pillaging precisely.
He appears to have been rather hostile, except for a few particular situations,
but he did not prove unduly severe towards pillagers.[5] He never succeeded in
eradicating pillaging from his armies. Pressed to move quickly, and deprived
of more than a few days' supply, they were always supposed 'to live off the
land' and exactions were proportional to the poverty of the regions they
crossed through: they reached a peak in the Peninsular War. It might even be
said that pillaging was generally indispensable to the mobility of Napoleon's
armies. Thus the following extract from the *Mémorial de Sainte-Hélène* must be
taken simply as an ideal to strive for, or even as a kind of covert regret:

Furthermore, he continued, politics is, happily, in full agreement with morality in opposing pillaging. I have thought a lot about this subject; I have often been put in the position of gratifying my soldiers; I would have done so if there had been advantages to be had. But nothing is more apt to disorganize and completely dissipate an army. Once he can pillage, a soldier no longer has any discipline; and if, by pillaging, he gets rich, he immediately becomes a bad soldier; he no longer wants to fight. Moreover, he went on, pillaging is not one of our French customs: the heart of our soldiers is not bad; when the first moments of fury have passed, he becomes himself again. It would be impossible for French soldiers to pillage for twenty-four hours: many would spend the last moments repairing the harm they had done in the first place. Later, in their barracks, they criticize one another for their excesses and brand those whose deeds were too odious with disapproval and scorn.[6]

The army must not attract bad or dangerous elements:

There is no need for brigands in Italy or France and putting bad elements in the troops that make up an army is a bad move. That's the method of Neapolitans and of countries that have no army [. . .]. I want all the regiments of my army to be good and suitably composed.[7]

You only need a few men per company to corrupt a whole regiment.[8]

Discipline does not do everything. An army is made up of men and they must be treated as such. In discussion with British officers on Saint Helena, Napoleon told them several times that he adjudged British discipline too harsh; that it was necessary to abolish caning, at least after a certain length of service:

that men are what one makes them and that their character is improved in accordance with the way people behave towards them.[9]

[. . .] to demoralize people is not the way to get the best out of them.[10]

What honour can a man possibly have who is flogged before his comrades. He loses all feeling, and would as soon fight against as for his country, if he were better paid by the opposite party. When the Austrians had possession of Italy, they vainly attempted to make soldiers of the Italians. They either deserted as fast as they raised them, or else, when compelled to advance against an enemy, they ran away on the first fire. It was impossible to keep together a single regiment. When I got Italy, and began to raise soldiers, the Austrians laughed at me, and said that it was in vain, that they had been trying for a long time, and that it was not in the nature of the Italians to fight or to make good soldiers. Notwithstanding this, I raised many thousands of Italians, who fought with a bravery equal to the French, and did not desert me even in my adversity. What was the cause? I abolished flogging and the stick, which the

Austrians had adopted. I promoted those amongst the soldiers who had talents, and made many of them generals. I substituted honour and emulation for terror and the lash.[11]

It is true that in Spain the Italian troops fought bravely and some soldiers performed exploits, notably at the siege of Tarragon.[12]

Honour, emulation, esprit de corps

No one has known how to speak to soldiers, and appeal to their sense of honour, better than Napoleon. Sometimes it was enough for the soldiers to know that they formed part of an army commanded by him in person:

> You know how much words do for soldiers: thus make it known to the various semi-brigades that they form the 2nd and 3rd division of the Army of Reserve.[13]

In 1807, the reserve divisions had to feel part of the Grande Armée:

> Take particular care over their food and have them given supplies that are at least as good as those for your troops, because they must not think that they are reject troops in the army corps. Men are what one wants them to be.[14]
>
> The Poles have officers who are willing. The way they will become worth something is to tell them so, to convince them of it. If headquarters is told every day that they are worthless, we shall get nothing from them.[15]

A fine uniform contributes to the soldier's sense of honour:

> Furthermore, the soldier must be concerned with his condition, must invest his inclinations, his honour in it. That is why fine uniforms, etc., do good. A trifle often gets people who would not stay to stand firm under fire.[16]

To stimulate emulation, honours are also required. This standpoint was fervently defended before the Conseil d'État in connection with the plan to establish the Légion d'honneur:

> I challenge anyone to show me an ancient or modern republic in which there are no honours. People call them *trifles*. Well, it is with trifles that one leads men. I would not say that to a tribune, but in a council of wise men and statesmen, one must say everything. I do not believe that the French people like *liberty* and *equality*; the French have not been changed by ten years of revolution; they are what the Gauls were—proud and fickle. They have but one sentiment: *honour*. It must therefore be nourished; they need honours. See

how the people prostrate themselves before the honours of foreigners: they have been surprised by them and they do not fail to wear them.

Voltaire called soldiers *five sous a day Alexanders*. He was right; that is what they are. Do you think that you would make men fight by analysis? Never. That is good only for the scholar in his study. Soldiers need glory, decorations, rewards. The armies of the Republic have done great things because they were composed of the sons of labourers and good farmers, not rabble; because the officers had taken the place of those of the *ancien régime*, but also out of a sense of honour. On the same principle, the armies of Louis XIV likewise did great things.[17]

Napoleon made honour the primary military quality, thus restoring a 'noble' value of the *ancien régime* to an army hailing from the Revolution.[18] By the number of its occurrences and related passages, the word honour occupies a central place in the *Mémorial de Sainte-Hélène*. With 242 occurrences, it is one of the primary words of Napoleonic discourse. Napoleon also employs the expression 'field of honour', as if to mask the harshness of war. For Didier Le Gall, this involves 'a premeditated discursive strategy that must ultimately diffuse new values, in order to mould the conduct of men and rally a large number of individuals around the Emperor's person'.[19]

Honour cannot be bought:

The troops must not become accustomed to receiving money for acts of courage; it is enough to write them letters of commendation.[20]

One does not pay for valour with money.[21]

As for the labour that the soldier will perform, it will not be waged and cannot be: that is to dishonour the soldier, who must do work of this kind [fortification] solely out of honour.[22]

It [the army] must serve without hoping for extraordinary wages, which would bring a stamp of venality to the noble duties of the soldier. One would fear creating some kind of interest in the duration of a lucrative service. The germs of such a debasement, which should be regarded as well-nigh impossible, must be carefully precluded.[23]

The sense of honour is also maintained by stigmatizing those who do not possess it:

They [the heads of corps] will take care to make an inventory of the dawdlers who, without good reason, have remained in the rear. They will recommend to the soldiers to put them to shame for it. For, in a French army, the heaviest punishment for those who have not been able to partake of the dangers and glory is the shame with which they are branded by their comrades.[24]

In your proclamation, I see that three soldiers let themselves be disarmed. Give orders that these soldiers must take part in parades with a staff rather than a musket for a month, and that their names are mentioned in dispatches.[25]

[. . .] it is with honour that one makes something of men.[26]

Emulation goes hand in hand with honour. To stimulate it, Napoleon multiplied elite units, to the point where the most elevated of them—the imperial guard—was fiercely envied by the rest of the army. In it, in each infantry regiment of the line, he advanced grenadiers and created light infantrymen:

> Lloyd abolished grenadiers. To abolish grenadiers is to abolish one of the main means of emulation in the army. [. . .] This battalion of five companies without grenadiers affords neither a way to create elite corps, nor a means to retain sufficient depth after having lost a third of one's number. This is what leads us to prefer a battalion of six companies, including one of grenadiers, of the stoutest men, and another of light infantry, of the smallest men. This provides the best means of emulation between men there is. Physical difference is perhaps greater than the difference in habits. Big men scorn small men and small men want to demonstrate by their daring and valour that they scorn big men.[27]
>
> Light infantrymen are excellent. Opposing big men to small ones is a new idea that belongs to me. The big man is naturally advantaged; size is an advantage. In convincing small men that they are worth the same as big ones, they are given the emulation to equal or surpass them. They are naturally squashed by big men; they are naturally jealous of them. This jealousy must be exploited. This is also a way of extending conscription to small men who did not used to serve.[28]

Not without cynicism, Napoleon thus exploited the rivalry between grenadiers and light infantrymen, flattering now the former and now the latter.[29] Emulation is bound up with esprit de corps, which must not be ruined by arrangements separating men from their leaders. Napoleon cautioned his brother Joseph, who was commanding in Naples, on the subject:

> Endeavour to keep the battalions together. In your position, there is no advantage in having the troops serve in squads or in forming battalions or strong detachments composed exclusively either of light infantry or grenadiers. This breaks up the corps and removes officers and soldiers from their principal leaders. The destruction of any administration and accountability is the inevitable sequel and everything ends up in disarray. It is a matter of principle that companies of light infantry and grenadiers should only be unified the night before an action. So look to keeping your battalions and

squadrons together, and not dividing them up. Otherwise your army will dissolve and be an utter shambles.[30]

You have regiments that have detachments in Gaeta, Naples, Abruzzo, and Calabria. So there is neither accountability, nor order, nor esprit de corps. The first task is to unify the battalions, without which one has no army.[31]

A warning of the same kind was sent to Jérôme in 1807:

It seems to me that you have 1,000 men from French depots. What aid do you anticipate from these 1,000 men, made up of soldiers without officers, belonging to different regiments? By contrast, if they were in their regiments, they would be of the utmost use.[32]

Napoleon also criticized Joseph for forming a royal guard in Naples by creaming off the best elements from units. Joseph wanted to imitate his brother's imperial guard, but did not seek to stimulate emulation. He cut corners by acting from authority—which had a very bad impact:

The King of Naples has taken the elite companies from regiments to form his guard: signal to him my discontent, as commander-in-chief of my army. All the regiments at Naples are doomed corps because they have been stripped of their cream.[33]

You have taken the elite cavalry companies to form your guard, so that these regiments no longer possess any nerve or perform any service. To isolate a small number of men in this way is to render a large number of men useless.[34]

You have disorganized my elite companies. Reflect on the fact that ten campaigns are needed to create the esprit de corps which is destroyed in an instant.[35]

The education of troops

The strength of the Grande Armée largely stemmed from the marriage of the Revolution's patriotic surge with the qualities of discipline and instruction adopted by Napoleon from the *ancien régime*. The camps of the Côtes de l'Océan were the great laboratory for the instruction of troops and Napoleon personally guided their philosophy. The orders given to generals testify to this:

Familiarize yourselves with the details of the major infantry manoeuvres. The season for drilling your troops will soon begin; and you will sense its full importance, especially in war, where the initial moments are the most intense and decisive. It is necessary to set the pace for the officers so that all see to it. [. . .]

See a lot of the soldier and see him in detail. The first time you arrive in the camp, mark out the row by battalion and see the soldiers one by one for eight hours in succession. Listen to their complaints, inspect their weapons, and assure yourself that they want for nothing. There are many advantages in doing these reviews from 7–8 o'clock. It gets the soldier used to remaining under arms, shows him that the leader does not indulge in dissolute behaviour and is wholly concerned with him—which is a great cause of confidence for the soldier. So let them think that, prior to embarking, I shall come to the camp and will watch them manœuvring and hand them flags.[36]

I recommend that you instruct your troops, and above all ensure that the staff officers and aides de camp have appropriate instruction. Many neglect knowledge of manoeuvres, which provides facilities for executing these manoeuvres and understanding them.[37]

The last order clearly indicates what was required of officers at the time in all armies worthy of the name: versatile knowledge, an ability to adapt to all the cogs of the machine. In a way, this prefigures interoperability, even if only staff officers are involved here. Plain infantrymen are also concerned:

These gunners will drill infantry soldiers in manoeuvres.[38]

Instruction is what makes the soldier:

[. . .] a combination of men does not make soldiers; drilling, instruction, and dexterity impart their true character to them.[39]

But soldiers, even large numbers of soldiers, are nothing if they are not well drilled; make them do manoeuvres; make them do target practice. Take care of their health.[40]

At Austerlitz, the Russians manoeuvred badly because their officers were not educated:

It is the ensemble of manoeuvres, of the education of officers, which forms a real army; it is also what shields civilized Europe from the ignorance and ferocious courage of the barbarians.[41]

Here the Emperor was thinking more of the Turks than the Russians.

What leaders should say to soldiers

In September 1804, Napoleon set out what is appropriate and what is useless to Fouché:

I see a *Lettre à l'armée*; it is by Barère. I have not read it, but I believe there is no need to speak to the army. It does not read the empty chatter of pamphlets and a word in the order of the day would do more than a hundred volumes by Cicero and Demosthenes. The soldiers can be inspired against England without speaking to them; to address a brochure to them is the height of absurdity. There is a whiff of intrigue and distrust about it; the army does not need it. Tell Barère, whose declamations and sophistries do not accord with his colossal reputation, not to get mixed up again in writing of this kind. He always thinks that the masses must be inspired. On the contrary, they must be led without realizing it. In sum, he is a man of little talent. If there is still time, do not allow his brochure to circulate and do not let it be sent to the army. It is not an authority. The only legal way of speaking to the army is the order of the day. All the rest is intrigue and faction.[42]

Harangues are useful only in particular circumstances. Otherwise, a few simple words suffice:

Discipline binds the troops to their flags. It is not harangues that make them brave in the midst of the shooting: old soldiers scarcely listen to them; young ones forget them at the first cannon shot. Not a single harangue in Livy is given by an army general, for not one possesses the features of[43] something improvised. A deed by a beloved general, who is esteemed by his troops, is worth as much as the best harangue. If harangues and arguments are useful, it is during campaigns, to destroy insinuations and false rumours, to maintain a good state of opinion in the camp, to supply material for chat in the bivouacs. The printed order of the day has many more benefits than the harangues of the ancients.

When Emperor Napoleon, riding along the ranks of his army amid the firing, said: 'Unfurl these flags! The moment has finally come!', the deed, the act, the move made French soldiers stamp their feet.[44]

When the leader's words have so much power over men, he must pay attention to what he says:

The 32nd [semi-brigade] would have got itself killed for me in its entirety because at Lonato I spoke these words: the 32nd was there, I was at ease.[45] The power of words over men is astonishing. There had been some acts of sedition in Toulouse. On my visit, I said to them [the mutineers]: so where are those who served with me in the 32nd, the 17th light? Are they all dead? (That was where they were recruited.) This brought everyone back to me. Provence was against me because, during the siege of Toulon, I said that the Provençals were bad soldiers. Princes must pay due attention to their words.[46]

A leader who does not know how to talk to his soldiers deprives himself of a major resource:

What is more popular in character than an army? The general who cannot move, electrify, is wanting in the most important of the qualities he needs.[47]

Officers must not only know how to speak to their men, but also be genuinely interested in them:

> I hear that some old sergeant-majors are stealing from conscripts and treating them badly. It is up to generals to deal with this abuse and ensure that conscripts are treated in such a way as to make their first steps in their military career easy.[48]
>
> A battalion leader must not take rest if he is not informed of all the details; he must even know the name and merit of the officers and soldiers of his battalion, when he has been commanding them for six months.
>
> As for captains, they must not only know the name of their soldiers, but even the regions they come from and everything that is of interest to them.[49]
>
> Nature formed all men equal.[50]

5
Boldness

Clausewitz devoted a chapter to boldness, for in war it is 'a principle in itself, separate and active'. It is a virtue required at all levels. Napoleon was conscious of what bold soldiers could do:

> The conduct at the Battle of Zela[1] of the 6th legion, which crushed everything before it, though composed only of 1,200 veterans, shows how influential a handful of brave men can be. Such influence was more marked among the ancients, just as it is more marked in the cavalry than the infantry among the moderns.[2]

After the Battle of Lodi, the entry into Milan, and the Austrian retreat in spring 1796, General Bonaparte extolled the merits of his soldiers:

> Nothing equals their boldness, except for the gaiety with which they perform the most forced of marches; they sing in turn of their homeland and love.[3]

The great victory at Ulm in 1805 was obtained through exhausting marches and a few engagements. It was not necessary to fight a major battle. Napoleon proclaimed:

> Soldiers, this success is due to your unbounded trust in your emperor, to your patience in putting up with fatigue and privations of every kind, to your uncommon boldness.[4]

In combat, officers must set an example:

> When the troops are demoralized, it is for the leaders and officers to restore their morale or perish. [...] Let no one counter me with an *if*, a *but*, or a *for*. I am an old soldier: you must defeat the enemy or die. I would have liked it if, at the first signal of the attack, the prince [royal of Bavaria] had taken himself to the outposts and restored morale in his division.[5]

Boldness normally becomes rarer in generals, because reason must predominate. Otherwise, the general no longer has the view of events corresponding to his rank:

Always first in the line of fire, Ney forgot the troops who were not before his eyes. The valour that must be displayed by a supreme commander is different from what a divisional general must have, just as the latter should not be the valour of a captain of grenadiers.[6]

But boldness is an additional asset when everything has been well calculated:

Intelligent and bold generals ensure the success of operations.[7]

Clausewitz joined Napoleon in denouncing pusillanimous generals. This was true of French generals in the Seven Years War:

On account of speechifying, trying to be witty, holding councils, French armies of the time experienced what has always occurred when this course is followed: they ended up adopting the worst course of action, which in war is nearly always the most pusillanimous, or, if you like, the most cautious. For a general true wisdom consists in vigorous determination.[8]

Napoleon bemoaned the fact that his admirals did not possess the boldness of his generals:

Our admirals need audacity so as not to take frigates for men of war and merchant shipping for fleets. It is necessary to be decisive in deliberations and, once the squadron has left, to head straight for the objective and not relax in ports or return.[9]

When France has two or three admirals who want to die, they [the British] will become petty indeed.[10]

6

Perseverance

At the start of this very brief chapter, Clausewitz waxes ironic: 'The reader expects to hear of strategic theory, of lines and angles, and instead of these denizens of the scientific world he finds himself encountering only creatures of everyday life.'[1] In a book devoted to strategy, it is indeed curious to find the warlike qualities of an army, rather than campaign moves and manoeuvres, being detailed. Clausewitz once again polemicizes against the geometrical mind-set of the theoreticians of his time, and distinguishes himself from them by stressing material to which they make insufficient reference. In war, the physical and moral being is always ready to give in. What is required is perseverance, 'steadfastness that will earn the admiration of the world and of posterity'. Napoleon had spoken of steadfastness:

> The primary qualities of the soldier are steadfastness and discipline; valour is only secondary.[2]
> If bravery is the soldier's primary quality, steadfastness is the second.[3]

Steadfastness and patience are qualities that are acquired gradually:

> Our armies are always worthy of their reputation: with the same valour and discipline, they have acquired the patience that awaits occasions without a murmur and trusts in the prudence and plans of the man who leads them.[4]

The result was evident in the stunning successes of October 1805, when the Grande Armée's marches ended up paralysing the Austrian army at Ulm:

> So much success is due to the army's patience, to its steadfastness in putting up with fatigue and privations: the soldier's first, most precious quality, because it is what makes it possible to do great things while sparing the shedding of blood.[5]
> [...] discipline and patience in putting up with fatigue and the travails of war are the main guarantees of victory.[6]

Steadfastness is a military virtue even aside from marches and engagements:

> Grenadier Gobain committed suicide for reasons of love; it was a very good theme. This is the second event of the kind to have happened to the corps in a month.
>
> The First Consul orders the following to be placed in the Guard's order:
>
> That a soldier must know how to overcome the grief and melancholy of the passions; that there is as much genuine courage in steadfastly enduring the pains of the soul as in standing firm under the fire of a battery.
>
> Abandoning oneself to sorrow without resistance, killing oneself to escape it, is to abandon the battlefield prior to being defeated.[7]

In Napoleon's strategy—his employment of military operations for the war's objectives—the psychological element was capital, both in his hold over his soldiers and his understanding of their mentality. He could distinguish between the griping of the 'old guard' and their real state of mind, between their mood and their sense of obedience. He knew that to make them go beyond the seeming limits of their endurance could be a source of elation. He understood that aspiring to honours and renown was one of the most powerful passions which could inspire men in combat. 'Napoleon grasped as few have done that esprit de corps and the glory that flowed from success in battle would, for many, prevail even over their feelings for their families, and that men might come to love him more than they loved their wives or their children.'[8] If Clausewitz elaborated on these aspects, it is because he had been deeply impressed by Napoleon's example.

7

Superiority of Numbers

'In tactics, as in strategy, superiority of numbers is the most common element in victory.'[1] The French Revolution proved capable of assembling much larger armies than those of the European monarchies; and this explains its victories. Napoleon frankly acknowledged this in January 1818:

> The Republic triumphed in the first instance solely by weight of numbers. It had enormous armies: 600,000 men under arms—which made for strength under arms [sic]. There were 80,000 men in Italy, 100,000 in Spain, 150,000 in garrisons and the interior, 350,000 on the Rhine and Sambre-et-Meuse—which makes more than 600,000 men. At Jemmapes the enemy had 19,000 men and we had 60,000. At Fleurus, we had double the enemy's strength.[2]

The last point is exaggerated. Most works situate French numbers at 70,000 to 77,000 men and those of the allies between 48,000 and 52,000. In the case of Jemmapes, the account is closer to what is generally agreed: 40,000 French against 13,000–14,000 Austrians.[3]

> War made no progress during the Revolution. People had mistaken ideas about it. The art of war regressed. We succeeded solely by mass and numerical superiority. The Battle of Jemmapes is shameful. It is to be noted that our generals are of the lowest rank. Having beaten the enemy generals, people concluded that one could succeed in war with bravery and without any need for instruction. I myself expressed this idea, always saying that I had no need of cultured generals, that what I needed was brave men, that fire in the belly was everything or almost everything.[4]

If he seeks to magnify his own merits, Napoleon is not wrong in recalling that at Jemmapes the French won thanks above all to their numerical superiority. With 40,000 men, Dumouriez had difficulty beating 14,000 Austrians. The French columns succeeded in absorbing the calibrated fire of their opponents solely by their mass and at the cost of heavy losses.[5]

Napoleon restored the value of the troops by improving training and reverting to certain practices of the *ancien régime* and its military virtues, as has been indicated above. To counteract the practices of the revolutionary era, he multiplied statements tending to relativize the numerical factor:

> It is not with a large number of troops, but with well-organized, highly disciplined troops, that one obtains success in war.[6]
>
> Consider the number of soldiers to be nothing and that it is only when officers and NCOs are aware they are manœuvring that one can expect something from them. It is the camps of Boulogne, where the corps have been constantly drilled for two years, which have vouchsafed me the successes of the Grande Armée.[7]
>
> It is not men that I count in my army, but men who have experience and valour.[8]
>
> Generals are always demanding; it is in the nature of things. There is not one of them who can be counted on for that. It is perfectly straightforward that someone who is charged exclusively with one task should think only of that; the more people he has, the more assurance he has for what he has to do.[9]
>
> It is not the number of soldiers that makes for the strength of armies, but their loyalty and their good disposition.[10]
>
> [. . .] the power of states consists in having good, loyal troops, as opposed to lots of troops.[11]
>
> Therefore you must strive to have not a large number of troops, but a small number of good troops whom you must gradually train.[12]

Reflecting on the influence of numbers in conflicts, Napoleon first of all notes that in antiquity a well-trained army devoted to its leader, like Hannibal's Carthaginians at Cannae, could prevail:

> This success was attributable to his cavalry and it is not surprising: they were good troops against bad ones. Number is not what matters most.[13]

At sea too, superiority of numbers was not decisive:

> Superiority at sea did not use to afford the same advantages as it does today; it did not prevent those who were inferior from crossing the seas, be it the Adriatic or the Mediterranean. Caesar and Antony crossed the Adriatic, from Brindisi to Epirus, in the face of superior fleets. Caesar crossed to Africa from Sicily and, although Pompey was well-nigh constantly master of the seas, he derived but little advantage from it. The fleets of the ancients would not have warranted the saying: 'The trident of Neptune is the sceptre of the world'—a maxim that holds true today.[14]

It will have been understood from this quotation that numerical superiority now prevailed on the high seas. This was because arms and equipment were approximately the same throughout Europe—which restored to numbers their full significance. The same applied to land:

> If it sometimes turns out that 17,000 men beat 25,000, this does not justify the temerity of the one who illegitimately exposed himself to the contest. When an army is expecting reinforcements that will triple its strength, it should not risk anything, lest it jeopardize certain success after the concentration of all its divisions.[15]

Clausewitz wrote that armies resemble one another and that the differences are to be sought in the military qualities of the troops and the skill of the commanding general. Unlike in antiquity, an army inferior in numbers is no longer seen to win completely. He refers to Marathon, not Cannae, but the same idea is being expressed.[16] Napoleon says this about his victories at Lonato and Castiglione:

> If I had just said that, with 30,000 men and 40 pieces of artillery, I had beaten Wurmser who, with 80,000 men and 200 artillery pieces, was arrayed in a good position, at a time when the Austrians were fighting well and almost matched us, I would be performing miracles. If I explain that these 80,000 men came from two or three different directions, and that I attacked them in succession with the same troops, who rapidly transferred from one corps to the next, I render the campaign intelligible and dispel the miracle.[17]

Gohier's *Mémoires* recount a dialogue between Generals Bonaparte and Moreau during their first meeting on 22 October 1799. To Bonaparte, back from Egypt, Moreau explained the French defeat at Novi, where they were overpowered by an overwhelming mass of enemies—'the large number always beats the small number':

> —You are right, said Bonaparte: the large number always beats the small number.
> —However, General, with small armies you have often beaten large ones, I [Gohier] say to Bonaparte.
> —Even in that case, he replied, it was always the small number that was beaten by the large.
> This prompted him to elaborate on his tactics for us:
> —When, with smaller forces, I was in the presence of a large army, concentrating mine rapidly, I fell like lightning on one of its wings and knocked it out. I then exploited the disarray this manoeuvre never failed to produce in the enemy army to attack it in a different part, still with my full

strength. I thus beat it thoroughly; and victory, which was the outcome, was (as you can see) still the triumph of the large number over the small.[18]

Clausewitz explains why a theoretician like Jomini was led to regard this as the basic principle of strategy, whereas Gohier, a man of the eighteenth century, spoke of tactics: 'We believe then that in our circumstances and all similar ones, a main factor is the possession of strength at the really vital point. Usually it is actually the most important factor. [. . .] Consequently, the forces available must be employed with such skill that even in the absence of absolute superiority, relative superiority is attained at the decisive point. To achieve this, the calculation of space and time appears as the most essential factor, and this has given rise to the belief that in strategy space and time cover practically everything concerning the use of the forces.'[19] Obtaining numerical superiority locally, even when one does not possess it generally—such was the procedure most often employed by Bonaparte in his first Italian campaign, and this from the outset:

> The French army only had 30,000 men present under arms and 30 artillery pieces; against it were 80,000 and 200 artillery pieces.[20] If it had had to fight in a general battle, doubtless its inferiority in numbers, its inferiority in artillery and cavalry, would not have enabled it to resist. It therefore had to compensate for numbers by the rapidity of its marching, for the lack of artillery by the nature of its manoeuvres, and for the inferiority of its cavalry by the selection of positions. For the morale of the French soldiers was excellent; they had reported and been toughened up on the rocks of the Alps and Pyrenees. Privations, poverty, misery—these are the school for good soldiers.[21]

It will have been noted that the army's military virtues are directly connected with the quality of its strategic manoeuvres—something that once again links Bonaparte and Clausewitz. According to the latter, the skilful concentration of forces at decisive points is not a kind of law that merely has to be applied, as Jomini put it. It is above all due to a correct assessment of these points, to an appropriate orientation of forces from the outset, and to 'the resolution needed to sacrifice nonessentials for the sake of essentials'.[22] This completely accords with the following comment by Napoleon on his reaction to Wurmser's offensive in late July and early August 1796:

> Wurmser's plan was thus exposed: he had taken the initiative and counted on retaining it. He assumed the army was fixed around Mantua and that, by identifying this fixed point, he would encircle the French army. To confound these plans, it was necessary to take the initiative oneself, render the army

mobile by lifting the siege of Mantua, sacrificing the trenches and siege equipment, to move rapidly with the whole unified army against one of the enemy's corps and then against the other two. The Austrians had two-and-a-half times as many soldiers. But if the three corps were attacked separately by the whole French army, the latter would have the numerical advantage on the battlefield.[23]

Clausewitz recognizes that Frederick II and Bonaparte were the modern generals who, thanks to accelerated marches, were more successful than others in routing several of their opponents with a single army. Conscious of his rivalry for glory with the king, the Emperor made this criticism of him in connection with the spring 1757 campaign:

> Frederick was numerically superior in the theatre of war and inferior on the battlefield: that is the biggest mistake a general could make. The great skill, when inferior in troops in the theatre of war[,] is to be superior on the battlefield. That is what I always did in my first campaign in Italy and in my last in France in 1814.[24]

8

Surprise

By achieving local numerical superiority thanks to rapid marches by his forces, Bonaparte surprised his opponents. Given its impact on morale, Clausewitz elevated surprise to the status of an autonomous principle. His theoretical approach becomes clearer: for him moral effects are what are most discernible and most real in strategy. This is what leads to decisions, abruptly alters them, and prompts the adoption of different ones. Secrecy and rapidity are the two factors in surprise. In strategy, surprise is all the more feasible if the measures to be taken approximate to the tactical sphere and concern a particular unit in the field. It is more difficult to create at the level of a whole army over a whole front. Clausewitz's reflection clearly describes the contrast between Napoleon's first and last campaigns.

In Italy, with a small army and in a succession of engagements over terrain that was initially mountainous, and then broken up by rivers, he surprised the Austrians with the unusual speed of his manoeuvres. Bonaparte was only a general at the time and his strategic decisions, given the number of men and the terrain, had a mainly tactical dimension and implications. He himself recounts this conversation, which occurred shortly after the Battle of Lodi in May 1796:

> Doing his night rounds, Napoleon came across a bivouac of prisoners where there was a talkative old Hungarian officer. He asked him how their affairs were going. The old captain could not deny that things were not going very well: 'But, he added, there's no way of understanding things any more. We're dealing with a young general who is sometimes ahead of us, sometimes on our tail, sometimes on our flanks. You never know where he is going to position himself. This way of waging war is intolerable and isn't the done thing.'[1]

In Egypt too, the forces were far from numerous—even fewer than in Italy. Against the mamluks, recourse to surprise could be made even more easily because they did not possess the discipline and organization of European armies. We are still at a tactical level in this recommendation to General

Desaix, who had set off with the advance guard in the direction of Cairo, shortly after disembarking at Alexandria:

> You will probably only encounter a few cavalry squads. Conceal your cavalry; only present infantry platoons to them: this will give them the confidence to remain within rifle range and will put you in a position to capture some of them. Do not use your light artillery. It must be saved for the big day when we shall have to fight four or five thousand horse.
>
> Do not make any use of your artillery, other than against houses.
>
> The art here consists in keeping all my extraordinary resources hidden, so as to employ them only when we shall have large forces to fight and surprise them all the more.[2]

Surprise is less easy at a strategic level. In 1815, Napoleon returned from the island of Elba, in defiance of the powers gathered at Vienna, and became French emperor once more. He wanted peace, but Europe armed against him and assembled armies on France's borders. Any decision he took as regards strategy would have profound political implications in the first instance. His plan was to surprise the forces of Blücher and Wellington billeted in Belgium. The move was well thought out and there was a certain effect of surprise. But the Prussians got themselves organized and the alarm was soon given. On Saint Helena, Napoleon continued to entertain illusions about the effect of surprise obtained on 15 June 1815.[3]

Nevertheless, he was a master of the art throughout his career and even in 1815 Wellington did not immediately believe in the reality of the French attack on Charleroi. But let us leave critical analysis of events there and turn to general reflections on the subject of surprise. There are not many. During disembarkations, one's whole army should descend in one go, not part of it, which would risk being crushed:

> The advantage of a squadron is that it creates surprise at the point where it presents itself, disembarking large forces that nothing can withstand at first. The art in this case is to disembark all at once, as I did at Marabout in dreadful weather.[4]

The French army then marched on Alexandria without intermission:

> If, in 1798, Napoleon had not appeared under the walls of Alexandria a mere thirteen days after dropping anchor at Marabout, he would not have succeeded: he would have found the walls crenelated and well armed, half of the mamluks already having arrived from Cairo, with an immense quantity of Arabs and janissaries. But he marched on Alexandria and mounted an attack

on its walls with only a handful of his men, without waiting for his guns, eighteen hours after his fleet had been reported. It is a principle of war that, when lightning can be employed, it should be preferred to cannons.[5]

On the level of 'the higher, and highest, realms of strategy', Clausewitz cites the crossing of the Alps by Bonaparte and the Army of Reserve in 1800 as an example of successful surprise. The instructions issued to General Masséna, commander-in-chief of the army of Italy at the time, are significant:

Citizen General, you are too well aware of the importance of the utmost secrecy in such circumstances for it to be necessary to commend it to you. You will employ all the displays and semblances of movement that you shall deem appropriate to trick the enemy as to the true objective of the plan of campaign, and persuade him that he is first of all to be attacked by you yourself. Thus, you will exaggerate your strength; you will announce enormous, imminent aid from the interior; finally, you will distance the enemy, as far as possible, from the real points of attack, which are the St. Gotthard and the Simplon.[6]

9
Cunning

Napoleon knew the value of secrecy, cunning, and deception. Some believe that he had a natural bent for them; that his temperament impelled him to Machiavellian schemes.[1] Clausewitz reckoned that 'it seems not unjust that the term "strategy" should be derived from "cunning" and that, for all the real and apparent changes that war has undergone since the days of ancient Greece, this term still indicates its essential nature'.[2] Yet he was scarcely under any illusions about the role of cunning in strategy. It can only be resorted to on the isolated occasions that present themselves. What a general needs above all is an accurate, incisive view.

Napoleon often used cunning in campaigns, but, as Clausewitz says, especially when he commanded a comparatively small force—that is, when he was only a general. At the very start of his first Italian campaign, Masséna was to attract the Italians to Voltri to cut them off from the Piedmontese:

> Do nothing that could create the impression that you want to evacuate this position, which must still be held for some time since we are occupying it. Keep your eyes open on Montenotte and always do what an enemy does when he wants to advance and believes himself stronger. Surveillance and boasting as required—all the methods common to war are invariably good and succeed.[3]

Before the island of Malta, General Bonaparte gave this order in 1798:

> In all cases, you must order that some sappers or labourers, whom you will take from the Marmont brigade and who will be paid, straight away move some earth so as to make the enemy believe that we are establishing batteries, by constructing a bit of a rampart with earth and others with barrels filled with earth. This will disturb the enemy and will have the double advantage of making him use his powder, if he is stupid enough to fire, and speeding up the negotiations that are under way.[4]

Before El-Arich in Egypt, the commanding general had mannequins made to create the impression that there were sentries.[5] During this campaign, he also employed a practice already adopted in Italy:

> In Egypt, Napoleon had agreed with all the heads of corps that, in the orders of the day, the actual quantity of provision of foodstuffs, weapons, clothing would be exaggerated by a third [. . .]. In the accounts of the Italian campaigns in 1796 and 1797, and since, the same methods have been used to convey an exaggerated idea of French strength.[6]

This practice was an important aspect of Napoleonic strategy:

> [. . .] the art of war consists in exaggerating one's own strength and diminishing the enemy's.[7]
> It would be a strange mistake to assume that all the conscriptions decreed have actually been carried out. It was a ruse used to impress foreigners; it was a means of exercising power.[8]

During his final campaigns, the Emperor's forces were decidedly inferior in numbers to those of the allies. It was necessary to compensate for this by exploiting his reputation. Thus, on several occasions he ordered that people should be led to believe he was present:

> A good ruse would be to have salvos fired in celebration of the victory won over the other army.[9] A ceremonial review should also be staged as if I was there and troops be made to exclaim: *Vive l'Empereur.*[10]
> [. . .] let [Macdonald] make all the arrangements required to get the troops to believe that I am present; in cases where he is in the presence of the enemy, to have them exclaim *Vive l'Empereur!* [. . .] and to spread rumours of my arrival among the soldiers and inhabitants tomorrow.[11]

Napoleon was in agreement with Clausewitz and 'the sword of Renaud'— doubtless a copy error for 'Roland'—is equivalent to 'direct action' in the following reflection, which relativizes the importance of cunning in strategy:

> The Austrians are very good at spreading false rumours, at creating false opinions among the inhabitants; they are great masters in sowing alarm in an army's rear. But if you unsheathe the sword of Renaud, the spell will be broken immediately.[12]

Nevertheless, the skill of the Austrians in spreading false rumours fooled the French, who were often unduly credulous, more than once, as is acknowledged by a bulletin of the Grande Armée in 1805:

A column of 4,000 Austrian infantry and a regiment of cuirassiers crossed our posts, which allowed them to pass because of a false rumour of suspension of hostilities that had been spread in our army. In the ease with which this was done one recognizes the character of the Frenchman, who, brave in the mêlée, is often unreasonably generous outside combat.[13]

10

Concentration of Forces in Space

The concentration of power in Napoleon's hands was accompanied by a strategy of concentrating military forces in a principal theatre. His undivided authority allowed him to construct a Grande Armée, similar in size to that of the Revolution, under his direct command, whereas the Republic's many armies had previously acted in un-coordinated fashion.[1]

The concentration of forces at the decisive point

For Clausewitz, the highest and most simple law of strategy consists in concentrating one's forces so as to be the strongest at the decisive point.[2] Here he drew on Napoleon, who added the specifications of someone who had been commander-in-chief:

> Only one army is needed, because unified command is utterly imperative in war. The army must be unified, concentrating the maximum strength on the battlefield and exploiting all opportunities, for fortune is a woman. If you miss out on her today, do not expect to find her again tomorrow.[3]

General Colin underscored the distinction between 'unifying' and 'concentrating' in this extract. The second term implies close congregation on the eve of an important action. By contrast, the army is unified 'as long as its different parts are sufficiently proximate to one another that the enemy cannot prevent their concentration or beat them separately. From this it follows that the various parts of the army must be close enough to be unified when in the enemy's vicinity. But when far removed from the enemy, they are unified while remaining very distant from one another.'[4] In the model campaigns—those of 1805 and especially 1806—the area covered by the Grande Armée was very extensive at the outset and then became progressively smaller. It was no more than 30 or 40 kilometres in size on the eve of battle. The transition from unification to concentration emerges from this letter to Marshal Soult at the start of the 1806 campaign:

> With this immense superiority of forces unified in such a small space, you sense that my wish is to risk nothing and to attack the enemy, wherever he chooses to stand, with double strength. [. . .]
> You definitely think that it would be a fine thing to move around this place [Dresden] in a battalion square of 200,000 men. But all this requires a little art and some events.[5]

The image of the 'battalion square' illustrates the concern for an army manoeuvre in which the corps proceed at a distance where they can support one another and be concentrated in less than twenty-four hours to confront the bulk of the enemy. In 1805, the various corps took parallel routes, at the risk of one of them, by distancing itself too much, being attacked in isolation. Another could have rapidly come to its aid, but in 1806 Napoleon sought to be more cautious. His deployment was narrowed to the extent of forming a gigantic square—which would make it possible to face up in any direction by rapidly altering the line of march.[6] When Napoleon wanted to attack, he did so with concentrated forces, 'en masse'. Before uniting the forces that he led from France with those commanded by Prince Eugène in Germany, at the end of April 1813—i.e. on the eve of the Battle of Lützen—he recalled this principle:

> You know that my principle is to debouch en masse; so it is en masse that I want to cross the Saale with 300,000 men.[7]

This was a principle valid for any military action, especially when it involved engaging the artillery. Forces should always be concentrated:

> The way I see the defence of Antwerp being organized exhibits little or no skill. Rather than positioning the batteries at 1,000 and 1,500 toises from one another, where they can only aid one another minimally, and have to fight separately against all the enemy's forces, this mass of guns should have been concentrated in a small space, so that they could defend themselves together and hit the same target. [. . .] When strewn all over the place, the greatest resources do not yield any result in artillery, cavalry or infantry, strongholds, and the whole military system.[8]

Staying unified when near the enemy

Separating one's army into different columns when near the enemy can prove very dangerous:

> Frederick was a great, audacious warrior, who had excellent troops. They were heroes. But he made big mistakes in his plans of campaign. He would not have made them with impunity in my presence. But as he was skilful, he soon corrected himself.
>
> In front of Prague, he marched along a bank of the Elbe and Marshal Schwerin arrived crossing two rivers by the other bank of the Elbe.[9] Jomini observes very well that Prince Charles[10] was to march towards these corps and beat them separately. It is clear that Frederick could not aid them. My art has always been that isolated corps could nevertheless communicate with one another and be aided. This is not the genius of war. I would not have done it at the age of twenty-five. [...] The Austrians fashioned their plan of campaign in Prussian style in Italy. They arrived in different columns that were not in communication and I beat them separately.[11]

At the start of August 1796, the Austrians made the mistake of attacking the French with two strong columns separated by Lake Garda. On account of the rapidity of his manoeuvres, Bonaparte could fight them separately at Lonato and then Castiglione:

> To operate in directions far removed from one another and without communications is one mistake that normally leads on to a second. The detached column only has orders for the first day; its operations for the second day depend on what happened to the main column. Either it wastes time awaiting orders, or it acts haphazardly. [...]
>
> It is therefore a matter of principle that an army must always keep all its columns unified, so that the enemy cannot insert itself between them. When, for whatever reason, one deviates from this principle, the detached corps must be independent in their operations and, in order to reunite, direct themselves to a fixed point, which they march towards without hesitation and without new orders, so as to be less exposed to being attacked in isolation.[12]

Field-Marshal Alvinzy's Austrians[13] repeated the same error in January 1797, at Rivoli:

> These dispositions were contrary to the major principle according to which an army is, *every hour of every day, in a state to fight*. Alvinzy was not in a condition to engage on his arrival in these mountains or during the time he required to

reach the Rivoli plateau. For in order for an army to be in a state to fight, it must be unified. But the 20 battalions skirting the Adige valley were separated and could only reunite after having taken the plateau of Rivoli. To fight, an army needs its cavalry and its artillery; yet the cavalry and artillery, which were under the command of Quasdanovich,[14] could only join the army via the Rivoli plateau. Alvinzy therefore assumed that he would not be obliged to fight from Corona until Rivoli, and that was not dependent on him. He had exposed 24 battalions, without cavalry or artillery, to attack by the whole French army, with 20,000 infantry, 2,000 horse, and 60 artillery pieces. This battle was not equal. [. . .]

In war, people are often mistaken about the strength of the enemy they are to fight. Prisoners only know their own corps; officers give very vague reports. This has led to the adoption of an axiom that cures all: *an army must be ready every day, every night, and every hour to mount all the resistance of which it is capable.* This requires that soldiers constantly have their weapons and munitions; that the infantry constantly has with it its artillery, cavalry, and generals; that the various divisions of the army are constantly in a position to support, back up, and protect one another [. . .].

A great commander must say to himself several times a day: if the enemy army appeared in front of me, to my right or my left, what would I do? And if he finds himself disconcerted, he is badly positioned, he is not in order; he must put it right. If Alvinzy had had asked himself this question: *If the French army seeks me out before my arrival at Rivoli, when I shall only be in a position to counter it with half of my infantry, no cavalry, and no artillery,* he would have answered: *I shall be beaten by forces inferior to mine.* How come the example of what happened at Lodi, Castiglione, Brenta, and Arcole did not make him more circumspect?[15]

The tradition of convergent attacking columns was so deep rooted in the Austrian general staff that Napoleon could still tell Masséna, commander-in-chief of the army of Italy in March 1800, how he would be attacked and what he should do:

> The enemy, Austrian fashion, will make three attacks: by Levant, by Novi, and by Montenotte. Refuse him two of these attacks and have all your forces ready for the third.[16]

'Incredible though it sounds', exclaims Clausewitz, 'it is a fact that armies have been divided and separated countless times, without the commander having any clear reason for it, simply because he vaguely felt that this was the way things ought to be done.'[17] For Napoleon, this was a major defect of the revolutionary armies. The reason for it was that the authorities in Paris did not want to concentrate too much power in the hands of a single general:

[. . .] notwithstanding Mathieu Dumas and all the military writers, I maintain that during the Revolution the art of war regressed rather than advancing.

[. . .] In the revolutionary wars, we had the false system of dispersing one's forces, of sending one column to the right and another to the left—which is worthless.[18]

Lecourbe was a pretty good general; the Directory counted on him a lot.[19] He had made a great name for himself in Switzerland, but I did not approve of [his campaign] in Engadine. With 25,000 men, he sought to beat Prince Charles, who had 80,000, and yet he placed 3,000 men here, another 3,000 [there], and finally all of them in small packs. This is to be ignorant of war. One must always have one's army at one's disposal, but that was the fashion. It was waging war in tune with the braggarts of the general staffs.[20]

When he acted differently from the other generals in his theatre of war in Italy in 1797, General Bonaparte explained himself as follows to an amateur critic:

> Perfection or the system of modern war consists, you claim, in throwing one army corps to the right, another to the left, leaving the enemy in the centre, even positioning yourself behind a line of strongholds. If these principles were taught to youth, they would put military science back four hundred years; and every time you behave thus, and are dealing with an active enemy with even a minimal knowledge of the snares of war, he will beat one of your corps and will cut off the other's retreat.[21]

If Napoleon won so many battles, it is precisely because he broke with the practice of the revolutionary wars:

> In the revolutionary wars, we had the system of extending ourselves, sending one column to the right and another to the left—which was worthless. To tell the truth, what led to my winning so many battles is that on their eve, instead of issuing orders to split up, I had all my forces converge on the point I wanted to force open, the mass. I mowed down what was in front of me, which of necessity was always weak. At Wagram, I even recalled Bernadotte, who was forty leagues away on the Danube. I concentrated all my forces. Thus I had 160,000 men and Prince Charles had left Jean at Presbourg.[22]

General Reynier got himself beaten by the British at Maida, in Calabria, on 4 July 1806:

> The army was dispersed; that is the way of not being able to resist properly anywhere with considerable forces. In dividing one's army by 40,000 men, as often happened, as the Directory did, one has no army.[23]

Next, an army should not be concentrated just anywhere. The situation of the rallying point is very important, as Napoleon stressed in connection with a manoeuvre by Turenne:

> The assembly point indicated for the quarters of the two armies was too close to the [enemy] army; that was a mistake: *an army's meeting point, in case of surprise, is always designated in the rear so that all the cantonments can arrive there before the enemy.*[24]

Napoleon made this an inviolable principle:

> The mistake was to have adopted an assembly point too close to the enemy. This is a principle of war that must never be violated. Turenne too violated it at Marienthal—which led to him losing the battle.[25]

II

Unification of Forces in Time

One of the strengths of Napoleonic strategy was to let army corps march according to different itineraries, which enabled them to resupply themselves more easily, and then to concentrate them rapidly in order to give battle. Compared with the *ancien régime* and the trial-and-error of the early revolutionary period, the procedure was new:

> Previously, people marched as they fought: in serried corps. Hence the need to carry their food supplies with them, in the reserves. People marched on all routes in divisions. It was easy for them to feed themselves, but in this way they were fighting at isolated points. What has been regarded as an advantage in the art of war was at bottom merely an inevitable result.
>
> Jomini notes this very well by saying that I always marched separately, so as to live, but [that] I arrived rapidly in order always to be able to be unified on the day of battle. Thus, in all battles I was unified; that is what was special about me.[1]
>
> General rule: when you want to fight a battle, assemble all your forces; do not neglect any of them. A battalion can sometimes decide the day.[2]
>
> [...] whenever one fights a battle, especially against the British, one must not divide one's forces, but unify them, present imposing masses. All the troops left in the rear run the risk of being thoroughly beaten or forced to abandon their positions.[3]
>
> The first principle of war is that battle should only be fought with all the troops one can concentrate on the field of operations.[4]

Clausewitz distinguished between concentration of forces in tactics and in strategy. During a battle, forces are used successively and it is always good to keep a reserve of fresh troops to be committed when the time comes. In strategy, a detachment cannot be allowed to operate far removed from the main forces. It must be there for the battle; otherwise, it is useless. Forces must be used simultaneously.[5]

While it is true that Napoleon was unrivalled in concentrating forces for battle, he could not foresee everything each time. At Marengo, he had imprudently detached Desaix. At Jena, he was mistaken about the distribution of the Prussian forces. At Eylau, Davout's corps only arrived during the battle and Ney's corps came much too late to take part in the action. The first two encounters, plus that of Auerstaedt where Davout confronted the Prussians alone, were nevertheless stunning victories. The same cannot be said of Eylau.[6]

Yet Napoleon was critical of detachments of troops on the eve of a battle:

> It is still my opinion that Charles XII was a great warrior, although Voltaire claimed that he did not know how to read a map and did not have any. He made some mistakes at Poltava. The biggest was a detachment on the eve of the battle of 15,000 men who did not arrive, as usually happens with detachments made on the night or several days before battle. The only good detachments are ones made at the time of the battle.[7]

Reality is often resistant to human volition, even that of a Napoleon. But his correspondence attests to his constant concern when on campaign to keep his corps unified. At the start of the 1805 campaign, he prescribed the following for Ney's corps:

> Once they have arrived at Stuttgart, all his divisions must be very close to one another, so that his whole army corps can be concentrated in less than two hours in line. I do not want partial affairs of divisions.[8]

To Davout, who commanded the Grande Armée's furthest advanced forces in Poland in November 1806, he recalled:

> In all this there is only one very important thing: it is that my three corps and my cavalry can be concentrated in a short space of time, if the Russians' manoeuvres require it.[9]

Some months later, he repeated this lecture to his brother Jérôme:

> General Lefebvre conducted himself well, but you were only able to come to his aid at 11 o'clock in the morning. It is a matter of principle in war that even a corps of 12,000 men cannot be more than one hour away from the bulk of the army. Had Lefebvre been beaten, you too would have been at 11 o'clock; thus you put yourself in jeopardy. Wage war properly. Be there with your 6,000 men unified.[10]
>
> I see you are on a false military track; I see you think that two columns, which place one-and-a-half in the middle, have the advantage. But that does not work in war, because the two columns do not act in concert and the enemy beats them one after the other. Doubtless the enemy must be turned; but first of all one must be unified.[11]

The finest set of lectures on the subject of unifying forces is given in a series
of letters sent to Joseph, when the latter departed to assume an independent
command to conquer Naples at the start of 1806:

> Your major task is to keep all your forces unified and to arrive as rapidly as
> possible in Naples with your full complement.[12]
>   You have five infantry divisions; always keep them unified.[13]
>   I repeat, concentrate all your forces, so that they are not so far apart that
> they cannot be unified in a day.[14]
>   Your army is too dispersed; it must always march in such a way as to be able
> to be unified in a single day on a battlefield. With 15,000 men, I would seek to
> beat your 36,000 and be superior everywhere on the day of battle.[15]

Whatever General Colin's view, we can see here, as earlier, that Napoleon
used the verbs 'unify' and 'concentrate' interchangeably. But the distinction
to be made in time between a relative distance between forces and their
rapid assembly is no less clearly reasserted. Once installed in Naples and
crowned king, Joseph had not done with hearing about it:

> I had warned you not to pay too much attention to Dumas, who is not used to
> war. It seems no one knows where your troops are, that they are dispersed
> everywhere and in force nowhere. [. . .] So march in force. So do not disperse
> your troops. [. . .] So finally adopt a vigorous course of action and keep your
> troops in hand, in echelons, so that you can concentrate 18,000 men on one
> point and crush your enemies. I do not see any unification of forces in your
> letter; all that does not seem clear to me.[16]
>   A long time ago, I told you that you were dispersing your troops unduly.
> Keep them concentrated and what has happened in France will happen to
> you: the British have disembarked several times, but they have been soundly
> beaten and no longer dare to disembark.[17]

Similar orders would be tirelessly repeated to Prince Eugène and the
marshals.[18] In a country like Spain,

> [. . .] you have to concentrate and not march in small packs; that is the general
> principle for all countries, but especially for a country where no communi-
> cations are possible.[19]

12

The Strategic Reserve

Napoleon denounced the idea of General Rogniat, who proposed, with an army of 180,000 men, to hold 60,000 of them in reserve, three, four, or five marches behind:

> [. . .] that is the way to have 180,000 men beaten by 140,000. This is not the secret of Caesar, Hannibal, or Frederick, but of the Soubises, the Clermonts,[1] etc. [. . .] In a word, if you have 180,000 men, enter enemy territory with this army; leave the lame, the hospitals, the convalescents, the depots, and minor garrisons in one or two strongholds. Be a conqueror: exploit the good will of fate and, whatever happens, don't count your chickens before they are hatched.[2]

In tactics, reserves are indispensable for continuing and renewing an engagement. Reserves can restore it if it is in jeopardy. In strategy, by contrast, once the decisive battle has been lost, there is no way to restore anything: 'All forces must be used to achieve [the decisive stage of the battle],' wrote Clausewitz, 'and any idea of reserves, of available combat units that are not meant to be used until after this decision, is an absurdity. Thus, while a tactical reserve is a means not only of meeting any unforeseen maneuver by the enemy but also of reversing the unpredictable outcome of combat when this becomes necessary, strategy must renounce this means, at least so far as the overall decision is concerned.'[3] And Clausewitz cites the 1806 campaign—a trauma from his youth—when the Prussian reserve of 20,000 remained uselessly in Brandenburg, while everything was being decided on the Saale at Jena and Auerstaedt. The example of Prussia in 1806 had also been cited by Napoleon:

> And this old army of Frederick's which had so many heroes at its head—the Brunswicks, Möllendorffs, Rüchels,[4] Blüchers, etc.—beaten at Jena, was unable to carry out a retreat. In a few days, 250,000 men laid down their arms. Yet they were not lacking reserve armies; they had one at Halle, one on

the Elbe, aided by strongholds; they were in the heart of their country, not far from their capital!

Give yourselves every chance of success when you plan to fight a major battle, especially if you are dealing with a great commander. For if you are beaten, should you be in the middle of your provinces, near your strongholds, woe to you![5]

13

Economy of Force

This notion refers not to a form of parsimony, but to an appropriate distribution of forces that does not leave any of them inactive. Everything must be conducive to military operations. Napoleon spoke out against the poor economy of force in the Convention's plan of campaign in 1793:

> At the time, France formed two large armies one of which operated on the enemy's right, the other on his left. This resulted in getting the enemy to recross the Rhine and recapturing Valenciennes, Landrecies, etc. It should not be concluded from this that the plan of campaign was a good one. On the contrary, we succeeded despite its defects and on account of our great superiority in troop numbers. The Austrian army in its entirety was concentrated at Fleurus, and the right wing of the army which the French had there, under Jourdan's command, was as strong as the whole Austrian army combined. Pichegru, who was under Dunkerque, had an inactive army. Had it been unified with Jourdan's, the enemy would have been overwhelmed by superior forces that would have burst like a torrent over his flanks and his rear, and we would have secured a great result without running any risks.[1]

The armies of the *ancien régime* frequently made this kind of mistake:

> The armies of Reserve, the corps not committed to facilitate retreats, are all too often ways of losing battles. It is with all our reserves, all our resources prepared for retreat, that we often lost battles and every campaign in the Seven Years War. This led to a regression in [the art] of war.[2]

The Austrians likewise often sinned against good economy of force. They did not know how to use all their troops or distributed them so badly that they got themselves beaten separately, as at Lonato and Castiglione:

> The Austrian general's plan, which might have succeeded in other circumstances, or against someone other than his opponent, was bound to have the fatal result that it had. And although at first glance the defeat of this great and fine army in such a short time seemed attributable exclusively to the skill of Napoleon, who constantly improvised his manoeuvres against a general plan

fixed in advance, it must be admitted that that plan rested on bad bases. It was a mistake to have corps that had no communication with one another operate separately against a centralized army whose communications were easy; the right could only communicate with the centre via Roveredo and Ledro. A second mistake was to sub-divide the corps of the right and assign different objectives to its various divisions.[3]

In 1809, Napoleon demanded of Fouché that the press should reveal the extent to which the British practised poor economy of force:

> Demonstrate the extravagance of ministers in exposing 30,000 British troops in the heart of Spain to 120,000 French, the best troops in the world, at the same time as they send 25,000 others to waste their time in the swamps of Holland, where their efforts result only in exciting the zeal of the national guards. Make palpable the ineptitude of their plans in thus dispersing their forces and the fact that small packs have always been the hallmark of idiots.[4]

From Paris, and sometimes without taking account of local constraints, Napoleon directed criticisms of the same kind at the marshals who were fighting in Spain—especially at Soult, commander-in-chief of the army of Andalusia, via the intermediary of Berthier:

> The Duke of Dalmatia has 60,000 men under his command; he could leave 30,000 of them under the command of the Duke of Bellune[5] and have more forces than he had at Badajoz. This way of seeking to hold all points at a difficult moment exposes us to great woes.
>
> The Emperor is unhappy that, while the siege of Cadiz risked being raised, the 12th, 32nd, 58th, and the 43rd, forming a division of more than 8,000 men, found themselves dispersed to what were then insignificant points. The six Polish battalions and Perreymond's[6] light cavalry were more than sufficient to remain in observation on this side. Consequently, the four French regiments and Count Milhaud's[7] cavalry division could have been available to support the siege of Cadiz. On the other side, General Godinot's[8] two regiments, forming six battalions, were doing nothing and were useless in their billets.
>
> The disposition of the troops is a general's primary merit and His Majesty notes with pain that the appropriate dispositions have not been made here.[9]

'If a segment of one's force is located where it is not sufficiently busy with the enemy, or if troops are on the march—that is, idle—while the enemy is fighting, then these forces are being managed uneconomically. In this sense, they are being wasted, which is even worse than using them inappropriately.'[10] Such is the meaning conferred by Clausewitz on the notion of economy of force.

14

The Character of
Modern Warfare

Warfare changed with 'Bonaparte's audacity and luck', writes Clause-witz, and also because it became national.[1] The Spanish, Russians, and Prussians ended up fighting with the force of their patriotic feelings. War could no longer be waged as it was at the time of the *ancien régime*'s standing armies. Napoleon does not offer such extended considerations as Clausewitz on this subject. But he extolled the merits of an army that had become national:

It is the soldier who founds republics; it is the soldier who maintains them. Without an army, without force, without discipline, there is no political independence or civil liberty.

When a whole people is armed and wants to defend its liberty, it is invincible.[2]

Soldiers themselves are but the children of citizens. The army is the nation.[3]

Only a national army can ensure the Republic tranquillity within and respect without.[4]

The establishment of paid troops would reduce your real strength, rather than increasing it. If you have a single regiment of paid troops, you must wave goodbye to energetic militias. As soon as the inhabitants see soldiers whom they are paying in place, they say: It is up to them to defend us. Paid troops destroy national energy; it would deprive you of the resources which, as and when necessary, you will find in the courage of your citizens.[5]

Without conscription [...], there can be neither national power nor national independence. All Europe is subject to conscription. Our success and the strength of our position stem from the fact that we have a national army; we must carefully endeavour to maintain this advantage.[6]

Conclusion to Book III

Accepting Jomini's definition, Napoleon only used the term 'strategy' on Saint Helena to refer to what he had hitherto called grand tactics—that is, the movements of armies preceding engagements. This anticipated Clausewitz's definition. This level of the art of war corresponds to what is today called operational art. Strategy is a question less of forms in conception than of moral forces in execution. Clausewitz stressed moral factors to demarcate himself from the geometrical mind-set of Jomini and other theoreticians, because that was precisely the lesson he drew from Napoleon. The Emperor not only waged war paying particular attention to opinion, but he proliferated sentences and advice on the subject. He maintained the morale of his men by inflating their numbers on paper, reducing those of their enemies, and controlling the press.

For Napoleon, military virtue was bound up with discipline. He had much to do in this connection when he became head of the army of Italy. He appealed to soldiers' sense of honour, emulation, and esprit de corps. Reintroducing these values from the *ancien régime* into an army of patriots enthused by the Revolution, he was the true source of the national army of today. Clausewitz clearly understood this and henceforth all armies regarded Napoleon as a kind of creator of the new military profession. This crucial development explains the 'Little Corporal's' immense popularity, despite the suffering occasioned by incessant warfare. To understand the acceleration imparted by Napoleon to the sense of equality, we should have to immerse ourselves in what French society was like under the *ancien régime*, with its privileged orders and social barriers. As an heir of the Enlightenment, he deemed all men equal by nature and his soldiers knew what he thought. In finding the words to speak to them, by regarding his soldiers as human beings, Napoleon was able to nurture in them steadfastness and perseverance, virtues deemed essential by Clausewitz.

With armies that had reached the same level of development, the numerical factor necessarily swung the balance. As we have seen, Napoleon became a master of the capacity to keep his army unified over a broad front and then to concentrate more troops at the decisive point than the enemy. He surprised his opponents with the rapidity of his manoeuvres. He could demand that from his men. We come back to the military virtues: they condition strategic moves. In moving rapidly with the mass of his army, Bonaparte was able to conquer the Austrian columns in Italy separately. He recommended never dividing one's forces in the enemy's presence. The concentration of forces was crucial in space, but also in time. In strategy, there was no point in retaining a reserve if it did not intervene at the decisive moment. Afterwards, all was lost. A good economy of force meant that the mass was concentrated for battle. Napoleon's essential contribution to the history of strategy, understood in the operational sense, consists in this ability to reconcile successively the advantages of dispersal and concentration, in the fusion of these two contradictory processes into a single war operation. Essentially, this is what made him a master in the art of directing a military campaign.[1]

BOOK IV

The Engagement

Clausewitz regards *Gefecht*, translated as 'engagement' here, as 'the essential military activity', whose effects are physical and psychological. Unless directly linked with the two forms of war that are attack and defence—the subjects of Books VI and VII of *Vom Kriege*—what Napoleon had to say about tactics in battles will feature below. We shall make several comparisons with Clausewitz's outline of a treatise on tactics. The only chapter that he completed was specifically concerned with engagement.[1] It is presented in the form of 604 numbered paragraphs, in an aphoristic style sometimes approximating to Napoleonic maxims.

I

The Nature of Battle Today

Napoleon distinguished between modern battles and the battles of antiquity:

> The task which the commander of an army has to acquit is more difficult in the case of modern armies than it was in those of antiquity. It is also true that his influence on the outcome of battles is greater. In ancient armies, the commanding general, at 80 or 100 toises' [150–200 metres] distance from the enemy, ran no danger. Yet he was appropriately positioned to direct all his army's manoeuvres. In modern armies, a supreme commander, positioned at 400–500 toises [800–1,000 metres], finds himself in the midst of the enemy batteries' fire and very exposed. Yet he is at such a distance that several of the enemy's manoeuvres elude him. There are no actions where he is not obliged to come within range of small arms fire. Modern weapons have even more impact if they are correctly positioned; an artillery battery that extends, dominates, and hits the enemy with a broadside can determine a victory. Modern battlefields are much more extensive; this demands a more practised, more penetrating *coup d'œil*. Much more experience and military genius are required to lead a modern army than were needed to lead an ancient army.[1]
>
> Today, the supreme commander is always obliged to advance in the direction of the cannon fire, often within range of grapeshot, and in all battles to go within musket range, so as to reconnoitre, view, and order. Vision has not been extended sufficiently for generals to be able to stay out of the range of bullets.[2]

For Clausewitz, major modern battles find two armies massing their troops, deploying only some of them, and letting them get worn down during long hours of firing, occasionally replacing one part of them by another. At the end of the day, each side draws up a balance sheet and a decision is then taken either to abandon the battlefield or to resume the engagement the next day. Battles took this form because European armies had reached virtually the same level organizationally and in the art of war.[3] Clausewitz accurately represents battles fought from 1809 onwards, when the Austrians began to organize army corps like the French. From 1809, Napoleon won his battles less decisively because his opponents began to fight like him.[4] Until then, he had defeated armies that were still organized like those of the eighteenth century.

Napoleon sometimes anticipated this description by Clausewitz when he pointed to certain details of modern battles. The following reflection on engagement of the infantry and skirmishers affords a glimpse of a war of attrition, which 'smoulders away, like damp gunpowder':[5]

> On an important day, a line passes completely to the skirmishers, sometimes even twice. They must be relieved every two hours, because they are weary, because their muskets are messed and clogged up.[6]

Infantry can do much, but is dominated by artillery fire and cannot hope to carry off the guns. For the artilleryman Bonaparte, the cannon dominates modern battles. He stressed this point more than the infantryman Clausewitz:

> To claim to rush the guns, snatch them with cold steel, or have the gunners killed by sharpshooters—these are chimerical ideas. It can sometimes happen and do we not have the example of strongholds captured by an improvised attack? As a general rule, however, there is no infantry, however brave, that could march with impunity for 500–600 toises [1,000–1,200 metres] without artillery against 17 or 24 well-placed artillery pieces, serviced by good gunners. Before they had covered two-thirds of the distance, they would have been killed, wounded, or scattered. Field artillery has acquired too much accuracy for one to be able to concur with Machiavelli, who, full of Greek and Roman ideas, wanted to fire a single salvo and then send the artillery back behind his line.
>
> Good infantry is doubtless the nerve of the army. But if it had to fight for a long time against superior artillery, it would become demoralized and be destroyed.
>
> In the initial campaigns of the revolutionary wars, what France always had that was best was artillery. I do not know of a single example where twenty guns, properly positioned, were carried off at bayonet point by infantry alone. At the action at Valmy, at the Battle of Jemmapes, at that of Nördlingen,[7] at that of Fleurus in 1794, we had superior artillery to the enemy's, though we often only had two pieces for every 1,000 men. But it was because our army was very large. It could be that a general who was more mobile, more skilful than his opponent, with better infantry at his disposal, would meet with success for part of the campaign, even if his artillery park was highly inferior. But on the decisive day of a general action, he would be painfully aware of his inferiority in artillery and risk losing everything in an instant.[8]

Very great man though he was, Frederick did not have a good understanding of artillery. The best generals are those who hail from artillery. Knowing how to position a battery well is regarded as nothing; but it is a lot. You form batteries behind the first line and suddenly unmask 60 or 80 artillery pieces on a point. You determine victory.[9]

In a note written in January 1789 in Auxonne, the young Lieutenant Bonaparte was already envisaging how artillery would decide things in a battle:

A battery of long guns, with a range of 1,000 toises, can concentrate all its fire on the section of the first line decided by the general, break that line, and sow confusion in the second and third lines. The infantry is then committed against a shaken army. During this, the direction is changed slightly and the horror is focused on a different section of the enemy army.

The battles of Raucoux, Dettingen, and Hastembeck furnish proof of all this.[10]

In November 1813, Prince Eugène, who was commander-in-chief in Italy, was told:

You must not leave Adige without fighting a major battle. Major battles are won with artillery; have plenty of 12-pounder guns.[11]

The Count of Las Cases explains:

He added that today artillery decided the real fate of armies and peoples; that one fought with cannon fire as with one's bare fists and that in battle, as in a siege, the art consisted in concentrating a great deal of shelling on a single point; that once the mêlée had begun, he who had the skill to have an unanticipated mass of artillery suddenly arrive, without the enemy knowing it, at one of his points, was certain to win. Such, he said, had been his great secret and great tactics.[12]

In other words, 'wherever it is the precondition for victory, a significant destruction of enemy forces is the plan's main objective'. Such, at any rate, is the commander's intention.[13]

2

The Engagement in General

The immediate destruction of enemy military forces is the decisive factor in any engagement. Tactics disposes of means to achieve this at a reduced cost. Generally speaking, Clausewitz reckoned that a determined, steadfast, and prompt opponent does not allow time for clever tactical ploys.[1] The lesson of the 1806 defeat always weighed on him. And he understood it well, inasmuch as Napoleon likewise did not encumber himself with undue tactical detail.

Predominance of firepower

One thing is for certain: firepower now prevails over shock:

> Firepower is everything; the rest is very little.[2]

Clausewitz uses a similar expression: 'In most cases, the destructive power of corps-to-corps fighting is insignificant, even nil.' The bulk of the destruction of enemy forces derives from firefights.[3] The predominance of firepower influences selection of the site of the engagement and deployment of the line of battle:

> The discovery of powder has changed the character of war: projectile weapons have become the main weapons; it is by firepower, not shock, that battles are decided today. The musket carries death 50, 100, or 200 toises; batteries positioned at 200, 400, and even 800 toises[4] have a direct, powerful influence on the success of engagements. All elevations and positions situated at this distance have therefore come to be regarded as forming part of the camp or battlefield. To establish their camp or their line of battle, modern generals have had to choose a terrain such they can: 1. bring the majority of their projectile weapons into play; 2. deploy their line in such a way as not to be compelled to keep masses under enemy fire; 3. position their infantry on slopes, walls, elevations, with a suitable command of enemy lines; 4. establish

their batteries in positions that dominate, overrun, extend, strike the enemy lines in the flank or head on, while at the same time positioning themselves so as not to be overrun or struck in the flank or head on by enemy batteries.

It follows from this: 1. that the Moderns have ranged their infantry in three ranks, because it is impossible to derive any advantage from the fire of the fourth, and the thinner the line is, the less of a target it offers to enemy fire; 2. that they have had to extend the line of their army so that it is not overrun, or extended or strung out, and so as to be able, on the contrary, to overrun, extend, string out the enemy, to occupy all the uneven ground that has an influence on the position, and finally so as to expose the smallest possible number of troops to enemy fire; 3. that they have abandoned, and have had to abandon, the custom of camping clustered, like the Romans, in a small square surface and instead camp in two lines, each made up of three ranks for the infantry and two for the cavalry, with the battlefield in front of the army.[5]

Battle dispositions

The disposition of troops for battle therefore obeys a few clear principles:

Duke Ferdinand's plan at the Battle of Crefeld[6] was against the rule that runs: *Never separate the wings of your army from one another, in such a way that your enemy can position himself in the gaps.* He divided his line of battle into three parts, separated from one another by empty spaces and narrow passes. He turned a whole army with a corps left exposed and unsupported, which went on to be surrounded and taken prisoner.[7]

As Bertrand recounts, Napoleon had asserted his distinctiveness in his battle dispositions:

I always had a different system from other commanders: I never sought to envelop the enemy army. On the contrary, I was often overrun and always occupied less space than the enemy, keeping my reserves in hand ready to deliver the decisive blow. At Wagram I occupied a third less [space] than the enemy.[8]

It was at Wagram that Napoleon led the largest number of men in a battle: more than 180,000 against around 140,000 Austrians. The front was enormous. For Clausewitz, the advantages of envelopment decline as fronts grow larger. The manoeuvre of envelopment disperses forces over an enormous space, reducing their effectiveness. They need much more time to cover a distance, whereas those who are enveloped have more facilities for employing their forces at different points. Overall unity is also weakened by the greater spaces that have to be covered by information and orders. When only a few battalions are involved, such disadvantages are well-nigh non-existent. But with large armies they become highly significant: 'For the difference between the radius and the circumference stays the same, while the absolute differences always increase with the expansion of fronts. It is precisely these absolute differences that have to be taken into account.'[9] The attacker has a tendency to choose the enveloping form and the defender the narrower form of front. But it can prove advantageous to the former to concentrate his forces against a point and adopt a more in-depth form of organization, allowing for use of his forces in succession.[10] At Austerlitz, Napoleon first allowed himself to be attacked by the Austro-Russians' enveloping manoeuvre. Inferior in numbers, he succeeded in beating them thanks to his dispositions:

> It is not extraordinary that the French army seemed immense in Russian eyes. They so dispersed their troops over the battlefield, and the French employed them so well, that the large Russian army seemed to be one division and the French army, which was smaller, seemed countless. Thus Emperor Alexander said to General Savary the following day: 'You were less numerous than me and yet you were everywhere stronger.' 'Such is the art of war', the general answered him. [...]
> No, the French did not spend the night positioning their troops near Pratzen. But, by a system quite unlike that of the Russians, they kept them concentrated so that these 65,000 men were at the Emperor's disposal like a battalion in the hands of a good major—ready for anything, even to withdraw if the enemy was wise.[11]

The balance of forces was to the advantage of the French at the main point. Clausewitz, paraphrasing Napoleon, also saw an advantage in this kind of disposition, 'a subjective reason for the director or leader of the war: having the main part of his forces more closely under his control.'[12]

Destroying the armed forces of the enemy means causing a reduction in their forces that is proportionally greater than the reduction suffered oneself. Yet experience shows that in battles the difference between victor and vanquished in terms of losses is often not very great. Napoleon recognized this before Clausewitz in two new comparisons with battles in antiquity:

> It is commonly held that the Ancients' wars were more sanguinary than those of the Moderns. Is this true? Modern armies fight every day, because cannons and rifles carry a long way; the front lines, the posts fire at one another and

often leave 500 or 600 men on the battlefield from both sides. Among the
Ancients, the engagements were rarer and less bloody. In modern battles, the
casualties caused by the two armies, which are almost equal as regards dead and
wounded, are higher than those of battles in antiquity, which only affected the
beaten army.[13]

At Pharsalus, Caesar only lost 200 men, while Pompey lost 15,000. We see
the same results in all the battles of antiquity. There is no example of this in
modern armies, where casualties in killed and wounded are more or less high,
but in a proportion of 1 to 3. The great difference between the victor's
casualties and those of the vanquished consists above all in prisoners: once
again, this results from the character of the weaponry. The Ancients' project-
iles generally did little damage. Armies first and foremost closed with one
another with cold steel. It was therefore natural that the loser lost many men
and the victor very few. Modern armies, when they close, do so only at the
end of the action and when much blood has already been shed. There is no
winner or loser for three-quarters of the day; the casualties caused by firearms
are pretty much equal on both sides. With its charges, the cavalry affords
something analogous to what occurred with armies in antiquity: the loser's
casualties are proportionately much greater than the victor's, because the
fleeing squadron is pursued and sabred, suffering great damage without
inflicting any.[14]

It is in retreat that the loser suffers his greatest losses. These are expressed
more in terms of moral than physical forces. The moral balance tilts
following loss of ground and as a result of the fresh reserves maintained
by one of the protagonists. An attempt must be made to force the enemy
to commit his reserve prematurely. Napoleon says that he acted thus at
Marengo:

> I saw that the Austrians had not employed their reserve; and, in such cases, the
> great point is to try to get the enemy to use all his forces, while sparing ours;
> and to engage him to attack us on the flanks as long as he does not realize his
> mistake. For the difficult thing is to compel him to employ his reserve.[15]

The one who then finds himself in a situation of moral inferiority loses his
order and unity. He beats a retreat. The victor must exploit this to harvest
prisoners and guns, which will be the veritable trophies of his victory.[16]
Battle arrangements must tend to this end. They concern tactics, but it is
clear (says Clausewitz) that '[t]he risk of having to fight on two fronts, and
the even greater risk of finding one's retreat cut off, tend to paralyse
movement and the ability to resist'.[17] Napoleon says virtually the same
thing when advocating supporting the flanks:

The important thing in battle organization and marching towards the enemy is to have the flanks supported. One is always undermined and defeated from the flanks. I therefore never lost sight of the flanks. They must always be supported by small or large squares that should sustain one another and resist, even when the line is broken. That is the main point which must never be lost sight of in any position in the face of the enemy.[18]

If the flanks are forced open, they jeopardize the line of retreat. 'Towards the end of the battle, consideration of the road of retreat becomes increasingly important and for that reason control of it becomes a valuable means of decision-making.' Clausewitz agrees with Napoleon in adding that 'as far as possible, the battle plan will be adjusted to this crucial point from the outset'.[19] In any engagement, there is a real instinct to preserve the security of one's own line of retreat at any price and to seek to capture the enemy's. It is ubiquitous and becomes the pivot of nearly all tactical and strategic manoeuvres. It is a fundamental expression of the instinct of self-preservation, as Napoleon implies in this passage from the *Mémorial de Sainte-Hélène*:

> He said that one could never make gunners fire on massed infantrymen when they found themselves under attack by an opposite battery. It was natural cowardice, he said gaily, a violent instinct for self-preservation. An artillery man [Gourgaud] among us rebelled against such an assertion. 'But that's how it is', continued the Emperor. 'You immediately take guard against whoever attacks you; you seek to destroy him so that he does not destroy you. You often stop firing so that he will leave you alone and return to the massed infantry, who are of a quite different relevance for the battle, etc.'[20]

Detailed tactics

As regards the detail of the tactical disposition of troops, no absolutes should be prescribed:

> Should the army be ranged in battle in several lines and what distance should be put between them? Should the cavalry be in reserve behind the infantry or positioned on the wings? Should all one's artillery be brought into action from the start of the battle, since each piece has enough to feed its firepower for twenty-four hours? Or should half of it be kept in reserve? The answer to all these questions depends on many circumstances: 1 on the number of troops that make up army, on the proportions between infantry, cavalry, and artillery; 2 on the relationship between the two armies; 3 on their morale; 4 on the objective one has set oneself; 5 on the nature of the battlefield; 6 on the

position occupied by the enemy army and the character of the leader commanding it. One cannot and must not prescribe anything absolute. There is no natural order of battle. Anything that might be prescribed would do more harm than good.[21]

There is no battle plan that guarantees victory. The forms are details:

It is ridiculous to attribute Frederick's successes to the details of his military organization of oblique formations, to the deployment of his column. How that is done matters little. In all ages, a battalion leader has positioned and deployed his battalion; a brigadier-general has positioned his brigade, the lieutenant-general his division. These details have but little influence on the success of a campaign and a battle. [. . .] Basically, all of that does nothing. All formations are good and, although there might be a better one, its influence on the war's outcome is limited. It is not these details that account for major successes.[22]

[. . .] since the invention of firearms, the way to occupy a position in order to camp or give battle depends on so many different circumstances that it varies according to these circumstances. There are even several ways of occupying a given position with the same army; the military *coup d'œil*, the experience or genius of the supreme commander determines it: that is the main thing.[23]

To those who rambled on about the most effective kind of infantry fire—by row or line—Napoleon responded:

In the face of the enemy, the only feasible kind of firing is at will, which starts on the right and the left of each platoon.[24]

This refers to firing in line at will.[25] As set down in the regulation of 1 August 1791, the first two rows fire together in a line, starting on the right and the left, with the third row confining itself to re-loading and passing the loaded weapon to the second. One should not seek to prevail by the complexity of one's schemes, but on the contrary by their simplicity. As we saw in Book III, Napoleon said and repeated this for all spheres of the art of war. One must adapt to circumstances. The French infantry synthesized the reflections of the *ancien régime* and the experience of the revolutionary wars. Under the Empire, it long possessed an advantage in flexibility over its rivals. It made the transition much more rapidly from columns or *ordre profond*, easier for marching, to line or *ordre mince*, which arrayed maximum firepower. This was a legacy from which Napoleon benefited and that often gave him the advantage in engagements:

Column order is a combat order when circumstances require it. That is why our tactics provide us with the means to make a rapid transition from thin to deep order. If one fears cavalry, one should march in columns, at a platoon's distance, so as to be able to form the battalion square by platoon on the right and left in battle.[26]

One does not have to be in columns or lines because a battle comprises an alternative between engagements and marches, but because the circumstances of attack or defence require one to be in columns or lines.[27]

Napoleon rarely prescribed a precise tactical disposition for his army corps and divisions, preferring to leave the initiative to his marshals and generals.[28] A plan could not determine the detail of a firefight because it lasted a long time and unforeseen events occurred in it. For Clausewitz, combat dispositions must afford an even greater margin of manoeuvre to large units than small ones.[29] At the start of the Prussian campaign in 1806, the Emperor nevertheless recalled a principle that facilitated proper exercise of command during an engagement:

In all your battle formations, make it a principle to position yourself in two or three lines, so that the same division forms the right of the two or three lines, another division the centre of the two or three lines, another division the left of the two or three lines. You saw the advantage of this formation at Austerlitz, because a lieutenant-general is at the centre of his division.[30]

When, in the spring of 1813, the Grande Armée lacked cavalry and the infantry was composed of young conscripts, it was indispensable for the latter in battle to be able to rapidly form itself into a square:

You will give orders that the manoeuvre of rapidly forming a battalion square is often performed, bending behind the last divisions of the battalion, at a platoon's distance and firing in lines. This manoeuvre is the most necessary one for colonels to know well, because the slightest hesitation can jeopardize the company.[31]

Clausewitz only had experience of land engagements between Europeans. In Egypt and Syria, Napoleon experienced what is today called a form of asymmetrical war. In particular, he drew the following reflection from it:

The Ottoman is generally dextrous, strong, brave, and a good shot; he defends himself perfectly behind a wall. In open country, however, the general lack of discipline and tactics renders him far from formidable. Individual efforts are unavailing against an overall movement.[32]

In going to Egypt, Napoleon also had the opportunity to appreciate the conditions of maritime engagements:

> Firearms, which have produced such a great revolution on land, have also brought about a very great one at sea. Battles there are decided by cannon fire and, as the impact of the gun depends on the position one occupies, the art of manœuvring and taking up position decides naval battles. The most steadfast troops cannot do anything in a kind of engagement where it is well-nigh impossible to come to grips. Victory is decided by 200 gun muzzles, which disorientate, shatter manoeuvres, sever masts, and spew forth death from afar. Naval tactics have therefore acquired a quite different importance. Maritime engagements no longer have anything in common with engagements on land. The art of the gunner is subject to the art of the manoeuvre that moves the vessel, provides batteries with enfilade positions, or presents it to the cannon-balls in the most advantageous manner. If, to the particular tactics of each vessel, you add the principle of general tactics that every vessel must man-oeuvre in the most appropriate way in the position, the circumstance, in which it finds itself to attack an enemy vessel, unleash on it as many cannonballs as possible, you will have the secret of naval victories.[33]

3
The Battle: Its Decision

The decisive moment and 'the event'

Every engagement 'reaches a point when it may be regarded as decided, so that to reopen it would constitute a new engagement rather than the continuation of the old one. The accurate perception of that point is very important in order to decide whether reinforcements would be profitably employed in renewing the action.'[1] From another standpoint, 'the decisive moment is the event that leads one of the two war leaders to make the decision to withdraw the troops'.[2] Napoleon alluded to this turning–point on several occasions:

> There comes a point in engagements when the smallest manoeuvre decides and confers superiority: it is the drop of water that creates an overflow.[3]
>
> The outcome of a battle, the Emperor used to say, is the result of an instant, a thought: one comes forward with various schemes, mingles, fights for a period of time, the decisive moment comes, *a moral spark* decides, and the smallest reserve does the job.[4]

This decisive moment is bound up with the combatants' psychological fluctuation:

> In every battle, a moment comes when the bravest soldiers, having made the greatest efforts, feel disposed to flee. This terror derives from a lack of confidence in their courage: nothing but a slight opportunity, a pretext, is required to give them this confidence; the great art is to create it.
>
> At Arcola, I won the battle with 25 cavalrymen. I seized on this moment of weariness in the two armies. I saw that the Austrians, even though they were old soldiers, would have asked for nothing better than to find themselves in their camp, and that our French, although brave, would have liked to be under their tents. All my forces had been committed; several times I had been obliged to reform them in battle. I had left no more than 25 guides. I sent them on the enemy's flanks with three trumpets that sounded the charge. A general

cry was heard in the Austrian ranks: 'The French cavalry are coming!' And they took to their heels. It is true that one must seize the moment. An instant sooner or later, and this attempt would have been unavailing; if I had sent 2,000 horse, the infantry would have made a *90 degree turn into line*; covered by its guns, it would have fired a good salvo and the cavalry would not even have attacked.

As you can see, two armies are two bodies that encounter and terrify one another; there is a moment of panic terror; it is necessary to know how to seize it. All this is merely the effect of a mechanical and moral principle: it requires nothing but habit. When you have been present at several actions, you discern this moment without any difficulty: it is as easy as adding up.[5]

At Marengo, the intervention of Kellermann's cavalry was decisive at the right moment because the Austrian cavalry

was half a league away: it needed a quarter of an hour to arrive and I have observed that it is always these quarter-hours that decide the outcome of battles.[6]

The report of the Battle of Lützen alluded to this moment:

His Majesty judged that the crisis moment which decides the winning or losing of battles had arrived: there was not a moment to lose.[7]

Marshal Gouvion Saint-Cyr provides some details on the decisive moment at Lützen, characterized by Napoleon as an 'event':

He replied to me that he did not have any preference for attack on the centre over attack on the wings; that he made it a principle to engage the enemy with the maximum resources; that the closest corps being committed, he allowed them to proceed, without worrying unduly about their good or bad fortune; that he only took great care not to succumb too readily to requests for help from their leaders. As an example he cited Lützen, where, he said, Ney had asked him for the most rapid reinforcements, still having two divisions that had not been committed. He assured me that in the same action another marshal had asked him for some even before he faced an enemy. He added that it was only towards the end of the day, when he saw that the weary enemy had risked the bulk of his resources, that he concentrated what he had been able to keep in reserve, launching onto the battlefield a powerful mass of infantry, cavalry, and artillery; that the enemy not having anticipated this, it caused what he called an *event*, and that he had nearly always secured victory thus.[8]

Chaptal confirms this account and adds that the event always occurs at a precise point:

He often said that a soldier who had fought for 4–6 hours asked only for a pretext to give up the fight, if he could with honour, and that the approach of a reserve corps, whatever the number, was nearly always a sufficient reason for him deciding to do so.

He added that when one had to fight superior forces, it was necessary to astound them with audacity; and that in this case he had always succeeded in concentrating his forces to bring them to bear impetuously on a point and create disarray among part of the enemy army. A skilful general, who knew how to energetically exploit this initial advantage, was sure of forcing his enemy to retreat. In the space of an hour, one loses all the people one would have lost without success in manoeuvres, marches, and counter-marches.[9]

The phases of the battle

Clausewitz comes to assess the moment of the decision. He estimates that, even if the engagement has been balanced, the two commanders-in-chief are relatively close to the final decision they must take when five-sixths of their troops have already tested their strength in the act of destruction. 'Nothing but a minor effort is required to force the decision. [. . .] When everything is still in the balance, success generally goes to the one who forces the decision, for the positive principle is of much greater weight at the point when a battle takes a decisive turn than at its start.'[10]

As in a tragedy, Napoleon identifies three major moments, three acts in a battle:

A battle is a dramatic action, with a beginning, middle, and end. The battle order adopted by the two armies, the initial moves made to come to blows, are the introduction; the counter-moves made by the army under attack constitute the nub; this induces new dispositions and brings on the crisis out of which the outcome or dénouement is generated.[11]

For Jean Guitton, our own situation and intention correspond to the thesis. The opponent's situation and intention correspond to the antithesis. The gravest moment is the synthesis, 'capable of creating an imbalance in spirit that is favourable to you. This most often comes about by overcoming negation.' Guitton observes that at Marengo and Austerlitz Napoleon initially experienced difficult moments; and that this attaches to the nature of human affairs, 'which has it that the most undoubted successes are lapses made good, negations taken up and sublimated—which is the essence of the dialectical process. But from the rectifier that is the great commander this demands a non-systematic mind capable of not being unhinged by a surprise, but instead, as it were, summoning the aleatory because his genius contains the faculty of the counter-aleatory. He knows that victory readily emerges from a semblance of defeat, because the opponent's most vulnerable moment is when he believes himself the victor.'[12]

For his part, Clausewitz breaks the engagement down into two acts: the destructive act and the decisive act.[13] He also perceives a whole wherein partial engagements combine to form an overall outcome. The decision resides in the latter. The general and army that have proved capable of conducting the engagement with the most economy of force will assert the psychological effect of a reserve, which is the guarantee of victory. 'In modern times the French must be credited with great mastery in this respect, particularly under the leadership of Bonaparte.'[14]

Used appropriately, cavalry makes it possible to decide certain phases of the engagement. It must not necessarily be kept back for a final decision. A major battle contains successive phases of fighting. Napoleon said this in connection with Waterloo:

> Cavalry charges are equally good at the beginning, in the middle, or at the end of a battle: they must be executed whenever they can be made on the flanks of the infantry, especially when the latter is taken and engaged frontally.
>
> The British general did very well to have a charge made on the flank of the French infantry, since the squadrons of cuirassiers who were due to support them were somewhat to the rear.
>
> General Milhaud did even better to have those two British regiments charged by his cuirassiers and to destroy them.[15]

4

Mutual Agreement to Fight

No engagement can occur without mutual agreement. However, in modern warfare those who really want to attack normally end up doing so, reckons Clausewitz. 'It is true that while the defender cannot nowadays decline an engagement, he can avoid it by abandoning his position and thereby his object in holding it. But this kind of success already constitutes the better part of victory for the attacker—the recognition of his provisional superiority.'[1] On 10 June 1807, the Grande Armée attacked the Russian army dug in at Heilsberg. The following day, it was ready to resume the engagement, but the Russians shied away. Napoleon observed that:

> [. . .] in such large armies, where it takes twenty-four hours to get every corps in position, you can only have partial actions when one of them is not disposed to end the dispute bravely in a general action.[2]

In a related line of thinking, Napoleon several times lambasted generals who gave battle lightly. Jourdan was referred to twice:

> Jourdan engaged in the Battle of Stockach unjustifiably; he claimed it had been insinuated that he should give battle.[3] A general should only decide to give battle when he has hopes of victory and what hopes could he entertain in setting 40,000 men against 65,000?[4]
>
> A battle is always a serious thing; victory depends on nothing, on a hair. One always runs a great risk in fighting it. It should not be started, unless one is forced into it, when the enemy has severed your line of operations.[5]

It might be thought that these reflections, formulated on Saint Helena, are those of a weary man whose career was at an end. But on the eve of Austerlitz, while preparing his trap for destabilizing the Austro-Russian army, Napoleon wrote to Talleyrand:

> Tomorrow there will probably be a very serious battle with the Russians. I have done a lot to avoid it, for it involves pointless bloodshed. [. . .] Do not

be alarmed: I am in a strong position; I regret what it will cost—and for virtually no objective.[6]

After Marshal Jourdan and King Joseph had lost the Battle of Talavera against Wellington in July 1809 in Spain, Napoleon proved even more explicit on the conditions in which battle can be given. He wrote to General Clarke, Minister of War, for him to communicate to Jourdan

> [. . .] that the enemy's position required prior reconnaissance and that my troops were led indiscriminately, well-nigh to butchery; that finally, having decided on battle, it was fought half-heartedly, since my arms suffered an affront and yet 12,000 men remained in reserve without firing; that battles must not be given if one cannot calculate a 70 per cent chance of success in one's favour; even that one must only give battle when no new opportunities can be hoped for, since by its very nature the outcome of a battle is always uncertain; but that once it has been decided, one must vanquish or perish, and that French eagles must bow in retreat only when all of them have made an equal effort.[7]

5

The Principal Battle:
Its Decision

Using one's aces

Napoleon thought it necessary to have the guts to do battle when the odds were in one's favour. In 1796, Archduke Charles of Austria did not have this spirit of decisiveness when confronting the French army of the Rhine:

> When two armies are doing battle with one another; and when one, like the French army, must effect its retreat over a bridge, while the other, like the Austrian army, can withdraw on all points of the semi-circumference, all the advantages lie with the latter. It is for it to be audacious, to strike major blows, to manoeuvre the flanks of its enemy. It holds the aces; it only remains for it to use them.[1]
>
> When one is within range to strike at the heart, one must not let oneself be distracted by contrary manoeuvres.[2]

All units must 'march towards the guns'. On 13 October 1806, an aide de camp had to leave at the gallop for the meeting of the cavalry generals d'Hautpoul, Klein, and Nansouty:

> He will make it known to them that, if they hear cannons in the vicinity of Jena, they must speed up their march and send officers to warn of their arrival.[3]

In Spain, Napoleon criticized Marshal Victor for having left one of his divisions to fight alone and for not having gone to its aid. He had Berthier write to him:

> You know that the first principle of war has it that when success is in doubt you go to the aid of one of your corps when it is under attack, because your salvation depends on that.[4]

Resolution and tenacity

When the decision is taken to do battle, the supreme commander must demonstrate determination and exhibit the utmost energy:

> Once ranged in battle, the supreme commander must reconnoitre the position of the enemy at daybreak, his movements during the night, and on that basis make his plan, send his orders, and direct his columns. [...]
>
> At the start of a campaign, whether to advance or not should be carefully considered. But once the offensive has been undertaken, it must be supported to the bitter end. For independently of the honour of arms and army morale, which is lost in a retreat, and the courage given one's enemy, retreats are more disastrous, more costly in men and material than the most bloody encounters—with the difference that in a battle the enemy loses almost as many men as you, whereas in a retreat you suffer casualties and he does not.[5]

Engaging involves the volition to stick to one's decision. When one gives battle, one engages in a process from which one cannot withdraw lightly. It is necessary to demonstrate firmness:

> The glory and honour of arms is the first duty that a general who gives battle must consider; the safety and preservation of his men is only secondary. But it is also in such audacity, such tenacity, that the men's safety and preservation are to be found.[6]

Reacting to what is said here of the secondary character of the men's safety, the British writer David Chandler regards it as symptomatic of a 'continental' great power with abundant reserves of manpower.[7] As we have said elsewhere, the Revolution in fact enabled France to create more massive armies and Napoleon was the first to find the means of profiting from this. Clausewitz, for Chandler another representative of a continental power, was in any event in agreement with the Emperor in magnifying the virtues of resolution in combat. He reckoned that it was the only way of 'controlling what is approximate. Resolution saves us from half-measures and is the most brilliant quality in conducting a major engagement.'[8]

Napoleon warned Marshal Marmont, with whose boasting he was familiar, when the latter bragged of being able to defeat Wellington in a major battle:

> But once the resolution has been made, it must be kept: there are no *ifs* or *buts*. You must select your position under Salamanca, be the victor, or perish with the French army on the battlefield you have chosen.[9]

Paradoxically, Marmont was to criticize the Emperor for his failure to respect this principle at the Battle of Borodino. In refusing to commit the Guard to finish off the Russians, he deprived himself of a great result and allowed them to save themselves:

> Thus Napoleon was unfaithful to one of his favourite principles, which I heard him repeat: 'it is those who retain fresh troops for the following day of a battle who are nearly always beaten'. He added: 'If it serves a purpose, the very last man must be committed, because the day after a complete success one encounters no obstacles. Perception alone ensures new triumphs for the victor.'[10]

Chaptal reports that Napoleon was convinced that tenacity alone often won battles. He heard him relate that he had fought with General Alvinzy at Arcola for five consecutive days, without either of them having the advantage:

> Because, he said, I was younger and more headstrong than him, I had no doubt that he would end up ceding the ground to me and I stuck by this conviction. On the fifth day, at 5 o'clock in the evening, he decided to order the retreat.[11]

Napoleon often said, adds Chaptal, that Alvinzy had been his best opponent and that was why he had never said anything about him in his reports, whether good or ill, whereas he had praised Beaulieu, Wurmser, and Archduke Charles, of whom he was not afraid. Fierce determination to win characterized Napoleon more than any other general. It was by persisting in engagements that he amazed his first adversaries in Italy.[12] In the sum total of partial results that forms the overall result in a battle, Clausewitz identified three essential elements: the rapidity with which troops are seen to melt away, the loss of ground, and the psychological strength of the leader's consciousness. Napoleon likewise takes the leader into account:

> I take only half the credit in the battles I won—and that's a lot. It is enough for the general to be named. The fact is that the army wins the battle.[13]

The principal battle is 'the true centre of gravity of the war', writes Clausewitz, and 'victory must be pursued so long as it lies within the realm of the possible'.[14]

6

The Principal Battle: The Use of the Battle

Seeking battle

Clausewitz characterizes as a 'dual law' that fact the 'destruction of the enemy's forces is generally accomplished by means of great battles and their results; and, the primary object of great battles must be the destruction of the enemy's forces'.[1] However, 'there have been minor engagements [...] in which favorable circumstances have resulted in the destruction of a disproportionate number of enemy forces'. He cites the Battle of Maxen, where on 20 November 1759 the Austrian Field-Marshal Daun encircled Finck's 15,000 Prussians and forced them to capitulate after a brief engagement. Similar in kind was the capture of Ulm, where 30,000 Austrians laid down their arms. Napoleon celebrated it in a proclamation to the Grande Armée:

> Soldiers, I had announced a great battle to you. But thanks to the enemy's poor ploys, I have been able to achieve the same success without running any risk.[2]

Two days before the Battle of Jena, Napoleon realized that his army was in the process of enveloping that of the Prussians. He wrote to Murat to attack the Prussian columns on the march and added:

> Two or three advantages of this kind will crush the Prussian army, possibly without any need for a general encounter.[3]

These two quotations have the effect of qualifying what one might imagine of a 'pugnacious' Napoleon. We saw above the reservations he expressed in this connection. For while a battle 'should not simply be considered as mutual murder', because 'its effect [...] is rather a killing of the enemy's spirit than of his men', it nevertheless always has blood as its price, and '[a]s a human being, the commander will recoil from it'.[4] The quotations in Chapter 5 indicate the human dimension to Napoleon. Perhaps Clausewitz did not sufficiently perceive this, when he wrote in connection with the victory of Ulm: 'The surrender at Ulm was a unique event, which would not have happened even to Bonaparte if he had not been willing to shed blood. It must in fact be looked upon as the aftermath of the victories that he had won in earlier campaigns.'[5] This description is accurate, since, as we have seen, Napoleon had initially announced a great battle to his soldiers. Clausewitz means that, more than any other military leader before him, Napoleon was determined to seek battle in order to finish things off as rapidly as possible. Marked to an extent that cannot be overstated by the defeat of his country in 1806, Clausewitz wrote this paragraph, which was to earn him a bad reputation, especially after the First World War: 'We are not interested in generals who win victories without bloodshed. The fact that slaughter is a horrifying spectacle must make us take war more seriously, but not provide an excuse for gradually blunting our swords in the name of humanity. Sooner or later someone will come along with a sharp sword and hack off our arms.'[6] For Clausewitz, Napoleon was the first to seek battle to this extent from the inception of each of his campaigns. The proclamation of Elchingen cited above proves it, as does his correspondence. But we must not forget the circumstances. On 10 October 1806, the Emperor felt that they were favourable to him, and that the enemy could not but make mistakes, when he wrote to Soult:

I greatly desire a battle.[7]

Seeking battle formed part of his strategy, bound up with his politics, insofar as a crisis had to be resolved as quickly as possible. This is attested to, for example, by his instructions to Ney and Marmont after the resumption of hostilities in August 1813. Napoleon knew that he must strike quickly, because his strength could not increase, unlike that of the allies:

It seems to me that the current campaign cannot result in a good outcome unless there is first a great battle.[8]

He reiterated the point to his Foreign Relations Minister:

Moreover, as we cannot arrive at any result without a battle, the best thing that can happen is for the enemy to march on Dresden, since then there will be a battle.[9]

Principal battle and campaign plan

For Napoleon, a battle was inherently bound up with his plan of campaign, with a 'system', as if it possessed no intrinsic value, but was merely one factor among others. The outcome did not have to be immediately legible or the success incontestably brilliant at once. The operational sequels to the battle were of even greater significance:

> At Marengo, the battle was nearly inconclusive. The enemy rallied his army in front of the Bormida.[10] If he had been in a natural position, he could have stayed there or withdrawn at pleasure. But it was necessary to cross the Po, which was guarded, his line of operations was cut, and he had either to capitulate or break through. After Eckmühl, the business at Landshut cut the archduke [Charles] off from Vienna.[11] I arrived there before him being close to Vienna. At Eylau, Bennigsen[12] acknowledged that, had he not intercepted the officer and the dispatch that informed him of my movements, he would have been lost.[13] He would not have been able to recover in time and restore his line of operations; he would have been thrown on Elbing and doomed.[14] These successes were due to my plan of campaign, my system. They are all different and Frederick has nothing that could be compared to that. At Austerlitz, the Russian army was cut off from its line of operations and if Davout had had made a more determined march, any retreat into Hungary would have been impossible.[15]

Napoleon was in fact the first to include the battle in a plan of campaign which, while developed in detail as events unfolded, nevertheless corresponded to a general idea, to a constant concern, in any event, to combine all the army corps' movements with what has since been called a system of operations. The operational level of war emerged with Napoleon, who combined the movements of his army corps on a vast scale and made them conducive to a single objective: stripping the enemy army of its capacity for action, whether by psychological effects, a series of engagements, an enveloping manoeuvre, a battle or, above all (as indicated above), by blocking its line of retreat.[16]

The link between the principal battle and the plan of campaign is clear in the case of Austerlitz:

> But the Battle of Austerlitz itself was only the result of the Moravian campaign plan. In an art as difficult as that of war, it is often in the system of campaign that one conceives the system of a battle: only highly experienced military

men will understand this. The people who were with the Emperor heard him
say a fortnight earlier on the heights on his return from reconnoitring
Wischau: 'Reconnoitre all these heights well; this is where you will fight
before two months are out.' At first they did not pay any attention to his
words, but the day after the battle they remembered them.[17]

In the Moravian campaign, the Emperor had understood that the Russians,
not having a top general, were bound to think that the French army's retreat
was to Vienna; were bound to attribute great importance to intercepting
them. But the army's retreat in the whole Moravian campaign was never
intended to be to Vienna. This single circumstance falsified all the enemy's
calculations and prompted manoeuvres from him that led him to his doom.

[. . .] He wanted the Russians to carry out false manoeuvres and make
mistakes, all resulting from his plan of campaign in Moravia—a plan that the
enemy neither must nor could penetrate.

Two days earlier, crossing the heights of Pratzen, the villages of Sokolnitz,
Telnitz, and Moenitz, the Emperor also said: 'If I wanted to prevent the
enemy from crossing, this is where I would position myself; but I would only
have a run-of-the-mill battle. If, on the other hand, I withhold my right wing
by withdrawing it towards Brünn, and the Russians abandon these heights,
even if there are 300,000 of them, they are caught and irretrievably lost.'[18]

The 'conception of the system of battle in the system of campaign' signifies
that tactics and strategy are interwoven on a tactico-strategic continuum.
Before Napoleon, the binary division of the art of war into strategic and
tactical levels involved a strict distinction between the two. The intermediate zone sensed by Guibert, and dubbed 'grand tactics' by him and Napoleon, heralded a third dimension: the operational dimension.[19] Kutuzov's
account of Austerlitz was so dishonest, as only the Russians knew how
(especially in the case of battles fought far from home), that Napoleon could
not resist reacting and he did so by confiding some crucial aspects of his
strategy. The 'run-of-the-mill battle' would perhaps not have been decisive.
It would have been limited to a tactical success. The attraction of the
Russians towards the withheld French right yielded greater results because
it completely destabilized their configuration. The French had seized the
heights of Pratzen, whence they could cut the Austro-Russian army in two.
The latter had fallen into the trap; it had sought to cut the French army off
from its supposed line of retreat, connected more with the campaign than
the battle. That is why the trap conceived by Napoleon was bound up, as he
put it, with this plan of campaign. At a higher level, his desire to fight not a
run-of-the-mill battle, but a decisive battle, stemmed from his campaign
plan to the degree that he sought to end the war by striking this great blow.

7
Strategic Means of Exploiting Victory

[...] winning is nothing; success has to be exploited.[1]

The conduct of the Austrian Marshal Daun, victorious over Frederick II at Kolin on 18 June 1757, illustrates what should not be done:

> The conduct of Marshal Daun, which is assumed to have been based on the resources that he knew existed in Prague, seemed good just after the Battle of Kolin. But he was guilty of not having exploited his victory: it was almost not worth winning! After twelve days of deliberating, he finally decided to make for Lusace. [...] The Austrian generals in this campaign were extremely timid. Although their troops fought with courage, their leaders did not show any confidence in them.[2]

Even Frederick, the most enterprising general of his age, scarcely pursued the enemy he had defeated. It was only after Leuthen (5 December 1757) that he organized a real pursuit of the Austrians. Even so, it was limited. Napoleon was right to say that he distinguished himself in this respect:

> Frederick took 5,000–6,000 prisoners and captured guns during the battle, but nothing the next day. I did the opposite. What I captured in my battles did not amount to much, but it was the next day, two days, three days, four days afterwards. That is the difference.
>
> The enemy might not have lost the Battle of Jena, but if he lost it, if he had a semi-success, he could no longer withdraw. That is my art. At Jena, basically, what I captured on the battlefield was not much—somewhere over sixty cannon—but the following day, at Weimar, Erfurt, etc. The enemy no longer had a line of retreat; he was doomed. At Ulm, the battle was run of the mill, but the following day we captured 33,000 men in the town, 10,000 at Memmingen.[3]

Let us take some orders issued during these exemplary campaigns. After the engagement at Elchingen (14 October 1805), and the flight of Archduke

Ferdinand, who escaped from Ulm into Bohemia, Murat was sent to pursue him and soon caught up with him. Napoleon wrote to Murat:

> I congratulate you on your success. But no resting: pursue the enemy with your sword in his back and cut him off from all communications.[4]

In mid-November, it was the turn of Kutuzov's Russians to be pursued, not always at the pace desired by the Emperor:

> If I was annoyed with someone yesterday, it was Walther,[5] because a cavalry general must always pursue the enemy with a sword in his back, especially during retreats; because I do not want them to spare the horses when they can capture men; and because I am aware that what is being done today could have been done yesterday.[6]

Physical and psychological overwork also affects the victor and, 'for purely human reasons less is achieved than was possible. What does get accomplished is due to the supreme commander's *ambition, energy,* and quite possibly his *callousness.*'[7] The Battle of Jena on 14 October 1806 was followed by the most terrible pursuit. Major results had already been achieved five days later:

> The campaign's primary objective has been accomplished. Saxony, Westphalia, and all the regions situated on the left bank of the Elbe have been freed of the presence of the Prussian army. This army, beaten and pursued with a sword in its back for more than fifty leagues, is today without artillery, baggage, officers, is reduced to less than two-thirds of what it was eight days ago and—worse still—has lost its morale and any confidence in itself.[8]

The pursuit became a veritable manhunt:

> The language of the Prussian officers had changed markedly; they sued for peace with great cries: 'Must your Emperor, they said to us, always pursue us with a sword in our backs? We have not had a moment's rest since the battle.' These gentlemen were doubtless accustomed to the manoeuvres of the Seven Years War.[9]

With this pursuit of the Prussians, the cavalry of the Grande Armée, Murat, Lasalle, and his 'infernal' brigade became the stuff of legend:

> The cavalry is the only way of exploiting victory. It is on account of the cavalry that the Prussian infantry was unable to cross the Oder after the Battle of Jena, which would otherwise have yielded paltry results.[10]
>
> It is for the cavalry to pursue the victory and prevent the defeated enemy from rallying.[11]

When the pursuit is conducted solely with an advance guard, it is necessary to beware the fury and offensive recovery of the beaten enemy, who seeks to break through. Should one not have enough troops on hand to dent the enemy, it is better to encourage his flight. This explains the following curious sentence about the engagement at Cerca (11 September 1796), when Masséna's advance guard was destroyed by Wurmser, who had been pursued after his defeat at Bassano:

> The fleeing enemy must be presented with a golden bridge or opposed with an iron barrier. We had to resolve to allow the escape of an enemy who, according to all calculations and probabilities, should have been forced to lay down his arms and render himself a prisoner that day.[12]

The 'golden bridge' is a reminiscence of Maurice de Saxe, who himself spoke in this connection of a 'proverb' that in fact dates back to Vegetius, an author of late Roman antiquity whose military writings enjoyed considerable success in the Middle Ages and the Renaissance.[13]

General Rogniat advocated more methodical, slower campaigns so as better to cater for the supply of provisions. He found himself reproved by the exile on Saint Helena:

> But if the battle is waged on the edge of your zone, forty leagues from your base, will you stop just at the border and allow your beaten enemy to escape, without exploiting the victory to pursue him with a sword in his back, ending up by scattering him, crushing him, and arriving with him in his capital or depots? Could your enemy give you more pernicious advice? Can you do anything more agreeable to him and more in his interests?[14]

'Little positive advantage would be gained in the normal course of events,' Clausewitz was to write, 'unless victory were consummated by pursuit on the first day.'[15]

8

Retreat after a Lost Battle

When an army has experienced defeats, unifying its detachments or relief forces and taking the offensive is the most delicate operation in war—the one that demands the most profound knowledge of the principles of the art from a general. It is especially then that their violation entails defeat and results in catastrophe.[1]

Clausewitz complements Napoleon: 'When a battle is lost, the strength of the army is broken—its moral even more than its physical strength. A second battle without the help of new and favorable factors would mean outright defeat, perhaps even absolute destruction. That is a military axiom.'[2]

The rules of war have it that an army division avoids fighting alone against a whole army that has already achieved success: that is to run the risk of losing irretrievably.[3]

To effect a retreat well, it is necessary to have several roads by which your army can retire in large bodies, and with celerity; and also able to defend themselves if attacked.[4]

The morale of the retreating army will not be restored with the arrival of young conscripts:

After the loss of a battle, the depots, the recruits cannot alter the fate of an army, since they cannot alter its morale; they can only worsen it and end up ruining everything.[5]

Remembering Waterloo, Napoleon made this observation about the French:

The French are the bravest one knows: in whatever position they are tested, they will fight. But they do not know how to withdraw in the face of a victorious enemy. If they suffer the least setback, they are no longer correctly behaved or disciplined: they will slip into your hands.[6]

We find the same observation from the pen of numerous French officers, such as Lieutenant-Colonel Lemonnier-Delafosse: the French show audacity and extreme impetuousness in attack, but if they fail, they flee shamefully and irresponsibly.[7] For Napoleon the British, once defeated, did not behave any better:

> Once confusion begins to enter into a retreating army corps, the sequels are incalculable, even more so for a British army than any other.[8]

Conclusion to Book IV

Modern combat is dominated by firepower—the firepower of riflemen and mass infantry, the firepower of artillery. Napoleon stresses the latter: it is what wins battles and an army's battle disposition must take account of it. The parts of the line of battle must be properly connected and the reserves at the disposal of the commander-in-chief concentrated and ready to be launched against a point. The flanks must be supported and the line of retreat ensured. The aim of an engagement is the immediate and significant destruction of enemy forces. Napoleon did not want to enter into the controversies, typical of the eighteenth century, about the advantage of some particular tactical disposition or way of firing. Clausewitz was to be of the same opinion. Circumstances require the infantry to be formed up in columns, lines, or squares. Brilliant operationally throughout his career, Napoleon displayed less and less subtlety at the level of battle tactics. A result had to be obtained by the application of brute force: making breaches in the enemy line with a mass of guns and then throwing columns of infantry and cavalry into them. The role of artillery increased from Friedland in 1807 and the casualty rate during battles traced an ascending curve. Perhaps one of the weaknesses of Napoleon's art, wrote Georges Lefebvre, was that it paid insufficient attention to detailed tactics and did not revitalize them by taking account of the coalition's tactics.[1]

The casualties suffered by contending armies are generally of the same order of magnitude for three-quarters of the day. When the balance of moral forces begins to swing to one side, and one of the adversaries initiates a withdrawal, preservation of one's line of retreat becomes vital, and it is by threatening it that the victor can take numerous prisoners. There is a moment in the engagement when the decisive moment occurs. Napoleon speaks in this connection of a 'moral spark'. Clausewitz acknowledges that the Emperor often managed more economy of force, enabling him at this point to create the psychological impact of a reserve of fresh troops, a guarantee of victory.

For Napoleon, battle should not be given lightly. Favourable conditions must exist: battle should be joined where one has a 70 per cent chance of success. If this is the case, all available troops must play their part and display energy and determination. One does not pull out of a battle without incurring a severe penalty, for retreat leads to heavy losses. In this sense, 'the glory and honour of arms' are the best guarantee of men's safety. Napoleon always began his campaigns by seeking battle, for he sought to finish them rapidly, but he knew how to await favourable circumstances. As at Ulm in 1805, these sometimes allowed him to secure victory without fighting.

For Napoleon, the battle was the result of a plan of campaign that he called a 'system' combining the manoeuvres of his army corps—what has since been called the operational level of war. In simpler terms, he gave battle while always envisaging the totality of the campaign. His mind refused to confine itself to the tactics of a closed field. He saw further. Pursuit of the defeated enemy completed the battle and made it possible to reap the benefits. If the battle had been lost, it was necessary to beat a retreat. The operation was not an easy one to conduct and Napoleon did not say much about it, because he was loath to envisage this type of situation. Despite the forced passages of the Berezina and Hanau, his retreats of 1812 and 1813 were catastrophic.

BOOK V

Military Forces

I

General Survey

On several occasions, Napoleon referred to the advantages of conscription:[1]

> Conscription creates citizen armies. Voluntary recruitment creates armies of vagabonds and bad elements. Honour guides the former; discipline alone commands the latter.[2]

In truth, the Napoleonic variety was among the most unjust forms of conscription. The poor accounted for the quasi-totality of recruits, because the wealthy could purchase a replacement.[3] In discussion with a British colonel on Saint Helena, Napoleon expressed his astonishment that soldiers so rarely became officers in the British army. The colonel replied that his compatriots were astonished at the reverse being the case in the French army:

> It is one of the major consequences of conscription, the Emperor observed: it had made the French army the most composite that had ever existed.[4]

Conscription yielded large armies. Napoleon was the first to be in a position to employ such masses. He realized the novelty of the phenomenon and sometimes was even frightened by it. Before a new war with Austria broke out in spring 1809, he wrote a long letter to General Caulaincourt, his ambassador in St Petersburg, in which he alluded to the prospect of a dismemberment of the Habsburg Empire so as finally to have peace in that quarter:

> When the last states have thus been split up, we shall be able to reduce the number of our troops, replace these general levies, which tend even to arm women, by a small number of regular troops, and thus change the system of large armies introduced by the late King of Prussia.[5] Barracks will become poorhouses and conscripts will stay at home to do the ploughing.[6]

2

Relationship between the Branches of the Military

For Clausewitz, an engagement had two components: the destructive principle of fire and hand-to-hand fighting or 'clash'.[1] The latter in turn was either offensive or defensive. Cavalry was suited only to attack. Artillery operated exclusively by firepower. Infantry alone combined offensive combat, defensive combat, and firepower, which made it the key element in armies. It was also the most independent of the three branches, while artillery was the least so. Yet an army was weakened less by an absence of cavalry than by an absence of artillery.

Proportion and relationship of branches

Staging a comparison with seventeenth-century armies, Napoleon indicated the ideal proportions of each arm of the services in his era:

> The armies of the time were at least half made up of cavalry; they had little artillery—one-and-a-half pieces per thousand men; the infantry was drawn up in four ranks—the fourth was armed with pikes.
>
> Today, an army is four-fifths infantry, a fifth at most cavalry, and has four artillery pieces per thousand men, a quarter of them howitzers; the infantry is arranged in three ranks; pikes and spontoons have been abolished.[2]
>
> In an army, infantry, cavalry and artillery are needed in the right proportions; these arms cannot be substituted by one another. We have witnessed occasions when the enemy would have won the battle: with a battery of 50–60 gun muzzles, he occupied a wonderful position; you would have attacked him in vain with 4,000 horse and 8,000 infantry more; what was required was a battery of equal power, under whose protection the attacking columns could advance and deploy. The proportions of the three arms have invariably been the subject of meditation by the great generals.

They have agreed that the following were required: 1 four artillery pieces per thousand men—which amounts to one-eighth of the army for artillery personnel; 2 cavalry equal to one-quarter of the infantry.[3]

The proportion of the cavalry would vary depending on the theatre of operations:

On the Rhine, an army's cavalry must be one-quarter of the infantry: one-twenty-fourth scouts, three-twenty-fourths chasseurs and light horsemen, two-twenty-fourths cavalry of the line and cuirassiers.[4]

The cavalry of an army in Flanders and Germany will be six-twenty-fourths of the infantry; in Italy and Spain, five-twenty-fourths; on the Alps and Pyrenees, one-fifteenth; and on the coasts, one-twenty-fourth. The size of France's cavalry will be one-sixth that of the infantry.[5]

In an army in Flanders or Germany, the cavalry, including the scouts, must be one-quarter of the infantry; on the Pyrenees and the Alps, one-twentieth; in Italy and Spain, one-sixth. [...]

You must have as much artillery as your enemy, reckon on four pieces per thousand infantry and cavalry. The better the infantry, the greater the need to conserve it and support it with good batteries.[6]

If the infantry was bad, a great deal of artillery was required:

The less good a troop, the more artillery it needs. There are army corps where I would only ask for one-third of the artillery I would require with other army corps.[7]

Accustomed exclusively to the plains of northern Europe, Clausewitz was of the same opinion as regards the relationship between cavalry and infantry. As for artillery, he differentiated depending on the point in time. At the start of a campaign, armies generally had two or three pieces per thousand men. Artillery was not lost at the same speed as infantry, so that at the close of a campaign the proportion rose to four or five pieces per thousand men.[8] Clausewitz declined to indicate the ideal proportions between the branches of the military, but established the consequences of pronounced superiority or inferiority in one or another. Infantry was the main branch. It was more difficult to do without artillery than cavalry. Artillery possessing greater destructive power, and cavalry less, it was always necessary to ask how much artillery one could have without disadvantage and what the minimum amount required was.[9]

In combat, liaison between the branches was indispensable. It often lay behind the success of Napoleon's armies. Even if he over-simplified

Wellington's dispositions in 1815, in order to discredit them, Napoleon put his finger on a defect in the army of the Low Countries at the time he launched his offensive in Belgium:

> The infantry, the cavalry, and the artillery of that army were billeted separately, so that the infantry was committed at Quatre-Bras without cavalry or artillery. This led to it suffering high casualties, because it was obliged to remain in close columns to confront the charges of the cuirassiers, under the hail of fifty gun muzzles.[10] These valiant men were thus destined for slaughter, without cavalry to protect them or artillery to avenge them. Since the three branches cannot dispense with one another for an instant, they must be billeted and positioned in such a way that they can always go to one another's aid.[11]

Artillery

Artillery is the most terrible of the branches. A surfeit of it imposes a defensive, passive character on operations, because it moves slowly. A lack of it makes it possible to prioritize the offensive, mobile principle. Increasing your artillery makes it possible to compensate for weakness in numbers, as in the case of Frederick II at the end of the Seven Years War.[12]

As is well known, organization of the artillery of the armies of the Revolution and Empire was attributable to Lieutenant-General de Gribeauval, who had lightened and standardized French guns under Louis XVI, at the expense of the old system of Vallière, an army inspector in 1732. There had been a controversy, but trial by fire had settled the matter. Gribeauval had strictly determined the minimum weight to which cannons and howitzers could be reduced, without diminishing their range or power. The result was that, in making field artillery lighter, it was rendered more powerful. You could now take cannon of 12 pounds and howitzers of 24 onto ground where you previously crawled along only with 8-pounders. Napoleon was the first great artillery general: he could commit and move guns on battlefields like no one else before him and, thanks to them, achieve decisive results.[13] He paid homage to Gribeauval and sought to go even further:

> [...] artillery is still too heavy, too complicated; it must be further simplified, standardized, reduced until we have attained the utmost simplicity.[14]

Las Cases confirms that for Napoleon 'nothing could be superior to the advantages of uniformity in all instruments and accessories'. For him, artillery must fire incessantly during battles, 'without counting the expenditure of cannon balls'.[15] He had started out in the artillery at Auxonne, where he certainly had in his hands *De l'usage de l'artillerie nouvelle dans la guerre de campagne*, a remarkable little book published in 1778 by Chevalier Jean du Teil, the younger brother of the general commanding the artillery school in Auxonne:[16]

> Old General du Teil was in command there. He was an excellent artillery officer who had sound ideas, who said that the first, principal, and greatest merit of an artillery officer was to place two pieces in a battery well; that the scholars of the corps, the officers of the arsenals and parks were nothing beside the officer who positioned his gun well, aimed his fire well, aptly and accurately, knew how to direct his cannons at the critical moment, halted a column; [...] that was the brilliant part of the profession. The genuine artillery officer is the one who positions eight artillery pieces and has them fire accurately and at the right time. This is the important, difficult, honourable, noble thing in the profession. The average artillery general is nothing. However, a general like Drouot, who can properly position, command, and direct thirty artillery pieces, counts for something;[17] such men are rare. It is from the light artillery that Sorbier, Dommartin,[18] and nearly all our genuine, good artillery officers came.[19]

The lesson was not wasted on Lieutenant Bonaparte. Horse artillery, combined with cavalry, enabled more rapid movements in order to direct fire at the crucial place:

> Gassendi[20] does not like horse artillery, especially ours, where the gunners are mounted. And yet, that alone has changed war. That is to say, putting artillery in a position always to follow the cavalry is a great change. With cavalry corps and batteries, we can now deploy in the rear of the enemy army, etc. After all, what is the cost of mounting a few artillery regiments on horse compared with the advantages afforded by this arm? [...] The fate of a battle, of a state sometimes attaches to the artillery making it through.[21]
>
> Horse artillery is the complement of the cavalry branch. 20,000 horse and 120 light artillery muzzles are equivalent to 60,000 infantrymen with 120 muzzles.[22]

Horse artillery was indeed a major factor in the superiority of the revolutionary and imperial French armies.[23] This quotation also discloses two major innovations by Napoleon: the creation of an artillery reserve and a cavalry reserve, both of them destined to play a decisive role on the battlefield. We shall not be surprised to find Napoleon, as a former artillery officer, extolling the importance of this arm:

> But if you only assign six artillery pieces to each division, that is not enough; you need twelve. It is with artillery that one wages war.[24]
> [. . .] it is only with guns that one wages war.[25]
> [. . .] wherever a regiment goes, artillery is needed.[26]
> The major battles are won with artillery.[27]

The quantity of artillery was bound up with the 'regular' character of the war. In other words, the more comparable the opponent was militarily to the French army, the greater the resources required to defeat him. Napoleon intimated this in connection with the Austrians in 1809.[28] On Saint Helena, Napoleon reckoned that all generals should spend some time in the artillery:

> A supreme commander should know artillery and engineering. Guibert was right to argue that any general officer should have been an artillery captain for a year or two. That is the most important thing for a supreme commander, who is otherwise held up at any moment by artillery problems that he could solve with a word. An artillery general answers: follow the road, when that cannot be done, because he does not understand the point of the general doing differently. Why would you go that way when you have a main road? It is easier and more convenient to say that you cannot get through and yet it is not true. Myself, when I was an artillery general, I thought like that.[29]
>
> But you need to know about artillery to know how it can be transported everywhere. So I believe that all officers should spend time in the artillery. It really is the arm capable of producing most good generals. They have personnel and materiel. Engineering is a good branch, but it is less applied than artillery.[30]

Firing artillery conformed to different principles on land and along coasts:

> On coasts, it is necessary to fire as far as possible and to fire continually. What are needed are projectiles that travel the furthest possible distance. In land artillery, all these principles change and it is rightly reckoned that to fire from a distance is to waste one's powder.[31]

Cavalry

Cavalry strengthened an army's mobility. A surfeit of it had only one negative consequence: potential supply problems. Plentiful cavalry facilitated possible major manoeuvres, but also major decisions. Bonaparte, says Clausewitz, knew how to employ his cavalry to deliver decisive blows.[32] Napoleon did indeed derive all possible advantage from his cavalry and did not spare them. He criticized his main opponents for doing the opposite:

> [...] the Germans do not know how to use their cavalry; they are afraid of jeopardizing it; they value it above its true value; they spare it unduly.[33]

Gourgaud reports similar words:

> His Majesty says that foreigners have never known how to make the most of their cavalry; that it is a highly advantageous arm: 'See what I did with it at Nangis, at Vauchamps,[34] etc., and if at Lützen the enemy had massed his cavalry on his left and, creating a breach, had come in the rear, what would have been the disorder!' [...] HM says that, when he charges, the cavalryman must have better morale than the infantryman, because he is on the receiving end of bullets and until the end of the charge the infantryman receives nothing, but that materially cavalry has a real advantage. One charges in columns by squadrons. The first echelon is felled, but the others win through.[35]

Cavalry must have as much order and discipline as infantry—even more. Napoleon criticized General Rogniat for saying the opposite:

> Cavalry needs more officers than infantry; it must be drilled more. It is not velocity alone that ensures its success; it is order, the totality, the good use of its reserves.[36]

An artilleryman, Napoleon used his cavalry a great deal and attributed much of his success to it:

> At Jena, the French infantry had won victory with only light cavalry; this victory was inconclusive. But the cavalry reserves arrived and then the Prussians could not rally. Demoralized, they were crushed on all sides and pursued with a sword in their backs. Of 200,000 men, not one crossed back over the Oder. Without cavalry, battles are inconclusive.
> [...] General Lloyd asks what use a great deal of cavalry is. For my part, I ask how it is possible to wage anything other than a defensive war, protecting oneself with entrenchments and natural obstacles, when one is not virtually on an equal footing with the enemy cavalry. Lose a battle, and your army is doomed.[37]
> Cavalry requires daring, skill, and, above all, not being dominated by the spirit of preservation and avarice. What can be done with great superiority in cavalry, well armed with dragoons' rifles, and with plentiful well-harnessed

light artillery, is incalculable. Of the three arms—cavalry, infantry, and
artillery—none is to be disdained. All are equally important. An army that is
superior in cavalry will always have the advantage of covering its moves well,
of being informed of its enemy's movements, and of only committing itself to
the extent it wishes. Its defeats will be of little moment and its efforts will be
decisive.[38]

For the Emperor, the cavalry must be directly at the disposal of the supreme
commander. He said as much to Eugène de Beauharnais, when the latter
commanded the remains of the Grande Armée in Germany at the start of
1813:

> It is appropriate for Generals Sébastiani and Latour-Maubourg to receive
> direct orders from you as frequently as possible; otherwise, the cavalry's spirit
> would be lost.[39] Infantry generals too often have the habit of crushing the
> cavalry and sacrificing it to the infantry. I therefore desire that you keep it in
> your hands and give it your orders directly.[40]

After Austerlitz, which was their first major battle, the Emperor delivered a
highly positive judgement on his cuirassiers:

> The science of the man on horseback must, to the highest degree, be in the
> heavy cavalry.[41]
> I cannot recommend drilling of my cuirassiers to you too highly. I imagine
> that they have schools. This arm, which has rendered me such important
> services, needs to be well drilled and it can be said that instruction does
> everything. The Russian cavalry is not wanting in courage, and yet has been
> almost completely massacred, while my Guard has lost no one.[42]
> You know that the cuirassiers are more useful than any other cavalry.[43]

Desirous of protecting them in their deployments, Napoleon strove to
equip them with a firearm:

> It is acknowledged that armoured cavalry can only use their rifles with
> difficulty. But it is completely absurd for 3,000–4,000 such brave chaps to
> be surprised in their billets or halted in their march by two companies of
> light infantrymen. It is therefore indispensable to arm them. [. . .] I cannot
> accustom myself to seeing 3,000 elite troops who in an uprising or surprise
> attack by light troops would be carried off by a partisan, or on a march
> would be halted by a few poor skirmishers behind a stream or a house: it is
> absurd. [. . .] War is made up of unforeseen events; to assume that 15,000
> heavy cavalry can always be kept in such a way as to be protected is to have
> no idea of it.[44]

An imperial decree of 18 June 1811 transformed six regiments of dragoons into regiments of light-horse lancers, specially intended to accompany the cuirassiers in their operations and to substitute for them on certain missions:

> Under no pretext can the cuirassiers be committed as orderlies. This function will be performed by lancers: the generals themselves will use lancers. The function of liaison, escort, and skirmishers will be performed by lancers.
>
> When the cuirassiers charge columns of infantry, the light horsemen must be positioned in the rear or on the flanks, to move into the gaps of the regiments and fall on the infantry when it is in disarray, or, if one is dealing with cavalry, on the cavalry and pursue it with a sword in its back.[45]

The cuirassiers represented the cavalry reserve par excellence. They must only exceptionally be exposed in marches:

> The cuirassiers will be specially placed in reserve to support the light cavalry and the dragoons. They will only ever be placed in the advance guards, rearguards, or wings when the latter need to be toughened up or to relieve the dragoons.[46]

The best types of cavalryman were categorized according to the various types of mission:

> The French cuirassiers were the best cavalry in the world *pour enfoncer l'infanterie*. Individually, there is no horseman superior, or perhaps equal, to the mamluk; but they cannot act in a body. The Cossacks excel as partisans and the Poles as lancers.[47]

Dragoons were particularly suited to pacifying an occupied region, if used well—that is to say, concentrated in a mobile mass. Joseph was told as much by his august brother when he reigned in Naples:

> Fools will tell you that cavalry is useless in Calabria; if so, it is no use anywhere. If Reynier had had 1,200 horses and used them well, he would have done the British dreadful damage, especially if he had had dragoons, which are armed with muskets and fight on foot.[48] [...] You have five regiments of dragoons dispersed; you must concentrate them and create a reserve out of them with four pieces of light artillery, harnessed. These 4,000 men, capable of doing 30 leagues in two days, can head for Naples and any other point that is threatened. What are you doing with 300 isolated dragoons, who will lose the spirit of their service arm and be of no use to you? [...] I reiterate the point: concentrate your dragoons, give them four or six pieces of light artillery, with caissons and cartridges. Regard them as infantry and organize them so that they can be anywhere smartly.[49]

In this desire to see the dragoons concentrated, the motive of preserving esprit de corps is evident. We can also divine what motivated the creation of a cavalry reserve on a larger scale. Napoleon made great use of his cavalry. This translated into heavy losses in horses. He acknowledged as much in 1809.[50]

Infantry, the French soldier

Refuting General Rogniat, Napoleon claimed that line infantry and light infantry did not have different missions. They performed the same function in war:

> In the 150 years since Vauban caused lances and pikes to disappear from all the armies of Europe, replacing them by the musket with bayonet, all infantry has been armed lightly, has been intended to skirmish, to scout out, to contain the enemy: there is now but one species of infantry.[51]

However, the First Consul was concerned that the make-up of light infantry regiments should be distinct from that of line regiments:

> In this picture [relative to conscription], insufficient attempts have been made to assign mountainous areas to the light infantry and flat areas to line infantry. For example, the 6th is assigned to the Allier, which is a completely flat region. It is essential to choose the 30 most mountainous localities so as to assign them to the recruitment of 30 demi-brigades of light infantry.[52]

Similarly, the role of a grenadier was not that of a fusilier. Grenadiers were the solid element of formation in battle order:

> [...] the function of grenadiers in wartime must be to remain with the mass of their battalion; they must never be skirmishers, never be placed in the principal guard, never be dispersed into guard corps. The company must always be able to be unified to take the head of a column or, in case of an alert, to march to support the principal guard and, by its composure, give the troops confidence. Sometimes, however, there are such crucial positions, such as a bridge, etc., that they must be guarded by the grenadiers.[53]

An infantryman must always carry his backpack, even in combat:

> There are five things from which a soldier must never be separated: his gun, his cartridges, his backpack, his victuals for at least four days, and his axe. Let his backpack be reduced to the smallest possible volume: let there be but one shirt, a pair of shoes, a collar, a handkerchief, a sabre—very good. But let him

always have it with him, for once separated from it, he will never see it again. Theory is not the practice of war. It was the custom in the Russian army that on the point of fighting the soldier put his backpack on the ground. One senses the advantages of this method: the ranks could close up more; the fire of the third rank could come into play; the men were more agile, freer, less tired; fear of losing his backpack, in which the soldier was accustomed to putting all his belongings, was such as to fix him to his position. At Austerlitz, all the Russian army's backpacks were found in battle order on the heights of Posoritz; they had been abandoned there during the rout. Despite all the specious reasons advanced for this custom, experience led to its abandonment by the Russians.[54]

Shortly before the Battle of Leipzig (16–19 October 1813), Napoleon ordered the infantry, in battle line, to stand in two ranks rather than three:

> My intention is that you should place your troops in two ranks rather than three: the third serves no purpose in firing; it serves still less when it comes to bayoneting. When you are in close columns, three divisions will form six ranks and three ranks of rearguards. You will see the advantage this will afford: your firepower will be greater; your forces will be increased by a third; and the enemy, accustomed to seeing you in three ranks, will adjudge our battalions to be a third stronger than they are.[55]

He maintained this position on Saint Helena:

> The fire of the third rank is acknowledged to be highly imperfect and even harmful to that of the first two; the first rank has been instructed to kneel in volley fire; and in firing at will the third rank loads the rifles of the second. This order is bad. Infantry should only line up in two ranks, because the rifle only makes it possible to fire in that order. It would have to be six feet long, and able to be loaded by the breech, for the third rank to be able to fire to advantage.[56]

On two occasions at least, Napoleon made the bayonet a favourite weapon of the French infantry:

> He [the supreme commander] has observed with regret the lack of bayonets occasioned by the negligence of a large number of soldiers. Yet it is the bayonet that has always been the weapon of the brave and the main instrument of victory. This is above all because it suits French soldiers.[57]
> The Emperor recommends that each man have his bayonet, which has always been the favourite weapon of the French soldier.[58]

The specificity of the French soldier is underscored on several occasions and analysed with a rare acuteness:

There is nothing [. . .] one cannot obtain from the French with the lure of danger; it is their Gallic heritage. The love of glory and valour are an instinct in the French, a kind of sixth sense. How many times, in the heat of battle, have I not seen our young conscripts throw themselves into the mêlée; they were sweating honour and courage at every pore.[59]

The primary talent of a general consists in knowing the soldier's spirit and gaining his trust. And, in these two respects, the French soldier is more difficult to lead than any another. He is not a machine to be got to move, but a reasonable being who must be led.

The French soldier has an impatient bravura and sense of honour that make him capable of the greatest endeavours. But he needs strict discipline and must not be left resting for long.

The French soldier is argumentative, because he is intelligent. He is a harsh judge of the talent and bravery of his officers. He discusses a plan of campaign and all the military manoeuvres. He can do anything when he approves of the operations and holds his leaders in high esteem. But when the converse is the case, one cannot count on success.

The French soldier is the only one in Europe who could fight on an empty stomach. However long the battle, he forgets to eat as long as there is danger. He is more demanding than any other when no longer faced with the enemy.

The French soldier is relentless when pursuing a retreating enemy. He can do 10–12 leagues a day and fight 2–3 hours in the evening. I often benefited from this disposition in my first Italian campaign.

A French soldier is more interested in winning a battle than a Russian officer. He always credits the corps he is attached to with the principal role in the victory.

The art of retreats is more difficult with the French than soldiers from the North. A lost battle strips him of his power and courage, undermining his trust in his leaders, and prompts him to insubordination.

Russian, Prussian, and German soldiers hold their position out of duty; the French soldier, out of honour. The former are almost indifferent to a defeat; the latter is humiliated by it.

Privations, bad roads, rain, wind—nothing discourages the French soldier when he pursues or has hopes of success.

The French soldier's sole motive is honour: punishments and rewards must be derived from this motive. If ever the punishments applied to Northern troops were established among us, the army would be ruined and it would soon cease to exist as a force.

A quip by the French soldier on his general, a song that depicts his state of misery for him, have often led to privations of every kind being forgotten and the greatest obstacles being overcome.

The French soldier is generous. He loots to spend, never to get rich. On this subject, I heard it recounted by General Lariboisière[60] that, at an inn in Germany, he found four French grenadiers in a berlin and one of them, charged with paying for the stage, asked the postilion what the Emperor gave him as a *guide*. When he replied that he gave him three francs a stage, he put six francs in his hand, observing that it was very easy to give his Emperor a lesson in generosity and that he would not fail to tell him when he next passed through.

The French soldier fights valiantly as soon as he has a uniform. He makes a trained soldier after two months' marching.[61]

The non-commissioned officer plays an essential role:

He is of the same class, the same stuff as the soldier, so that while commanding him, he sympathizes with him, persuades him. He exercises a moral influence over him that not only gets him to obey, but trains him. He knows what must be said to him and never shocks him, because he is his equal. Why has the French army today become the most formidable in the world? It is because with the officers having gone into emigration, they were replaced by non-commissioned officers who have subsequently become the generals and marshals of France. It is with non-commissioned officers that one leads the mass of the soldiers; they can inspire them for the simple reason that they share the same backgrounds, they are part and parcel of them.[62]

The engineering corps

Passing over the old dispute between artillery and engineering, General Bonaparte deplored the loss in Egypt of an officer of engineers whom he greatly esteemed. He said that the latter mastered especially well

[. . .] that difficult science in which the smallest blunders have such an impact on the outcome of campaigns and the destiny of a state.[63]

Napoleon acknowledged that engineering was a more learned branch than artillery, inasmuch as the studies were longer at the specialist school at Metz:

One appreciates that two years are required for engineering; a year of schooling is enough for artillery.[64]

The engineering corps must enjoy autonomy:

Engineering should not be combined with artillery, but [have] sappers and pontoniers.[65]

As is well known, pontoniers theoretically formed part of the artillery. In some campaigns, such as that of 1809, Napoleon placed them at the disposal of the engineers. While he appreciated the technical knowledge of his engineers, he did not see them readily commanding an army:

> The knowledge of a supreme commander and that of an engineering officer are different and rarely converge. Myself, I had no clear, sharp ideas on the possibility of digging in in twelve hours or twenty-four hours. I have never believed that or seen a very important idea here.[66]

Engineering works swallowed up much money and took time. Napoleon feared the waste of both and distrusted the calculations of engineers:

> I am discontent with the fact that in its calculations the engineering corps today ignores money, which should be the basis of its surveys, and time, of which I am not the master. I shall repeat to you here my adage: every time one spends 100,000 écus in work on a site, an additional degree of strength must be imparted to it. This is what did not occur in the last instance, for, having spent eight or ten million in Italy, these positions were none the stronger. When an engineer requests several years, his plan is badly drafted; what he can be granted is the time of a campaign, but one is not always master of it.[67]

Engineer-geographers did not form part of the engineering corps, but came directly under the War Ministry. The only military thing about them was their uniform. Their task consisted in drawing maps of marches, camps, positions, battlefields, and occupied countries. Following reflection on them, Napoleon preferred engineering officers to them:

> Engineer-geographers are a bad institution: they had no esprit de corps or military spirit. Rather than carrying out reconnaissance, they constructed ridiculous plans of campaign, like people who are not military men. A reconnaissance must say positive things: the size of a river, the depth, the quality of the roads, the nature of the country. This was also the mistake of military engineers, but to a lesser extent; and ultimately they were military men and there was an esprit de corps. These duties should have been entrusted to them.[68]

The medical service

Relative to medical progress, the conditions in which the sick and wounded were treated were atrocious in the Napoleonic era. Could Napoleon have been more concerned with his armies' medical services? Only in 1813 did he become aware of the need to aid the wounded on the battlefield immediately. In fact, doctors could only intervene once the fighting was over. But nothing changed before the reign of his nephew, Napoleon III. The impossibility of providing first aid was the primary obstacle to better work by surgeons.[69] The situation in the British army was no better. Napoleon questioned a British doctor on Saint Helena about the administration of hospitals. In Great Britain, doctors signed the supply inventory for hospitals. This had been proposed and discussed in France, but ultimately rejected so that doctors would remain exclusively medical professionals, interested solely in the soldier's well-being:

> Today, they are the tribunes of the people; they are always concerned with the soldier's interests and complain about everything: the running of the hospitals, the supply of the hospitals, the poor quality of the food, the wine, the broth, the remedies provided. They could be corrupted, or be suspected of so being, if they were charged with administration.[70]

3
The Army's Order of Battle

Clausewitz agreed that the notion of order of battle pertained to tactics, rather than strategy. This, he said, was because formerly 'battle . . . used to be the whole of war and will always be its main element'. The larger armies had grown, the more integrated they had become and the more a kind of interaction had become established between tactics and strategy. This was most apparent at the junction of the two, 'where general deployment of armies passes into actual dispositions for battle'.[1] Clausewitz thus understood order of battle as the organization and disposition of military forces, which remained more or less the same throughout a campaign.

The organization of the army

The armed forces must be organized in accordance with the number of men. Large masses were obviously more difficult to handle:

> When Turenne says that an army must not exceed 50,000 men, we need to understand what he means by army. In his time, the army was not organized by divisions. The supreme commander had to organize everything, appoint the generals to command a particular corps, and so we can appreciate that, having to see to everything himself, there would have been utter confusion with more than 50,000 men. But he does not say that with 50,000 he would have the advantage over an army of 200,000. So he would create several armies. That is our divisions, our army corps. I will go further: 30,000–40,000 men are all that is required for an army corps of three divisions. You can conduct it well, feed it well.[2]

The example of Turenne anticipates what Clausewitz would have to say about seventeenth- and eighteenth-century armies in this chapter. Then an army's order of battle took the form of a compact, indivisible mass of infantry, with the cavalry on the wings.[3] Armies began to be integrated and to contain more flexibility in the late eighteenth century, with the emergence of divisions:

> The great advantage of today's armies is their division into divisions. Like the Roman legion, each division can be self-sufficient.[4]
>
> The current organization into divisions is excellent; each division has its own complete organization. It is like a legion. If the French army had taken this form at Fontenoy, the French manoeuvres would not have been partial as they were.[5]

A division essentially comprised infantry, but also some cavalry, artillery, and engineers. The pooling of two or three infantry divisions with a brigade of light cavalry, and artillery and engineering elements, yielded the army corps. 'Units could be easily detached and reattached without disturbing the order of battle. This was the beginning of corps made up of all arms.'[6] Clausewitz forgot to attribute its paternity to Napoleon. On 25 January 1800, the First Consul contemplated organizing the army of Reserve intended to serve in Italy. The inter-service character and critical size of the army corps were fixed in this letter to Berthier, which is a milestone in the history of war, since this model was to be universally followed down to the two world wars:

> My intention, Citizen Minister, is to organize an army of Reserve whose command will be reserved for the First Consul. It will be divided into right, centre, and left. Each of these three great corps will be commanded by a lieutenant of the supreme commander. In addition there will be a cavalry division, likewise commanded by a lieutenant of the supreme commander.
>
> Each of these great corps will be divided into two divisions, each of them commanded by a major general and two brigadier-generals; and in addition each of the great corps will have a senior artillery officer.
>
> Each lieutenant will have a brigadier-general as the head of his general staff; each major general will have an adjutant general.
>
> Each of these corps will be made up of 18,000–20,000 men, including two regiments of hussars or chasseurs, and 16 artillery pieces, 12 of them serviced by companies on foot and four by mounted companies.[7]

In 1805, within the Grande Armée, the corps were numbered, with a marshal at the head of each of them, with the exception of the 2nd corps, which was commanded by General Marmont. The army corps were needed to deal with larger masses of men, but could cease to exist if the army was smaller:

It is good that the army corps are not all equal, that there are some with four divisions, some with three divisions, and others with two. At least five army corps of infantry are required in a large army.

When the army's infantry amounts to no more than 60,000 men, it is better only to have divisions and lieutenant-generals to command the wings and detachments.[8]

The extent to which Clausewitz complements and elucidates Napoleon is striking. For the Prussian, nothing was more difficult to deal with than an army divided into three, except for one divided into two. By contrast, it was possible to go up to eight parts. It might be thought that a commander-in-chief would find his task simplified if he only had three or four major subordinates. But this would have two drawbacks. The longer the chain it had to pass along, the more an order lost in promptness, vigour, and precision. This is what happened when commanders of army corps came between the supreme commander and division leaders. 'A general's personal power and effectiveness diminishes in proportion to the increase in the sphere of actions of his closest subordinates. A general can make his authority over 100,000 men felt more strongly if he commands by means of eight divisions than by means of three divisions. There are various reasons for that; the most important being that a subordinate commander thinks he has a kind of proprietary right over every part of his corps, and will almost invariably object to any part being withdrawn for however short a time. Anyone with experience of war will be able to understand this.'[9] We could not encounter a better explanation of Napoleon's last sentence quoted above.

Furthermore, the number of parts must not be too large, on pain of disarray. If the divisions or brigades became too numerous, commanders of army corps had to be introduced. But, let us not forget, warned Clausewitz, that 'this adds another *power* to the chain of command'. Army corps operated autonomously within the framework of an overall strategy. This modular structure devolved the execution of manoeuvres, but strategy remained centralized. There was more flexibility in operations, but unity of command was preserved.[10]

Unity of command

After his victory at Lodi on 10 May 1796, and on the eve of entering Milan, General Bonaparte learnt that the Directory intended to send him to Rome and to entrust the command in northern Italy to General Kellermann.[11] While his arguments scarcely masked the disappointment of his growing ambition, Bonaparte explained how such a decision would contravene unity of command in the Italian theatre:

> I believe it very impolitic to divide the army of Italy in two; it is also contrary to the interests of the Republic to install two different generals.
>
> The expedition to Livorno, Rome, and Naples is a trifling matter; it must be carried out by divisions in echelons, so that, by retreating, we find ourselves in a position of strength against the Austrians and can threaten to envelop them if they make the slightest move.
>
> For that what is required is not only a single general, but also that he should not be obstructed by anything in his marching and operations. I have conducted the campaign without consulting anyone; I would have done no good if I had had to reconcile myself to someone else's way of seeing things. [...]
>
> If you weaken your resources by dividing your forces, if you rupture the unity of military thinking in Italy, I tell you with regret that you will have squandered the most wonderful opportunity to impose laws on Italy.[12]

The same day he wrote to Carnot:

> I cannot readily serve with a man who considers himself the first general of Europe; and moreover, I believe that it would be better to have one bad general than two good ones. War is like government; it is a matter of tact.[13]

The same idea was expressed to the same correspondent a few months later:

> If Prince Charles commands the two armies of the Rhine and of Italy, it is imperative that, when we are in Germany, there should be unity of command among us.[14]

At sea, on board a vessel, there should only be one leader. Apparently, this was not really the case at the time of the Directory, since General Bonaparte made this remark from Egypt:

> Juries, councils, assemblies must be banned. On board a vessel, there must only be one authority—that of the captain—which must be more absolute than that of the consuls in the Roman armies.[15]

The principle of unity also assumes a political dimension in the following order given to General Bernadotte in the first months of the Consulate:

> All the generals in the army of the West need to be brought round to the unity which is the basis of an army. Contain those who would like to act solely as they see fit.[16]

The principle was recalled on Saint Helena:

> There must only be one army, because unity of command is of the utmost necessity in war.[17]
>
> Unity of command is the most important thing in war. Two armies must never be placed in the same theatre.[18]

Napoleon stressed this point, because the Revolution's armies had been multiplied to the point of absurdity by the Convention and the Directory. There was one for each front. By its very name, the creation of the Grande Armée in 1805 signified that unity of command was henceforth a reality. This did not prevent the survival, for example, of an army of Italy. But the latter's moves were subject to an overall manoeuvre directed by the Emperor.

4
General Disposition
of the Army

When the armies of the *ancien régime* billeted in the field, they did not really feel themselves to be in a state of war. Frederick II had already changed that and thereafter the concern to fight pervaded everything. An army in the field must simultaneously subsist and be capable of fighting in its entirety without impediment. Certain dispositions facilitated this.[1] Napoleon indicated the following:

> A key thing for an army is to properly support its wings, which are the weak points. The wings are supported by a river, a mountain chain, a line of neutrality. If you cannot support one of your wings, that is a negative to be corrected. The two wings are the two weak points. But if, instead of having a single army, you have two, then you have four wings. You therefore have four weak points and if each of these armies is then divided into two further corps, which are separated and each of which has two wings, then you have 6, 8, 10, 12 wings or weak points, against which the enemy can throw significant forces. This is a capital error.[2]

In 1796, the armies of the Rhine and Sambre-et-Meuse made the mistake of forming three separate masses:

> Thus, in this march the French formed three separate corps, with nothing in common, having three lines of operation and six flanks, five of them exposed. Since the flanks are the weak part, it is necessary to support them and, when that cannot be done, to have as few of them as possible.[3]

In his desire to provoke and to challenge certain received ideas, Clausewitz said something different. According to him, the wings were not an army's weak points, because the enemy army likewise had them and could not threaten them without exposing its own. It was only when the enemy army was superior, and had better communications, that the wings became vulnerable. The wings were important, because flanking movements made resistance there more complicated than at the battlefront. The wings must therefore be well defended against the enemy's initiatives and this 'can be done by placing stronger forces on the wings than are needed simply to observe the enemy'.[4] These ideas coincide with those of the Emperor.

Clausewitz then raises the question of the distances between separate corps intended to support one another and fight together. There was no absolutely valid answer: 'So much depends on absolute and relative strength, on weapons and terrain, that no iron-clad regulation can be formulated; only a general one, a sort of average measure.' Clausewitz provided a few indications, those about the advance guard being the easiest to determine.[5] Napoleon had similar ideas:

> The distances which army corps must put between themselves on marches depend on localities, circumstances, and objective. Either the terrain is viable everywhere, and then why march on a front of 10–15 leagues? Or it is viable only on a certain number of pavements or vicinal paths, and then one accepts the law of localities.[6]

The increased size of armies had significant consequences for the dispositions to be made. It was no longer possible to proceed as in Turenne's time:

> Armies then were small. Posts with small armies play a big role. There is no position that can stop 200,000–300,000 men, whereas favourable positions for armies of 20,000–30,000 are to be found all over. An occupied village was a significant point then. Its importance diminishes with the strength of the army. An army of 25,000 men opposed to one of 20,000 is not in the same proportion as one of 250,000 faced with one of 350,000. Armies are not in geometrical proportion, but arithmetical proportion. For example, an army of 25,000 only disposes of 5,000 to create a detachment. It will have even more difficulty in hiding [it] from the enemy and, moreover, 5,000 can do nothing. The smallest post, the smallest position will stop them, whereas an army of 250,000 men can create a detachment of 50,000, who can subjugate a region, take positions, etc. on their own; and the enemy will find it difficult to make out whether he is confronted with only 200,000, rather than the 250,000 who were there prior to the detachment.[7]

It is in passages like this that Napoleon emerges as the inventor of mass warfare. It cannot be repeated often enough that this was the primary sign of war under the Revolution and the Empire (see Book III, Chapter 8). Napoleon was the first to be capable of handling great masses of men and to become intoxicated with the novelty of such power, as this passage indicates.

5

Advance Guards and Outposts

Here we enter a sphere 'where the threads of tactics and strategy are interwoven'.[1] Initially, outposts pertained to billeted troops and advance guards to troops on the march. But when the advance guard halted at the end of the day, it was not transformed into a mere outpost. What happened in outposts was of capital interest to the supreme commander. Napoleon had the habit of sending adjutants and general staff officers there, who were answerable directly to him:

> This was a great advantage. These sorts of detail have a major influence on war. This is what makes a lot of generals have contempt for books, because practice [of the profession] is everything.[2]

Outposts must be the object of every effort on the part of generals and colonels:

> Before the break of day, generals and colonels must be at their outposts and the line must remain under arms until the return of the reconnaissance teams. It must always be assumed that the enemy has manoeuvred during the night so as to attack at dawn.[3]

Unlike Frederick II, Napoleon had his movements preceded by a strong advance guard. His army no longer took the form of a single bloc, but was organized into corps. He also managed much larger numbers of men:

> The art of a general of the advance guard or rearguard is, without jeopardizing himself, to pursue the enemy or draw him away, to contain him, to delay him, to compel him to take three or four hours to do one league. Tactics alone supplies the means to achieve these major results; it is needed more by cavalry than infantry, by the advance guard or rearguard than any other position. [...]
> The movements of an advance guard or rearguard do not consist in advancing or withdrawing at a gallop, but in manœuvring, and for that it requires good light cavalry, good reserves of line cavalry, excellent infantry battalions, good light batteries. These troops must be well drilled; and the

generals, officers, and soldiers alike must know their tactics well, each in accordance with the exigencies of his rank and his branch of the service.[4]

An advance guard was therefore an important corps, made up of three branches. It needed artillery, as the First Consul recalled in connection with the army of Reserve commanded by General Brune[5] after Marengo:

> The advance guard of this army is on the march to Switzerland; but an advance guard without artillery is absolutely nothing [...].[6]

In 1809, Eugène de Beauharnais only placed one infantry regiment in the advance guard of his army of Italy. The result was disastrous and the Emperor told him why:

> It seems that the 35th of line was isolated and surrounded by the enemy. It is a matter of principle in war that an advance guard must be composed of 10,000–12,000 men.[7]

Light cavalry was destined for the advance guard and outposts. Its reconnaissance mission was essential and its leaders must be constantly alert:

> [...] a colonel of chasseurs or hussars who, rather than spending the night in the bivouac and in continuous liaison with his guard corps, goes to bed, deserves death.[8]
>
> It is necessary to reiterate the order to the light troops never to spend the night in a town: they must bivouac and change bivouac in the evening, so as to lie half a league or a league from where they were at sunset. That is the way never to be taken by surprise [...]. 200–300 light cavalrymen must not take up position like an infantry corps; their objective is to scout and not to fight. [...] It must be made known that the penalty for the patrol commanders of light troops who spend the night in a town is death.[9]

Dragoons were particularly useful for the advance guard, as in retreats:

> Dragoons are necessary to support the light cavalry in the advance guard, the rearguard, and on the wings of an army. Cuirassiers are less suitable than them for this duty on account of their body armour. It is necessary to have them in the advance guard, but only so as to accustom them to war and keep them in suspense. A division of 1,600 dragoons moves rapidly to a point with 1,500 horse of light cavalry, dismounts to defend a bridge there, the top of a pass, a height, awaits the infantry's arrival. In a retreat, what advantage doesn't this branch possess?[10]

6

Operational Use of
Advanced Corps

The role of an advance guard was to observe the enemy and delay his advance. Corps flanking the bulk of the army could also play this role. Clausewitz devotes a chapter to the role of these advanced corps. He explains why they do not necessarily suffer significant casualties if they are attacked by the enemy. For the enemy was likewise preceded by an advance guard, did not advance in full strength, and did not know how far off his opponent's reinforcements were. This probing and caution afforded the advanced corps the possibility of withdrawing before any real danger arose. A division of 10,000–12,000 men reinforced by a cavalry detachment, and positioned in advance a day's march away on ordinary terrain that was not especially hard, could hold the enemy, retreat included, around one and a half times the time required for retreat. The enemy would thus find it difficult to commence his attack on our army the same day that he repulsed our advance guard. The advance corps were not intended to halt the enemy's movements, but to moderate them, to regulate them 'like the weight of a pendulum . . . so as to make them calculable'.[1]

We can compare these considerations with the following extract from a letter by Napoleon to his brother Joseph:

> Assuming that the British had a lot of forces in Calabria and seriously wanted to sustain such a disproportionate war, with an advance guard at Cassano, supported at a distance of a few marches by two or three brigades, you would be reinforced in three days by 9,000 men. And if, ultimately, they did not think themselves sufficiently strong, they would withdraw the distance of one march and be joined by 3,000 men. This is how you wage war, when you have several points to hold and do not know via which one the enemy will attack you.[2]

Further details were given in June 1809 to Eugène de Beauharnais, who commanded the army of Italy:

You must therefore all march properly concentrated and not in small packets. Here is the general principle in war: a corps of 25,000–30,000 men can be isolated; well led, it can fight or avoid battle and manoeuvre in line with circumstances, without anything bad happening to it, because it cannot be forced into an engagement and ultimately must fight for a long time. A division of 9,000–12,000 men can be left isolated for an hour without any drawbacks. It will contain the enemy, however numerous, and will give the army time to arrive. Also it is customary not to form an advance guard of less than 9,000 men, to have its infantry camp properly concentrated, and to position it at most at an hour's distance from the army. You lost the 35th because you ignored this principle:[3] you created a rearguard composed of a single regiment, which was turned. Had there been four regiments, they would have formed a mass of resistance such that the army would have come to their aid in time. There is no doubt that in observation corps, as Lauriston was,[4] you can place an infantry detachment with a lot of cavalry. But this is because you assume that the enemy is definitely not on regulated operations, you are going in search of him, and finally this infantry will be capable of impressing the enemy cavalry, the peasants, and a few companies of enemy chasseurs. In general, in flat lands, cavalry must be on its own, because on its own, unless it is a question of a bridge, a pass, or a given position, it will be able to withdraw before the enemy infantry can arrive.

Today, you are going to embark on regulated operations. You must march with an advance guard composed of a large force of cavalry, a dozen artillery pieces, and a good infantry division. All the rest of your corps must bivouac an hour behind, with the light cavalry, as one might expect, protecting as much as possible. You must reckon that it is in the mind of Colonel Nugent, who directs Prince Jean,[5] that as soon as he sees you march on him from one side, and Macdonald from the other, he will march on one of you and, since he possesses the advantage of having the locals with him, he will march in concentrated formation, without having scouting done by his light cavalry, and can fall on you unawares. Consequently, you must organize your march well; the artillery must be in the divisions and everyone at his post, when marching and when bivouacking alike; you must bivouac as in wartime and in such a way as to shoulder arms and fight at the crack of dawn.[6]

The beginning of this extract reveals Napoleon's conception of the army corps: with a critical mass of 25,000–30,000 men, it had such firepower that it could engage greatly superior forces on its own and resist for a whole day, or the time it took for aid to arrive. Each corps commander must be capable of confronting the unforeseen for a dozen hours, in the knowledge that the Emperor would manoeuvre the other corps to reinforce it.[7] When only one division was involved, Napoleon added artillery to it and did not envisage it being further away than an hour's march. He too evoked 'regulated operations', the only ones envisaged by Clausewitz in this chapter. In 1813, Napoleon used the phrase 'regulated warfare' in the same sense, to characterize operations where the balance of forces operated in strict fashion as regards numbers of men, horses, and guns—where it was more difficult, in other words, to count on a factor other than numerical superiority alone:

> Finally, in my position, any plan where I am not personally at the centre is inadmissible. Any plan that removes me determines a regulated war, where enemy superiority in cavalry, number, and even generals would lead me to utter perdition.[8]

7

Camps

Camps are contrasted with 'quarters', which armies in the field take with the inhabitants. Whether composed of tents, huts, or bivouacs, their disposition conditioned combat strategically.[1] Napoleon said as much before Clausewitz:

> The art of setting up camp in a position is nothing other than the art of adopting a line of battle in this position. All projectile machinery should be in play there and advantageously positioned; the position adopted must not be dominated, extended, or enveloped, but, on the contrary, must dominate, extend, and envelop the opposed position as much as possible.[2]

The various camps of the same army could not be positioned indiscriminately. During the 1762 campaign, Prince Henry of Prussia[3] was mistaken in this respect:

> During this campaign, that prince breached the principle that the same army's camps must be positioned in such a way as to be able to support one another.[4]

Until the French Revolution, recalls Clausewitz, armies were always billeted under tents in fine weather. In winter, they took quarters. Since then tents had been abandoned, because they required too much baggage. The latter had proved too cumbersome for rapid, extensive movements by armies. With an army of 100,000, the preference was for 5,000 more cavalry or some artillery pieces, rather than the 6,000 horses required to transport tents. Clausewitz stressed that the abandonment of tents had led to higher casualties among military forces and greater devastation of regions. However feeble the protection offered by a tent, in the long run troops suffered from their absence and illnesses ensued. The abandonment of tents was an indication of the greater violence of war. Clausewitz doubted whether the use of tents would return, because '[t]he bounds of military operations have been extended so far that a return to the old narrow limitations can only occur briefly, sporadically, and under special conditions. The true nature of war will break through again and again with overwhelming force.'[5] In the Iberian Peninsula, Wellington ended up reintroducing the use of tents and his troops' health improved.[6]

Refuting General Rogniat, who was close to Clausewitz on this point, Napoleon vehemently rejected the idea of reverting to tents:

> Tents are not healthy. It is better for the soldier to bivouac because he sleeps with his feet to the fire, shelters from the wind with a few planks or some straw, and the proximity of the fire rapidly dries the ground he is sleeping on. Tents are necessary for leaders, who need to write, read, and consult the map. A tent for the battalion leader, one for the general, would be useful; it would make it possible to impose on them the obligation never to sleep in a house—a dire abuse that is the source of so many disasters. Following the French example, all the nations of Europe have abandoned tents; and if they are still in use in leisure camps, it is because they are economical, conserve forests, thatched roofs, and villages. The shade of a tree against the sun and heat, the most meagre shelter against the rain, are preferable to a tent. Transporting tents would use five horses per battalion, which are better employed carrying food. Tents are an object of observation for retainers and enemy staff officers. They provide intelligence about your strength and position. This is a permanent, constant drawback. An army drawn up in two or three bivouac lines only affords a distant glimpse of smoke which the enemy confuses with atmospheric mist. It is impossible to count the number of fires; it is very easy to count tents.[7]

These reflections were not born out of a desire to contest Rogniat's statements. In 1808, Napoleon did not want Prince Eugène to use tents to camp his troops in Italy:

> You must not consider tents; they are good only for inducing illness. It often rains in Italy.[8]

In support of these statements, it is accepted that tents played a significant role in the Prussian defeat in 1806. Not only did transporting them slow the Prussians' movements, but they were indeed harmful to the soldiers' health because they dispensed with supplying them with a greatcoat and were lost with the baggage at the start of the campaign. As a result, the Prussian soldiers, beaten at Jena and Auerstedt, had to disperse into villages to shelter from the bad weather. It took a long of time to reassemble them and the French, who were in pursuit, fell on them.[9] Napoleon's concern for the health of soldiers was genuine. A few weeks after the previous letter, when the Spanish spring began to get hot, he wrote to Murat, his lieutenant-general in Madrid:

In every camp you must take care to have installed horizontal tents, like canopies, which are attached either to trees or stakes and provide a good deal of fresh air. Tents intercept the sun, not the wind. The bottom of them must be watered frequently by men doing fatigues. The soldier can have tables and chairs in them and will not suffer from the heat. This is the way to compensate for the lack of trees and this is what Arabs do in the desert.[10]

The men's health reappears in several letters extolling the merits of camps with a different argument—that of drilling. Napoleon obviously had in mind the camps of Boulogne and the Atlantic coast, where the Grande Armée of Ulm and Austerlitz drilled. Napoleon was in no doubt:

It is the camps of Boulogne, where the corps were constantly drilled for two years, which won me the Grande Armée's successes.[11]

In March 1806, Napoleon wanted Prince Eugène to set up two camps in Istria:

I would like two camps to be established there, almost like in Boulogne, except that they should be square and located in important positions. In this way, discipline would be maintained. These corps would get instruction and contain the country.[12]

A different role was therefore added in occupied territory.[13] This applied to Portugal, where General Junot[14] was installed in late 1807. Napoleon told him how, and for which troops, camps might prove useful:

It grieves me to see that you have placed the first division in Lisbon. Your depots are sufficient to hold the forts. All the troops must be camped in squares and available straight away. You have made your second division, which is the worst, camp; it is exactly the opposite. It is not enough to have troops to throw around a few marches from Lisbon. They must be available to head anywhere without being spotted. That is the advantage of camps.[15]

The least good troops sufficed to guard the capital's forts. The best must be camped in such a way as to be ready for any eventuality.

8

Marches

A large army that one wanted to concentrate at a particular point, in order to give battle, had to march divided into several columns taking different routes. 'If 100,000 men were to march in a single column on one road without breaks in time, the tail of the column could never reach its destination on the same day as the head. Progress would either be extremely slow, or the column would, like a falling jet of water, break up into drops. Such a dispersion, together with the extra effort that the length of the column would impose on those at the tail, would soon result in general disorder.'[1] With General Rogniat elevating this need to march in several columns into a sort of general rule, Napoleon responded as follows:

> There are instances where an army must march in a single column and there are others where it must march in several. An army does not normally advance in a defile that is twelve feet in width; the roads are four or six toises and make it possible to march in two ranks of carriages and fifteen or twenty men abreast. One can nearly always advance on the right and the left of roads. We have seen armies of 120,000 men, marching in a single column, go into battle in the space of six hours.[2]

Napoleon's response dispenses with unnecessary detail and is not convincing. Without saying so, he was probably trying to justify his march on Moscow, where the main mass of the Grande Armée took a single route—something criticized by Clausewitz.[3] An opponent of all dogmatism, the latter implicitly agreed with the Emperor, however, when he recalled that 'the conditions of either the disposition or the march may predominate'. For him, the development of the art of war had imparted an organic distribution to armies and that is what enabled a corps or division to march and fight separately, for a certain period of time at least. Thanks to this, 'marches almost organize themselves; at least they do not call for complicated planning'.[4] In his first Italian campaign, General Bonaparte was the first to demonstrate all these new possibilities. After the event, he did not really build a theory of these marches which brought him such great success. We find just a few details, like this comment on the Piacenza manoeuvre in May 1796, which forced Beaulieu's Austrian army to retreat and fight a hard rearguard action at Lodi:

> In such marches, one must avoid exposing one's flank to the enemy and, when one cannot avoid it, they must be made as short and rapid as possible.[5]

It was necessary to get one's army moving. Even when active operations were temporarily suspended, prospects for the sequel could be enhanced and it could be secured by appropriate marches. Commenting on his operations in Italy, Napoleon uttered this famous sentence where he compared an army to a mechanical force, its speed multiplied by its mass increasing its force:

> The Battle of Borghetto was fought on 30 May [1796]; Wurmser's attack was on 1 August. In the intervening sixty days, part of the army crossed the Po, captured the legations of Ferrara and Bologna, Fort Urban, the citadel of Ferrara, Livorno, and disarmed these provinces. The troops were back on the Adige before Wurmser was in a position to begin his operation; that is using one's time well. The force of an army, like the quantity of movement in mechanics, is measured by the mass multiplied by the speed. Far from weakening the army, this march increased its materiel and morale, augmented its means of victory.[6]

The art of war was supposedly subject to the laws of mechanics: increase in the speed of an army augmented its force and imparted kinetic energy to it, with a more powerful impact on the enemy set-up. Speed could therefore substitute for number.[7] In fact, Napoleon confused force with acceleration or momentum. The latter, the product of the body's mass and speed, measures its movement, not its force. The conception articulated by Napoleon was predominant in the nineteenth century. Since the momentum of the body given impetus described the force imparted by the initiator of the movement, force was measured in kilograms–metres per second and the body's impact velocity was its movement multiplied by its mass. Momentum was referred to as force.[8] Napoleon was a member of the mechanics section of the Institute's class of sciences and mathematics.[9] Like Napoleon, Clausewitz was a follower of Newton and derived some of his main concepts—centre of gravity, equilibrium, friction—from mechanics. This allowed him to show that they were inherent in war, extraneous to human manipulations.[10]

Napoleon celebrated rapid marching—conduct that often afforded him victory, but which was also to reveal its limits. We shall return to it. The role of marches in operations emerges from the correspondence, especially relating to the victorious campaigns. In September 1805, with the next campaign in Italy in mind, Napoleon wrote to Prince Eugène to have 50,000 pairs of shoes made in Milan with the utmost possible discretion— real shoes, 'not cardboard, as is customary in Italy'. These pairs were to be added to what was already planned:

> Some good cooking pots and camping tools will be useful to you, in reserve; have this done with the least possible noise, without the corps knowing about it, so as not to stop them making do with what they have and authorizing them to count on this resource. In war, it is always shoes that are lacking.[11]

A month later, when the seven 'torrents' of the Grande Armée had poured into Germany, the future General Lejeune[12] reported to the Emperor on his mission, which consisted in transporting 300,000 pairs of shoes. He complained about such an unexciting task, which prevented him from taking part in the start of the campaign, and received the following reply:

> Child that you are, you do not understand the full importance of the task you have just performed: shoes facilitate marches and marches will win battles.[13]

In a striking letter, Napoleon explained to Marshal Soult the result that should be obtained by his army corps' marches:

> I recommend that you have your horses exhausted by your aides de camp and deputies. Position them in relay on the Weissenhorn road so that I can have your news rapidly. It is not a question of beating the enemy; not a single one of them must escape. Assemble your generals and corps leaders when you are in Memmingen and, if the enemy has done nothing to escape the hammer blow that is going to fall on him, make it known to them that in this important circumstance I am counting on the fact that nothing which can render our success complete and absolute will be spared; that this day must be ten times more famous than that of Marengo; that, in the most distant centuries to come, posterity will know in detail what everyone did; that, if I had only wanted to beat the enemy, I would not have needed so many marches and fatigues, but that I wanted to capture him; and that, of this army which was the first to break the peace and caused us to fail in our plans for maritime war, there should not remain a single man to carry the news to Vienna; and that the perfidious Court corrupted by British gold must learn it only when we are under its walls.[14]

What emerge from this letter are a frenzied desire to conquer and the wish to obtain 'complete and absolute success'. The second adjective makes it possible to establish a certain connection with the notion of absolute war in Clausewitz. The 6th bulletin of the Grande Armée, written at the time of the surrender of Mack's Austrians at Ulm, set out the results:

> The Emperor crossed the Rhine on 9 vendémiaire, the Danube on the 14th at 5 o'clock in the morning; the Lech the same day, at 3 o'clock in the afternoon; his troops entered Munich on the 20th. His outposts arrived on the Inn on the 23rd. The same day, he was master of Memmingen and on the 25th of Ulm.
>
> In the engagements of Wertingen, Günzburg, and Elchingen, the days of Memmingen and Ulm, and the engagements of Albeck, Langenau, and Neresheim, he had captured from the enemy 40,000 men, infantry and cavalry alike, more than 40 flags, and a very large number of artillery pieces, baggage, coaches, etc.[15] And, to achieve these great results, he had required nothing more than a few marches and manoeuvres. [. . .] Thus, the soldier often says: 'The Emperor has found a new way of waging war; he only uses our legs, not our bayonets.'[16]

For Josephine, Napoleon further amplified these results, but the essentials were tersely stated:

> I have carried out my design; I have destroyed the Austrian army by simple marches; [. . .].[17]

The outcome was spectacular, but had not been foreseen as Napoleon would have liked people to think. Until the night of 12 or 13 October, he was mistaken about Mack's intentions and was not thinking about trapping him in Ulm.[18] Nevertheless, his success was essentially obtained through marches.

For Napoleon, being commander-in-chief involved knowing how to direct a 'marching war'. Some of his generals were capable of understanding something of this, but not all. Joseph Bonaparte, placed in 1808 on the Spanish throne where he had difficulty maintaining himself, needed a leader who understood marching by his side, as his brother wrote to him:

> Savary is a very good man for secondary operations, but does not have enough experience and calculation to be at the head of such a large machine.[19] He understands nothing of this marching war. I very much hope that Jourdan has arrived with you.[20] The habit of commander-in-chief, which creates the habit of calculations and ploys, cannot be replaced by anything.[21]

In 1814, Napoleon again surprised his opponents with the rapidity of his marching. He took advantage of the support of the population to march at night:

[...] night time marches are especially advantageous when one has the country for oneself; we must therefore make the most of them and take the enemy's positions, since we can have understandings with the inhabitants, who will tell us the number of men the enemy has and will lead us to his rear, where he has a lot of baggage and pieces bogged down.[22]

Clausewitz examined the destructive effect of marches on fighting forces. Disproportionate efforts destroyed men, animals, vehicles, and clothing. He accepted that '[t]ools are there to be used', but was 'opposed to bombastic theories that hold that the most overwhelming surprise, the fastest movement or the most relentless activity cost nothing; that they are rich mines which lie unused because of the generals' indolence'.[23] When one sought to wage a war full of movements, large-scale destruction of one's own forces was to be anticipated. In Napoleon's armies, never was this destruction greater than in Russia, on the way out as well as the way back. Napoleon ventured very few general reflections on this. At the start of the Russian campaign, he made this observation to Marshal Davout:

The King of Naples has hitherto perhaps been too quick.[24] The army is only in the process of regrouping and one should not march against an entire army in the same way that one marches against a defeated army.[25]

In late October 1806, a way of rounding up stragglers was suggested to Marshal Lannes:

In forced marches, the course to adopt is every day to create a rearguard of 400 men out of the loiterers with whom you leave a good staff officer, who will be charged with getting it to catch up. By this means, you will prevent it from creating disturbances and the soldiers from tiring [themselves] unduly.[26]

In connection with Spain, where the rebellion had just broken out in July 1808, the concern to economize men was asserted from the outset, at a time when the Emperor was not yet intervening personally and the guerrilla warfare was only just starting:

In a war of this kind, you need sang-froid, patience, and calculation; and you must not exhaust the troops with mistaken marches and counter-marches. When you have performed a mistaken march of 3–4 days, you should not believe that you can correct it with a counter-march. Normally, that constitutes two errors instead of one.[27]

9

Supply

The issue of supplies had assumed major importance in the modern age, as a result of the expansion of armies and their constant availability for waging war. In what was the poor country of Poland in early 1807, Napoleon registered the fact:

> My position would be very good if I had provisions; the lack of provisions makes it mediocre.[1]
>
> My position here is excellent, militarily speaking; it is bad when I have no food.[2]

'Ability to endure privation is one of the soldier's finest qualities', wrote Clausewitz: 'without it an army cannot be filled with genuine military spirit. But privation must be temporary; it must be imposed by circumstances and not by an inefficient system or a niggardly abstract calculation of the smallest ration that will keep a man alive.'[3] Bodies responsible for war administration, which were independent organizations, had been charged with supplying the armies of the *ancien régime* by establishing 'depots' that combined requisitioning, field ovens, and flour rations in fortified sites. The armies of revolutionary France dispensed with them, taking what they needed from the lands they crossed through. With Napoleon, France remained between two systems. Clausewitz reckoned that supplies took four different forms. The first was supplies at the expense of the inhabitants or municipality. In a territory of average density, an army of 150,000 could be supplied thus for one or two days. The French armies advanced 'from the Adige to the lower Danube and from the Rhine all the way to the Vistula without substantial means of provisioning other than living off the land, and never suffered want'.[4]

When Joseph Bonaparte, having embarked on the conquest of his kingdom of Naples, prohibited his soldiers from demanding food from their hosts, Napoleon criticized him for taking 'unduly restrictive measures':

Understand clearly that, if circumstances have not required you to carry out major manoeuvres, there remains to you the glory of proving capable of feeding your army and drawing resources of every kind from the land where you are. This forms a large part of the art of war.[5]

Requisitioning carried out by troops—the second form of supply according to Clausewitz—was insufficient for a large army, especially since the men sent on the mission only ended up harvesting part of what actually existed. The formula only suited a division of 8,000–10,000 men. The third form of supplies—'regular' requisitioning—involved the collaboration of the local authorities. From the initial campaign of the Revolutionary wars, notes Clausewitz, the requisitioning system was always basic for French armies. 'Their enemies were forced to adopt it as well, and one can hardly expect it ever to be abandoned. [. . .] war has gained the utmost liberty by these arrangements.'[6]

To Eugène de Beauharnais in 1809, and then to his brother Jérôme in 1813, both of whom were in charge of a distant command but unaccustomed to war, the Emperor advocated requisitioning, brushing their scruples aside.[7] In Spain, it was necessary to go even further. War 'must feed war':

> Repeat the order to him [General Suchet] to slap a tax of several millions on Lerida, in order to procure the resources for feeding, paying, and clothing his army in the country. You will get him to understand that the war in Spain requires such an expansion of forces that it is no longer possible for me to send money; that war must feed war.[8]

Would supply by means of depots—the fourth mode of supply according to Clausewitz—return once more? Clearly perceiving the advantage of the requisitioning system and living off the land, he nevertheless recognized that this was not enough for Napoleon, who also had to organize a system of depots, with the construction of ovens in the vicinity of his armies.

The need to have wheat ground and bread baked for his troops was a major headache for Napoleon in his campaign plans and major expeditions,[9] to the extent that he imagined the soldiers grinding their own wheat and baking their own bread:

> The great improvement to be made in war, the only important one that remains to be made, which will confer a great advantage on the first to do it, who will inevitably be followed by others, is to accustom the soldier to carry his food, to make his flour and bread, so that he is sure of having bread without

recourse to the administration and always has his food provisions on him. We find wheat everywhere, but cannot grind it. The Romans distributed wheat, not flour. The soldier saw to things back then. If the government would only give the army wheat, then it would no longer have any administration. The outcome would be tremendous.

I had mills made for the Russian campaign, but they were too heavy. Since then some simple but excellent ones weighing eight pounds have been made. This is a major development.[10]

Modern troops have no more need of bread and biscuit than the Romans: give them flour or rice or vegetables during marches and they will not suffer. It is a mistake to assume that generals in antiquity did not pay great attention to their depots. In Caesar's commentaries, we can see the extent to which this important concern preoccupied him in several of his campaigns. They had only discovered the art of not becoming slaves to it and not depending on their purveyors. This art was that of all the great commanders.[11]

The organization of depots, reckoned Clausewitz, must not be presented as a 'perfecting of warfare simply because it is more humane. War itself is anything but humane.'[12] Furthermore, were war to be waged with the unbridled violence that constituted its peculiarity with 'the craving and need for battle and decision, then feeding the troops, though important, is a secondary matter'. Clausewitz claimed that Napoleon did not wish to have people to speak to him about foodstuffs. The citations given above contradict this assertion, which is nevertheless in tune with this extract from a letter dealing with the march of Junot's corps into Portugal:

> I do not intend his march to be delayed by a day on the pretext of a lack of food: that is a good reason only for men who do not want to do anything. 20,000 men can live anywhere, even in the desert.[13]

For Clausewitz, Napoleon was a 'passionate gambler', who frequently took 'reckless risks', especially in Russia, but he dispelled a massive prejudice about supply. He showed that it must be regarded not as an objective, but only as a means.[14] We can see just how far the power of Napoleon's example went for the Prussian thinker. History has subsequently established that supply was the principal weakness of the Napoleonic system of war.[15]

10

Lines of Communication and Lines of Operations

Napoleon's armies lived off the land. In addition to food, they needed reinforcements in munitions, men, horses, and materiel. They had wounded and sick men and prisoners to evacuate. This did not require daily contact with the rear, but from time to time it was necessary to organize convoys circulating on what Napoleon called the army's line of communication—that is to say, a 'post road, of stages':[1]

> In any operation, the first concern is to establish one's line properly.[2]
> The orders to be given to trace one's line of communication are one of a general's major concerns.[3]

The line of communication did not have to be protected at all times. To devote the fewest men possible to it, Napoleon endeavoured to have a stronghold on this line every five or six marches or, where there was not one, he had the medieval outer wall still possessed by most towns strengthened and then established '*field posts*', which were less strong but shielded from the enemy's light troops and raiding parties:

> In the last century, people asked whether fortifications were of any use.[4] There are sovereigns who have deemed them useless and who, as a result, dismantled their posts.[5] As for me, I would invert the question and ask if it is possible to plot war without strongholds and I declare no. Without depot posts, one cannot establish good campaign plans and without what I call field posts—i.e. sheltered from hussars and raiding parties—one cannot wage offensive war. Thus several generals who, in their wisdom, did not want strongholds, ended up concluding that you cannot wage a war of invasion.[6]

The distinction between field posts and depot posts does not revolve only around their strength and capacity for resistance. Depot posts, it will be understood from this extract, make it possible to 'plot war' and establish 'plans of campaign'. In other words, they are more directly bound up with operations. Speaking in the third person like Caesar, in 1805 Napoleon indicated how he relied on depot posts as he advanced:

> In 1805, having taken out the whole Austrian army, 80,000 strong, at Ulm, he headed for the Lech, had the old ramparts of Augsburg rebuilt, had them armed, constructed a strong bridgehead on the Lech,[7] and made this great city, which supplied him with so many resources, his depot post. He wanted to restore Ulm, but the fortifications had been razed and the locality was too poor. From Augsburg he headed for the Inn and seized Braunau. This stronghold ensured him a bridge over the river. It was a second depot post that enabled him to go as far as Vienna; the capital was itself placed out of reach of an improvised attack. After this, he headed for Moravia, seized the citadel of Brünn, which was immediately armed and supplied. Located 40 leagues from Vienna, it became his fulcrum for manœuvring in Moravia. One march from there he fought the Battle of Austerlitz. From that battlefield he could withdraw to Vienna and cross back over the Danube, or go via the left bank to Linz so as to cross the river on its bridge and reach Braunau.[8]

Here it is no longer simply a question of ensuring the army's line of communication, but of making it possible to advance further, of serving as a 'fulcrum for manœuvring'. In October 1806, the Grande Armée headed en masse for the Prussians' line of retreat. Compacted, it could no longer live off the land, as it had during the initial marches. Napoleon created a centre of operations where, by means of requisitioning carried out over a large radius by his cavalry, he established food depots, where he left large stocks and organized his hospitals.[9] He used the term line of operations to refer to the fairly short road linking this centre to his army and enabling it to operate against the enemy—that is, to seek battle with it. Napoleon expounded this manoeuvre to his brother Louis on 30 September 1806:

> My intention is to concentrate all my forces on my extreme right, while leaving the whole area between the Rhine and Bamberg entirely empty, so as to have almost 200,000 men concentrated on the same battlefield. If the enemy makes breakthroughs between Mainz and Bamberg, I shall not be concerned, because my line of communication will be established on Forch-heim, which is a small stronghold, and from there on Würzburg. [...] The nature of the events that might occur is incalculable, because the enemy, who supposes my left to be on the Rhine and my right to be in Bohemia, and who believes that my line of operations is parallel to my battle front, could have a great interest in outflanking my left, and because in that case I could cast him into the Rhine.[10]

Here the line of communication is slightly altered to take account of the concentration of forces on the right, but it still links the Grande Armée to France. For its part, the line of operations is completely different from what the enemy supposes it to be. The Grande Armée was not going to operate— i.e. combine its movements—in the way expected by the Prussians. On Saint Helena, Napoleon credited himself with inventing this modus operandi:

> The great art of battles is to change one's line of operations during the action. This is an idea of mine and is completely new. It is what won me Marengo. The enemy attacked my line of operations, in order to sever it. But I had changed it and he found himself cut off.[11]

While he did indeed surprise the Prussian army in 1806 by partially altering the orientation of his line of operations, Napoleon had not had this intention at Marengo. There it was he who was surprised by the Austrians. He was inches away from losing the battle and this passage forms part of the construction orchestrated by him to credit himself with exclusive responsibility for the victory, for the purposes of political propaganda.[12] Retreat was camouflaged as a rotation of the army, the change of front corresponding to a so-called alteration in the line of operations.[13] Moreover, Napoleon was not the first to have this idea. In Bourcet, he had read that 'there are circumstances in war when you must know how to abandon one of your communications in order to support the other'.[14]

> What most distinguishes Frederick is not the skill of his manoeuvres, but his daring. He did what I never dared to do; he abandoned his line of operations and often acted as if he had no knowledge of the military art.[15]

Normally, one did not abandon one's line of operations:

> There are two principles of war that are not violated with impunity: the first, *Do not carry out flanking marches in front of an army that is in position*; the second, *Carefully maintain, and never willingly abandon, your line of operations.*[16]

If we do not confuse 'abandon' with 'change', a line of operations could therefore be altered in the course of a campaign. This could be done in order to fool the enemy, as Frederick II did at the Battle of Leuthen:

> Nor did he violate a second, no less sacred principle, *that of not abandoning one's line of operations.* But he changed it—something regarded as the most skilful manoeuvre taught by the art of war. In fact, an army that changes its line of

operations fools the enemy, which no longer knows where its rear is and the sensitive points where it can threaten it.[17]

There was one point that admitted of no exceptions: a country was not invaded with a dual line of operations, contrary to Frederick II at the start of the Seven Years War, even if fortune smiled on him.[18]

In September 1808, Napoleon offered a useful clarification of the difference between a line of communication and a line of operations. His brother Joseph had evacuated his capital, Madrid, which had rebelled against him. Wishing to follow the Emperor's reiterated advice on concentrating forces, he proposed to assemble all his troops in a mass of 50,000 men and then march on the capital. This concentration of forces would be such that he would momentarily be obliged to interrupt communications with France, pending the arrival of the first troops from the Grande Armée from Germany. During this period, Napoleon would have no news of his brother's army and his brother would have none of the Grande Armée. Napoleon cautioned him and stressed the need for a veritable line of operations linked to a centre of operations:

> The military art is an art that possesses principles which it is never permissible to violate. To change one's line of operations is an operation of genius; to lose it is such a serious matter that it renders the general culpable of it criminal. Thus, retaining one's line of operations is necessary in order to reach a depot where one can evacuate prisoners, the wounded, and the sick, find some food, and rally.
>
> If, being at Madrid, we had concentrated our forces on the town, and regarded the Retiro[19] as a meeting point for hospitals, prisoners, and a way of containing a large city and conserving the resources it offered, that would have involved losing our communications with France, but ensuring our line of operations, especially if we took advantage of the time to bring together a large quantity of food and munitions, and had organized at one or two marches, on the main openings, like the citadel of Segovia, etc., support points and sentinels for the divisions. But today, when we have shut ourselves up in the interior of Spain without having any organized centre, any depot, and find ourselves in the situation of having enemy armies in the rear and on the flanks, it would be such a great folly that it would be unparalleled in the history of the world.
>
> If, before taking Madrid, and organizing the army there, with depots for 8–10 days, having sufficient munitions, we ended up being beaten, what would become of the army? Where would it rally? Where would it evacuate its wounded? Where would it draw its war munitions from, since we have but

a simple supply of provisions? We shall say no more of it. Those who dare to advise such a measure would be the first to lose their head as soon as events had exposed the folly of their operation.

When you are in a place under siege, you have lost your line of communication, but not your line of operations, because the line of operations is of the buffer zone in the centre of the place, where the hospitals, depots, and means of subsistence are to be found. You have been beaten outside? You rally in the buffer zones and you have three or four days to patch up the troops and restore their morale.

With an army made up completely of men like those of the Guard, and commanded by the most skilful general, Alexander or Caesar, if they could do such stupid things, one could not answer anything, a fortiori in the circumstances in which the army of Spain finds itself.

We must renounce this course of action condemned by the laws of war. The general who undertook such a military operation would be criminal.[20]

The message is clear: you can momentarily interrupt communications with your country of origin, and the friendly forces arriving as reinforcements, but you must ensure that the operations to be undertaken will be connected to a centre with hospitals and provisions in foodstuffs and munitions. This centre, where you can recuperate and resupply, has to be organized: that is what having a line of operations consists in. Yet Joseph planned nothing. He tried to be clever, seeking to get his brother to understand that he had taken on board the principle of marching as a unit. But he forgot another principle, which was just as essential.

A few days later, Napoleon returned to the same arguments, but did not use exactly the same terms, which demonstrates that he attached little or no importance to the theoreticians' definitions. What had to be done was very clear in his mind. He did not hesitate for a moment. But it must be acknowledged that the nature of his power, meaning he constantly had to take decisions on a large number of matters, and his disinclination for theory could render his orders confused, because he initially referred to the line of communication in the sense of a line of operations and then in the sense of a road to France that might be momentarily cut. The end establishes the difference clearly, by introducing the notion of centre of operations:

According to the laws of war, any general who loses his line of communication deserves to die. By line of communication I mean the line where the hospitals, aid for the sick, war munitions, foodstuffs are, where the army can rebuild, recuperate, and, in two days of rest, recover its morale, sometimes lost due to an unforeseen accident. Losing one's line of communication does not mean it being disturbed by barbets, miquelets,[21] rebellious peasants, and, in general, by what is called partisan warfare. That stops couriers, a few isolated men who always break through, some course of action, but it is not in a position to confront an advance guard or a rearguard. So that is nothing. The line of communication is organized on the principle that everything would be redeployed to Madrid. For that, everything had to be concentrated at Retiro—war munitions, foodstuff, etc.—and we could have concentrated a larger number of troops there in a few days if necessary. This is very different from operating with a system fixed on an organized centre, or proceeding randomly and losing one's communications without having an organized centre of operations.[22]

The notion of line of operations had been proposed by Lloyd, an author read by Napoleon. By contrast, the latter did not adopt the expression 'base of operations' popularized by Bülow and taken up by Jomini and Clausewitz.

II

Region and Terrain

Land and soil are closely and constantly related to military activity. They have a decisive influence on engagements—their preparation, their unfolding, and their exploitation. Clausewitz used the French term *'terrain'* to represent this problematic.[1] In 1805, Napoleon pressed Prince Eugène to familiarize himself with the terrain where he was going to be in command:

> Study the land; local knowledge is precious knowledge that one rediscovers sooner or later.[2]

For Clausewitz, the role of terrain corresponded above all to three properties: its value as an obstacle to an approach, as an obstacle to an overall view, and the protection it afforded against the impact of firearms. The value of an obstacle to an approach emerges from this extract from a letter to Prince Eugène once again:

> In plains like Hungary, it is necessary to manoeuvre differently from in the gorges of Carinthia and Styria.[3] In the gorges of Styria and Carinthia, if you catch up with the enemy on an intersection, like Saint-Michel, for example, you cut off an enemy column. But in Hungary, by contrast, as soon as he has been caught up at one point, the enemy will turn to another. Thus I assume that the enemy is heading for Raab and that you will arrive there before him. Learning this en route, the enemy will change direction and head for Pesth.[4]

Mediterranean and artilleryman, Bonaparte saw the terrain of his origins and the conditions of actions of his service forming a happy combination. In *Le Souper de Beaucaire*, written in 1793, he did not believe that Marseille would be able to defend itself. The hills covered with olive trees surrounding it would impede the defenders and give an advantage to the attackers:

> For it is on broken up land that the good artilleryman has superiority, by dint of the speed of his movements, the precision of the unit, and the accuracy of the assessment of distances.[5]

Napoleon's geographical knowledge was extensive. He engaged in some reflections of a geopolitical kind in connection with Italy:

> Secluded in its natural confines, separated by the sea and very high mountains from the rest of Europe, Italy seems to be called upon to form a great and powerful nation. But in its geographical configuration it possesses a crucial vice, which may be regarded as the cause of the misfortunes it has suffered and the division of this beautiful country into several independent monarchies or republics: its length is out of proportion to its width. Had Italy been bounded by Monte Velino—that is to say, almost at the height of Rome—and all the terrain between Monte Velino and the Ionian Sea, including Sicily, been discarded among Sardinia, Corsica, Genoa, and Tuscany, it would have had a centre near to all the points of the circumference; it would have had a unity of rivers, climate, and local interests. But, on the one side, the three large islands, which are one-third of its surface, and which have interests, positions, and are in secluded circumstances; on the other, the part of the peninsula south of Monte Velino which forms the kingdom of Naples—these are foreign to the interests, the climate, and the needs of the whole Po valley. [...]
>
> But although southern Italy is, by its situation, separated from the north, Italy is a single nation. The unity of language, customs, and literature must, in a more or less distant future, finally unite its inhabitants in a single government. The precondition for this monarchy existing is for it to be a maritime power, in order to maintain supremacy over its islands and to be able to defend its coasts. [...]
>
> With its population and its wealth, Italy can maintain 400,000 men of all branches of the armed services, quite apart from the navy. The Italian war requires less cavalry than the German; 30,000 horses would be enough. The artillery arm should be sizeable, in order to provide for the defence of the coasts and all the maritime establishments.[6]

Conclusion to Book V

Napoleon reckoned that on average an army must have four artillery pieces per thousand men and that the cavalry must amount to one-quarter of the infantry. Compared with Clausewitz, he tended to reckon cavalry less costly. Artillery was obviously essential for him: it is what won battles. It must be all the more plentiful in a regular war, faced with an opponent of the same calibre. Cavalry was also very useful in all circumstances to deliver blows and complete victory. It must not be spared. Napoleon greatly appreciated his cuirassiers. More difficult to replace, they must not be unduly exposed on marches. Dragoons were highly suitable for pacifying an occupied territory. Line infantry and light infantry had the same duties, with the second recruiting more men in mountainous regions. Engineering was the most scholarly of the service branches.

An army was to be organized into corps and divisions in accordance with its mass. Unity of command was vital. Advance guard and rearguard must manoeuvre with great tactical skill. They must be composed of the three branches, thus possess artillery, and number at least 10,000 men. Arranged in echelons, they could fall back on reinforcements if they were too hard pressed by the enemy. Napoleon's considerations on these issues, and especially the mastery he displayed in 1805 and 1806, have gone down in history. 'Until then people had not known how to get 100,000 men to act together.'[1]

Concerned for the well-being of his soldiers, Napoleon nevertheless rejected tents, which were too burdensome to transport. He waged war with the legs of his soldiers, believing that an army, like the quantity of motion in mechanics, was assessed by mass multiplied by speed. If he was preoccupied with provisions, had what he needed requisitioned, and organized depots, he did not hit upon an ideal formula and, on Saint Helena, still imagined a system where soldiers ground their wheat and cooked their bread. The army lived off the land, but a line of communication must make it possible to send reinforcements in men, munitions, materiel, and equipment. The army could momentarily abandon this line to manoeuvre against the enemy. But then it constituted a centre of operations for recuperating and re-stocking. It was connected by a line of operations. The latter could therefore be changed and it was even a stroke of genius to do so opportunely. But it must always have one. For General Riley, notwithstanding its deficiencies, the logistics of Napoleon's armies were the best of the age and would not be superseded before the great battles of 1916 on the western front. Napoleon established the regimental transport system. Even today, in a way that is certainly more respectful of the population, armies complement their provisions by recourse to local resources. Given the state of the road network, the modes of transport and organization, the way that Napoleon's armies were supplied, represented a masterwork of forecasting, planning, and implementation. What made possible the victories of Ulm, Austerlitz, and Jena was superior logistical organization. However, the system was fragile and could not withstand extended operations or the largest and poorest theatres, such as those of Spain and especially Russia.[2]

BOOK VI

Defence

Napoleon had to fight defensively as well as offensively, even if he had a marked preference for the latter. His observations on defence are frequently very concrete, sometimes of the order of detailed tactics. He engaged in a protracted strategic defence only from 1812, compelled and constrained as he was, and always took the initiative again whenever possible. Marked by the defeat of Prussia in 1806, Clausewitz had to show patience and had the time to reflect on defence. The 1812 campaign in Russia, which he took part in, marked him deeply. He developed an extended theoretical reflection on defence. Book VI is at once the least organized and the longest in *Vom Kriege*: it makes up one-quarter of the book and contains numerous repetitions.

I

Attack and Defence

For Clausewitz, defence signified the 'parrying of a blow'. It was characterized by anticipation of the blow. But the latter would have to be delivered and, in the context of defence itself, waging war therefore involved undertaking an offensive operation at a given point. To put it differently, one could fight offensively during a defensive campaign.[1] The alternation of defensive and offensive modes characterized the activity of war. What Napoleon had to say was no different:

> Defensive warfare does not exclude attack, just as offensive warfare does not exclude defence, although its objective is to breach the frontier and invade enemy territory.[2]
> [...] the art of sometimes being audacious and sometimes very prudent is the art of succeeding.[3]

The objective of defence was to conserve—something easier than acquiring. Defence was resorted to when in a situation of weakness. What was expected from it was an increase in strength. The defensive form of war is therefore in itself stronger than the offensive form, notes Clausewitz. 'Experience shows that, given two theaters of operations, it is practically unknown for the weaker army to attack and the stronger stay on the defensive. The opposite has always happened everywhere, and amply proves that commanders accept defense as the stronger form, even when they personally would rather attack.'[4]

Similar ideas can be found in a set of orders given by Napoleon during the British landing at Walcheren in 1809. When he was still in Austria after his victory at Wagram, he told War Minister Clarke how to employ the national guards and the new recruits raised to defend Zeeland. He repeated that only hardened, disciplined troops were suitable for the tactical offensive and that the national guards must be confined to defence—something that implicitly anticipates Clausewitz's argument:

No offensive, no attack, no audacity. Nothing can succeed with bad or new troops; if one attacks Vlissingen, one jeopardizes them.[5]

[...] do not go onto the offensive anywhere, unless you are four to one and have a lot of field artillery. Once again, the current British expedition cannot produce any result; and the only way of conferring one on it would be to go and imprudently attack them, because then our national guards would be demoralized and the effects would make themselves felt on all those in reserve.

[...] I cannot repeat it too often: you must act prudently, not jeopardize poor troops, and not have the folly to believe, like a lot of people, that a man is a soldier. Troops of the kind you have are the ones that require the most redoubts, works, and artillery. [...] The less good a unit, the more artillery it needs.[6]

Clarke having written to Bernadotte that it was necessary to risk a battle, Napoleon resumed his arguments:

I fear you have not fully understood my idea. I said that in no instance should a battle be risked, unless it was to save Antwerp, or unless we were four to one and in a good position covered by redoubts and batteries. [...]

To have real success against the British, we must have patience and look to the weather, which will destroy and repel their army, allow the equinox to come, which will leave them with no recourse but to slip away in capitulation. As a matter of principle, we should move to new positions but avoid general actions.[7]

On Saint Helena, commenting on operations in 1798, Napoleon criticized the way that the Austrian General Mack commanded the Neapolitans. The latter were two or three times more numerous than the French:

But the Neapolitans were not trained troops; he should not have used them in attacks, but to wage a war of position that obliged the French to attack. Military men are sharply divided over the question of whether there is more advantage in carrying out an attack or being on the receiving end of one. The issue is not in doubt when on one side there are tough, mobile troops with little artillery, while on the other is a much more numerous army, with a good deal of artillery in its retinue, but whose officers and soldiers are not battle-hardened.[8]

2

The Relationship between Attack and Defence in Tactics

Only three factors could have a decisive importance in tactics, according to Clausewitz: surprise, advantage of terrain, and concentric attack. In particular, the last factor made fire more effective. The author of *Vom Kriege* estimated that only a small part of surprise and advantage of terrain favoured attack, whereas for the most part they and the second factor in its entirety served defence.[1]

On this point, Napoleon came down more in favour of the offensive. He wondered if the increase in firepower favoured attack or defence. The issue was still posed in history:

> Guibert claimed that the invention of powder was favourable to defence. [...] Guibert based his view on the fact that it was difficult to take a battery position that was vomiting cannon balls and grapeshot. But he did not realize that the great advantage of artillery is to envelop the point one wishes to attack with a ring of fire. One begins by overwhelming and extinguishing the firing from this point. Positions that formerly resisted for a very long time are rapidly ground down today. Plentiful artillery and underground war are great means of destruction.[2]

Responding to Lloyd this time, the Emperor affirmed unhesitatingly:

> [...] firearms are more appropriate to offensives than defensives.[3]

With Napoleon, however, nothing was absolute in tactical matters. He wished above all to counter the excessive character of certain assertions by Guibert and Lloyd. As we shall see later, he never advocated attack in all circumstances. Circumstances determined everything. The following passage indicates that he was perfectly aware of the advantages that tactical defensive might possess:

Prince Charles is certainly one of the best Austrian generals, or simply the best, of recent times. He was less unfortunate confronted with me than the others. But this stems from the kind of war that is waged. It has wrongly been said that in Italy I always waged an offensive war. The opposite is true: I always waged a defensive war because of the mountains. I waited for the enemy to make a push. Except that Prince Charles was attacked by me at Essling.[4]

3

The Relationship between
Attack and Defence in Strategy

Depending on whether an offensive or a defensive strategy is adopted, different troops are needed:

> When you take the war into an enemy country, you must only lead trained soldiers there. The same is not true when you are the one invaded. So as not to weaken the army, you can shut up depots, recruits, and national guards in your strongholds. Everything is useful for holding a fortified position, if there are also good staff, artillery, and engineering officers. In this instance, finally, you do not have the initiative.[1]

When General Bonaparte came to take command of the army of Italy in Nice, in late March 1796, he decided to adopt an offensive strategy by skirting the Alps:

> In seeking to skirt the Alps and enter Italy by the Cadibona Pass, it was necessary for the whole army to be assembled on its extreme right—a dangerous operation if the snows had not covered the outlets of the Alps. The transition from defensive order to offensive order is one of the most delicate operations.[2]

From 10 October until 15 November 1799, while Masséna, at the head of the main army in Switzerland, remained on the defensive after his great victory at Zurich, General Lecourbe, with a force inferior to the enemy facing him, conducted offensives beyond the Rhine and 'fought for fighting's sake'. He would have done better to stay on the left bank of the Rhine. Had he wanted to create a diversion, he should have established himself firmly on a point of the right bank, in a defensive camp:

> When the main corps is immobile, a separate, secondary corps must not create a diversion by an active movement or invasion. It must conform to the posture of the main corps and influence the theatre of operations with a defensive posture, taking a position that is, by its very nature, threatening.[3]

4

The Convergent Character of Attack and the Divergent Character of Defence

According to Clausewitz, the imagination spontaneously attributes a convergent character to attack and a divergent character to defence. In reality, this was not always the case. If the front to be defended ran in a straight line from one sea to the next or from one neutral territory to the next, then it was impossible to launch a convergent attack and freedom of choice was limited. If the defender mobilized, and his available forces were closer to one another, he could exploit the advantage of internal lines and achieve a greater concentration of forces. The advantage of internal lines increased with the spaces in which these lines figured. For Clausewitz, the convergent form led to brilliant results, but those obtained by the divergent form were more reliable.[1]

These considerations clarify those of the young General Bonaparte, who, having initially proved himself at the siege of Toulon in late 1793, turned his gaze towards the Italian border. In this period, he multiplied reflections tending to demonstrate that the French army had no interest in remaining on the defensive and that, on the contrary, only an offensive would bring benefits:

> The border of Piedmont forms a semicircle, with the two armies of the Alps and Italy occupying the circumference and the King of Sardinia occupying the diameter.
>
> The circumference that we occupy is full of difficult passes and mountains.
>
> The diameter occupied by the King of Sardinia is an easy, fertile plain, where he can shift the same troops from one extremity of the diameter to the other in a few days.

The defensive system is therefore always to the advantage of the King of Sardinia. We need double the number of troops the enemy has to be equal in strength.

These observations are of the greatest significance. It would be easy to prove this by a detailed description of the borders of Spain and Piedmont and through an analysis of the different wars. It would be clearly demonstrated that every time we stayed on the defensive on the borders of Piedmont, we needed a lot of troops and in minor actions were always inferior.[2]

These reflections directly register the impact of the reading of Feuquière and Bourcet. They did not refer to circumference and diameter, but curve and cord.[3]

From the Saint Bernard to Vado, the Alps, which our army occupies, form a circumference of 95 leagues. We therefore cannot shift our troops from left to right in less than two or three weeks, whereas the enemy holds the diameter and communicates in three or four days. This topographical circumstance alone renders any defence disadvantageous, more murderous for our army, more destructive for our cartage, and more burdensome for the public Treasury than the most active campaign.[4]

However, the Piedmontese army, encamped in the plains and on the hillocks at the foot of the Alps, was in the utmost abundance. It recovered from its fatigue and made good its losses. It was reinforced every day with the arrival of new Austrian battalions, whereas the French armies, encamped on the ridges of the upper chain of the Alps, on a semi-circumference 60 leagues in length, from Mont Blanc to the sources of the Tanaro, perished from miseries and illnesses. Communications were difficult, food scarce and very costly; the horses suffered greatly, as well as all the army's materiel. The air and unboiled water of these elevated regions caused many illnesses. The losses suffered by the army in hospitals every three months would have been sufficient for the consummation of a major battle. This defensive was more of a burden on the finances, and more dangerous for the men, than an offensive campaign. In addition to these disadvantages, the defence of the Alps had others that pertained to the nature of the topography. The various army corps camped on its summits could not aid one another; they were isolated. To go from right to left, it took 20 days, whereas the army defending Piedmont was positioned in beautiful plains, occupied the diameter, and in the space of a few days could be concentrated in strength on the point it wanted to attack.[5]

The offensive was the best way of depriving the Piedmontese and Austrians of their advantage of operating on internal lines. The terrain made it possible for the French to make a convergent kind of attack. In other circumstances,

Napoleon naturally realized all the advantage of a defensive on internal lines. In 1808, he advocated this in the event of the French army in Spain being beaten and having to retreat:

> If he [Marshal Bessières] has been defeated and soundly beaten, all the troops must be assembled and concentrated in the circle 7–8 days from Madrid and we must study the dispositions in different directions, to know where to position the advance guards so as to exploit the advantage we have of being in the middle, so as to successively crush the enemy's various corps with all our strength.[6]

5

The Character of
Strategic Defence

In February 1797, General Bonaparte gave a veritable lecture on strategic
defence to one of his lieutenants, General Joubert,[1] who was charged
with guarding the outlets of the Tyrol:

> I urge you to reflect and to observe the localities more. For I cannot conceive
> that, with your line at Lavis broken and your retreat carried out during the
> night, you have not adopted an intermediate position, as close as possible to
> the first, where you can remain all day, putting together your troops, and
> receiving the scattered men or the corps that could not join in the night, the
> following night start marching again, if necessary, retake the line of Mori and
> Torbole, and keep the enemy in check there for several days; finally arrive in
> Corona, in the defensive camp of Castel Nuovo, or finally under the walls of
> Mantua or Verona. To act otherwise would be no longer to wage war, the art
> of which consists in gaining time when one has inferior forces.[2]

Clausewitz could complement this advice by recalling that defence is 'simply
the more effective form of war: a means to win a victory that enables one to
take the offensive after superiority has been gained; that is, to proceed to the
active object of the war'. Counter-attack was a key factor in strategic defence.
'A sudden powerful transition to the offensive—the flashing sword of
vengeance—is the greatest moment in defensive war. If it is not in the
commander's mind from the start, or rather if it is not an integral part of his
idea of defence, he will never be persuaded of the superiority of the defensive
form.'[3]

During the summer of 1806, Joseph Bonaparte was installed as king in
Naples, but the British landed in Calabria and beat General Reynier's
exposed corps at Maida (or Sant'Eufemia) on 4 July. Napoleon criticized
his brother for his poor dispositions. Reynier could not be aided because the
other French forces were too far off. The way to organize a strategic defence
was clearly set out in this letter:

Once General Reynier has pulled out and combined with your reinforcements, you must keep your troops in echelons, by brigades, a day's distance between them from Naples to Cassano, in such a way that in three days four brigades of 10 or 12,000 men can be concentrated. [...] With this positioning in echelons, you are on the defensive, sheltered from all events, in that, when next you want to take the offensive for a specific objective, the enemy cannot know it because he has seen you engaged in a formidable defensive operation, and because, before the changes have occurred on the defensive, the ten or twelve days of operations will be over. I do not know if you will understand some of what I am saying here. Major errors have been committed in the defensive; they are never made with impunity; the experienced man spots them at once. But the effects will still be felt two months later. Since the two important points were Gaeta and Reggio, and you have 38,000 men, you should have had brigades in echelons forming five divisions and which, positioned a day or two apart if needs be, could liaise. The enemy would have found you in such a position that he would not have dared to move,[4] for in an instant you could have concentrated your troops in Gaeta, Reggio, Sant'Eufemia, without a day being lost. Here are the dispositions that must be adopted for the Sicilian expedition. You must set off with a defensive order so formidable that the enemy does not dare attack you, and abandon any position behind you, other than defensive dispositions for your capital, and be wholly on the offensive against the enemy who, the raid having been made, cannot try anything. This is the art of war. You will see a lot of people who fight well and none who knows how to apply this principle. Had there been a brigade of 3,000–4,000 men at Cassano, nothing that occurred would have come to pass. It would have been at Sant'Eufemia at the same time as General Reynier, and the British would have been beaten or, rather, they would not have disembarked. What emboldened them was the incorrect position of your defensive.[5]

6

The Defensive Battle

In a defensive battle, one waits for the attack in a chosen, prepared position. This position is held in depth. A substantial mass—one-quarter or one-third of the total—is kept in the rear, to be thrown at the assailant when the latter has fully exposed his plan and used most of his troops.[1]

Napoleon always looked to keep back this '*masse de rupture*', whether he was the defender or attacker. What is less well known is that he also advocated measures of tactical defence. At the height of his power, in 1806, he recommended recourse to various fieldworks to protect the infantry:

> It seems that what is most to be feared from the Prussians is their cavalry. But with the infantry you have, and by always staying in a position to form yourself into squares, you have little to fear. However, no means of war must be neglected. Ensure that 3,000–5,000 pioneer axes always march with your divisions so as to construct a redoubt or even a simple ditch when required.[2]

7

Fortresses

Clausewitz referred to fortresses from the standpoint of a state's defensive strategy. They were depots for provisions, protected the wealth of major towns and cities, served as barriers in blocking roads and rivers, were tactical fulcra, stages, and refuges for weak or defeated troop corps, and shields against enemy attack.[1]

The usefulness of fortresses

Napoleon's reflections were very varied. In particular, he envisaged the defensive possibilities of fortresses in the context of a war of movement of an offensive kind. The first task of fortification, whether permanent or temporary, was to enable field armies to rest, recuperate, or compensate for a situation of numerical inferiority:

> The natural positions one ordinarily finds cannot shield an army from a stronger army without the aid of human art.
>
> There are military men who ask what use strongholds, defensive camps, and the engineer's art are. But in our turn we shall ask them, how is it possible to manoeuvre with inferior or equal forces without the aid of posts, fortifications, and all the supplementary means of art?[2]

Fortresses could serve many purposes:

> Strongholds are good for sheltering munitions depots, etc., holding soldiers who in the open would flee in the face of a few hussars and who, in these places, are trained during the campaign and can then act offensively and procure new resources for the active army, disrupt the enemy's rear if he advances, and oblige him to leave corps to conceal it. Another advantage of posts is to shorten the line of operations. When I marched on Vienna, Würzburg and Braunau were highly advantageous to me.[3]

Strongholds are useful for defensive and offensive war alike. Doubtless on their own they cannot stand in for an army. But they are the only means there is of holding up, impeding, weakening, disrupting a victorious enemy.[4]

As the current system of war requires very large active armies, an invaded country cannot fill the garrisons of all its posts, when they are numerous and vast, without significantly impoverishing its active force and without rendering it incapable of disputing the ground with an offensive army.

There is a way of using all of a border's strongholds without their defence absorbing too many men. This is to complete the garrisons and supply of the small number of genuinely important places, those in the front line, whose garrisons could unsettle an enemy that risked embarking on the siege of second-rank places, whose fortifications are in a less respectable state and where all that is left is a garrison capable of withdrawing into the citadel and defending it all-out, when the guard corps, having resisted sufficiently to force the enemy to prepare a siege, has been captured.[5]

Depot posts and field posts

For Napoleon, there were different kinds of strongholds. The most important were depot posts:

Here we must establish our ideas on the usefulness of strongholds. There are strongholds that defend a gorge and which, solely by virtue of that, have a specific character. There are depot strongholds which, capable of containing large garrisons and resisting for a long time, enable an inferior army to be reinforced, to rebuild and take new risks. In the first case, a stronghold or small post can be indicated; in the second, a large place where neither money nor work should be spared.[6]

Field posts are more modest:

Should it be asked what a permanently fortified field post means, let us take a look at the events that occurred last vendémiaire [October 1805]; let us see what use that poor castle of Verona was:[7] perhaps it had an incalculable influence on events. That castle secured the Adige, which immediately conferred a different physiognomy on all the actions of the campaign. [...] Yet, during all the time an army manoeuvres, evacuates one wing to head for a different wing, pulls back a little to combine with its relief or reinforcements [...], during all these manoeuvres, the enemy has neither the time nor the means to conduct a siege. He blocks the fortifications, fires a few shells, some salvos of field artillery: that is just the degree of strength a field post must have.[8]

However poor, field posts could support the operations of an army if they only required a small garrison, as Napoleon stated in connection with Osoppo in Italy:[9]

> [...] this original position fulfils the two indicated conditions on its own: it can offer protection to a division, contain its depots, and be defended by a handful of men; and it is never any trouble, for strongholds are also often very troublesome, undermine an army, and are the cause of the loss of a battle and a campaign.[10]

Borders and capitals

Following places defending a gorge and depot posts, Napoleon perceived a third instance of the usefulness of strongholds in the complete fortification of a border:

> Thus, the border from Dunkirk to Maubeuge contains a large number of fortified places of different size and value, placed in a chequered pattern in three lines, so that it is physically impossible to pass without having captured several of them. In this instance, a small post has the purpose of withstanding the flood that proceeds from one place to the next or plugging a re-entrant. In the midst of all these places, another kind of war is instituted. The carrying off of a convoy, a surprise attack on a depot, give a highly inferior army the advantage, without confrontation or any risk, of getting a siege lifted, of causing an operation to fail. In a few words, this is the action at Denain, one of little value in itself, but which obviously saved France from the greatest catastrophes.[11]

Vauban's system made possible Marshal de Villars's victory at Denain in 1712. As for capitals, they must be fortified:

> If Vienna had held out [in 1805 or 1809], it would have changed operations, but the inhabitants in a capital that can be bombarded with shells have a great influence. Once I was master of Vienna, 8,000–10,000 men were enough for me for its defence. I was quite sure that the enemies would not want to burn it and the mere threat of having it burnt by the garrison kept the inhabitants in order. Paris had to be and is to be fortified. In our day, armies are so large that border strongholds do not stop a victorious army and it is of the utmost consequence to let the enemy, following a victory, march on the capital and take it. But for that its outer wall should prevent the impact of bombardment. My intention had always been to do so.[12]

In fortifying capitals, generals have at their disposal all their resources, all their wealth, all their influence. In them they find basements, public buildings, which serve to hold the army's depots. These towns, having nearly all formerly had fortifications, still have ramparts in stonework or floodgates, etc.—which is useful; whereas earthen posts are not shielded from an improvised attack, unless you station a garrison in them that is as numerous as in a defensive camp.[13]

A large capital is the homeland of the nation's elite; all the great and good have their homes and families there. It is the centre of opinion, the depot of everything. It is the greatest of contradictions and inconsistencies to leave such an important point without direct defence. Back from the Austerlitz campaign, the Emperor often talked of it and had several plans drafted to fortify the heights of Paris. The fear of upsetting the inhabitants, and the events that succeeded one another with incredible rapidity, prevented him from following up on this plan.

How, he was to say, can you claim to fortify towns that have 12,000–15,000 toises of perimeter? You will need 80 or 100 fronts, 50,000–60,000 garrison soldiers, 800–1,000 artillery pieces in a battery. But 60,000 soldiers are an army: would it not be better to use them in line? This objection is generally lodged against large strongholds, but it is false in that it confuses a soldier with a man. Without doubt you need 50,000–60,000 men to defend a large capital, but not 50,000–60,000 soldiers. In times of misfortunes and great calamities, states may lack soldiers, but they never lack men for their internal defence. Fifty thousand men, including 2,000–3,000 gunners, will defend a capital and will deny entry to an army of 300,000–400,000 men, whereas 50,000 men, in open countryside, if they are not soldiers forged and commanded by experienced officers, are thrown into disarray by a charge of 3,000 cavalry. Moreover, all the great capitals are capable of protecting part of their outer wall with floods, because they are all situated near large rivers and ditches can be filled with water, either by natural means or by fire pumps. Such large posts, which contain such sizeable garrisons, have a certain number of dominant positions, without control of which it is impossible to risk entering the town.[14]

Posts near borders must delay invading armies by forcing them to undertake long sieges. Should the field army protect the capital? No doubt that is the wisest course of action, but the question cannot be settled in the abstract:

It should not be possible to cross the border without undertaking sieges or blockading places. Prince Eugène did not dare do it and lost two campaigns in sieges.[15] The Prince of Coburg did not dare to do so in 1793.[16] In 1814, the allies were sufficiently afraid of the Flanders border to avoid it, violate Swiss neutrality, and adopt their line of operations on this side. In 1815, before crossing the Somme, they were obliged to undertake sieges and await

coalition armies. It was the revolt of the Chambers that made them decide to
bear down on Paris and to demand my abdication.

The wisest thing is to cover the capital, but this issue cannot be posed thus
in the abstract. What one must do depends on the flight of a fly. Dumouriez
covered Paris. Kutuzov covered Moscow in 1812. I covered Paris. Turenne,
manœuvring on the flanks of the Archduke[17] and the Prince of Condé, did
not allow himself to be cut off or surrounded. By covering the capital and
withdrawing, one often succeeds. A great commander will not answer this
question. In examining what they have done, we can find reasons to justify
their conduct without deducing general principles from it.[18]

Defending yourself until the last moment

Fortresses have a mission to accomplish within a plan of campaign: they
must hold out for as long as possible. We have seen (Book I, Chapter 2) that
Napoleon feared and denounced premature surrender. Loyal to the rules of
the *ancien régime*, he repeated that a place could only surrender with honour
when the breach had been made by the besieging force. He went so far as to
lay down this principle:

> To be honourable, a capitulation must stipulate bad terms for the garrison.
> There is always a negative presumption against a garrison that leaves a place on
> a golden bridge.[19]

The officers defending a place must not believe the calculations of engineers
as regards the duration of resistance. It could last longer. Napoleon was
anxious that this was clearly stated in a work that was to help train artillery
and engineering officers at the Metz School:

> On this occasion, it is necessary to protest against the mania of engineering
> officers for believing that a site can only be defended for so many days; to bring
> home how absurd this is, and to cite well-known examples of sieges where,
> rather than the number of days that it was calculated would have to be spent
> routing the parallels, people have been forced to spend much more time on it,
> either by sorties or by crossfire or by any other kind of delay that the defence
> of the position has given rise to; and to bring out, when a breach exists, all the
> resources that still remain if the counterscarp has not exploded, if all the fires
> have not been extinguished, and how the very assault on the breach can fail if
> one has dug in behind. [. . .] he [the author of the book] must teach shrugging
> off the false rumours that the enemy may spread, and make it a matter of
> principle that the commander of a place under siege must not engage in any

kind of reasoning foreign to that with which he is charged; that he must regard himself as isolated from everything; that he must finally have no other idea than defending his position, rightly or wrongly, until the very last minute, in accordance with what is prescribed by the orders of Louis XIV and the example of brave men.[20]

To command in a town under siege involved novel responsibilities for an officer:

When a town is under a state of siege, it seems to me that a soldier becomes a kind of magistrate and must conduct himself with the moderation and tact required by the circumstances, and must not be a tool of factions, an advance guard officer.[21]

A post must be defended to the last:

One must hold out until the last moment, without calculating if one will be aided or not. To surrender a day earlier is a military crime. A commander must be oblivious of everything but his post and prolong its defence without seeking any political justification.[22]

In order for a place to resist for a long time, it must be armour plated—that is, its superstructures must be shielded from the bombs launched by howitzers:

The defence of a place is not organized when some military installations are not shielded from bombs. The result is that at the first accident the commander summons his council of war and proposes to surrender the place. This is how with a few mortars and howitzers the enemy seizes a place which, well organized, could have withstood a long siege. [...] At the point of the investment, it is too late to make armour, unless the commander is a man of firm character—something that cannot always be counted on. It is necessary to work on the assumption that he will be a mediocre man.[23]

A fortification must be 'actively' defended. Napoleon explains what he meant by this in connection with the defence of Corfu:

The first consideration in an active defence is that the enemy must not be allowed to establish himself in any point that has domination. [...] The art consists in keeping the enemy far removed from Mount Pantokrator [which dominates everything], in engaging him in a war that is disadvantageous to him, because we are masters of the higher position and that is incorrigible for the enemy; in harassing him, killing some of his men, tiring him, because the advantage lies with the one under siege, because he has chosen the position and it is for him.[24]

Already in *Le Souper de Beaucaire*, written in 1793, Bonaparte flung this at his audience:

> [...] it is an axiom in military art that he who remains in his defensive positions is beaten: experience and theory are in agreement on this point [...].[25]

For Charles of Lorraine, who let himself be holed up in Prague by Frederick II at the start of the Seven Years War, Napoleon has a sentence that might seem to be intolerably cynical:

> Prince Charles should have made sortie upon sortie. Troops are there to get killed.[26]

This was the major difference introduced by conscription: the generals of the French Revolution were able to dispose of a virtually unlimited reservoir of men and could conceive war in a more intense, more audacious way. We saw above that Napoleon pressed his generals not to attack, but to bypass fortified positions in order to spare their men. This did not prevent him embodying a more vigorous, but also more sanguinary, way of waging war—something underscored by Clausewitz when he said that war had approximated to its absolute form.

Another way of improving the defence of a fortress was that men should be in charge of a specific sector for good, rather than moving from one point to the next:

> This duty must not be performed by guards who relieve one another, but by battalions or companies that remain in a fixed post in bastions, demi-lunes, and contre-gardes; this is the Turkish method; it is the best for defence and the most economical in men. [...] You must therefore issue instructions that, were the siege to occur, the defence of each front will have to be entrusted to an officer and a portion of the garrison which will barrack there; the soldiers will always remain there.[27]

This system made the soldiers more responsible and they would defend a sector better if they were accustomed to it. An order of the same kind was given to Joseph Bonaparte in March 1814:

> Give orders that duties should be performed in Turkish fashion in posts—that is, the same men always remain charged with defence of the same bastion and sleep there as in a barracks. With this method, guarding a place requires only one-quarter of the troops necessary for our usual procedure.[28]

8

Defensive Positions

Troop mobility increased during the revolutionary wars and all combat plans now more or less sought to circumvent and envelop enemy positions. Circumventing entailed cutting lines of retreat and communications. Outflanking a position meant not bothering with it and pursuing one's objective by advancing on another route. The art of entrenchment seconded nature to confer a defensive character on a position.[1]

Although Napoleon had an attacking mind-set, he was also very sensitive to the advantage of a good position. According to him, choosing it was much more important now than in antiquity:

> A Roman camp was positioned regardless of the locality: all localities were good for armies whose strength consisted exclusively in weapons with blades. Neither *coup d'œil* nor military genius was required for setting up camp appropriately, whereas choice of position, the manner of occupying it and positioning the various arms, exploiting circumstances of terrain, is an art that forms part of the genius of the modern commander.
>
> The tactics of modern armies are based on two principles: 1 they must occupy a front that enables them to activate their projectile weapons advantageously; 2 they must especially prefer the advantage of occupying positions that dominate, extend, and enfilade the enemy lines to the advantage of being protected by a ditch, parapet, or some other item of field fortification.[2]
>
> In war, in any movement, the objective must be to gain a good position.[3]

In Palestine, Bonaparte prescribed to Murat:

> It is a matter of military principle that any detached corps should entrench itself and it is one of the first concerns people should have when occupying a position.[4]

The extent of the position depended on the kind of force involved. In 1813, lacking cavalry, Napoleon prescribed stretching sufficiently to enable the infantry to deploy its firepower:

[. . .] we must avoid the drawback of taking up too enclosed a position, which prevents weapons from being deployed and confers a great advantage on the enemy cavalry.[5]

On Saint Helena, Napoleon stressed the usefulness of field fortifications, as if the memory of the losses experienced in the major battles haunted him and he now wished to spare soldiers more by shielding them. He reckoned that the custom of an army digging in had been abandoned without much justification.[6] He confessed that had he known that an entrenchment could be constructed in twelve hours, even six or three, it would have had a major influence on his way of waging war:[7]

> The principles of field fortifications need to be improved. This part of the art of war is capable of making great progress.[8]

The Emperor concerned himself with field fortifications for three months, from the end of August 1818 until November, focusing his attention on such dry subjects 'as if it still involved the fate of the world'.[9] He had the great Marshal Bertrand, a general in the engineers, undertake a number of works and projects, to the point where Bertrand could not keep his diary as regularly:

> The problem can be solved: the principles of field fortification need to be improved; this important part of the art of war has made no progress since antiquity. Today, it is even inferior to what it was two thousand years ago. Engineers must be encouraged to perfect these principles, to raise this part of their art to the level of the other parts. It is doubtless easier to proscribe and condemn in a dogmatic tone from one's cabinet; moreover, you can be certain of flattering the troops' spirit of laziness. Officers and men find wielding pickaxes and shovels repugnant. So they incessantly repeat and claim: field fortifications do more harm than good; they should not be constructed. Victory goes to those who march, advance, manoeuvre: there is no need to work— doesn't war impose enough fatigue? . . . Flattering, but contemptible talk.[10]

This preoccupation with field fortification was not born on Saint Helena. Before starting the 1806 campaign against Prussia, Marshal Berthier had to see to equipping each division with 400–500 tools, in addition to 1,500 for each army corps.[11] In the course of the 1806 and 1807 campaigns, Marshal Soult received several dispatches to similar effect:[12]

> It is highly appropriate to move earth. This is the case with redoubts and field fortifications which, regardless of their real value, possess an advantage in perceptions.[13]

Earth must be moved and wood cut in order to create palisades. That is the way to spare the infantry and to have no fear of incursions by cavalry.[14]

On the eve of the 1809 campaign against Austria, General Bertrand, commanding the engineers of the army of Germany, had to reserve tools for battles.[15] It was above all during the 1813 campaign in Germany that Napoleon had important battlefield works constructed, principally around the city of Dresden, the pivot of his operations from the summer onwards. They bought him time, with much smaller forces, to stop the army of Schwarzenberg. The blow of 26 August and victory of the 27th were possible thanks to them. The supervisor was the commander-in-chief of the Grande Armée's engineers, General Rogniat, of whom Napoleon approved prior to reading his *Considérations sur l'art de la guerre*.[16] Still, in April 1819, the Emperor made some calculations about the depth of trenches, the height required to cover a man, and so forth. It is astonishing that, having read Rogniat and criticized him, Napoleon went so far in stressing the usefulness of engineering works in the field, which the former advocated with all his might. Bertrand, likewise a general in the engineers who had good relations with Rogniat, revealed the Emperor's innovative proposals in detail. The pickaxe would become a true adjunct of the musket.[17]

An important, frequently ignored aspect of Napoleon's thinking about war consists in this research on Saint Helena into better ways of ensuring defensive positions for troops thanks to field fortification. Rogniat's critique doubtless stimulated Napoleon's reflections on this and naturally it was General Bertrand who gave the best account of it. Anticipating a mid-nineteenth-century development,[18] Napoleon realized the importance of light field fortification for increased firepower:

> The infantry's power consists in its fire and that is all the more formidable when it is ranged over a larger number of ranks. Infantry is positioned in battle in three ranks, because it is acknowledged to be impossible for the fourth rank to fire. The third rank's fire is itself of very limited impact, more harmful than useful. However, with a simple fieldwork constructed with a small amount of labour, the terrain can be arranged in such a way that third, fourth, fifth, and sixth ranks can fire without harming one another—something that doubles and triples the depth and power of the infantry. [...]
>
> The peculiar thing about the clash of infantry and cavalry is that, if the infantry is knocked down, it is doomed, whereas, if the charge fails, no disadvantage results for the cavalry: hence the concern of all military men to

increase the power of the infantry. With a small amount of labour, it can be protected by a ditch and a wall of sandbags. Our ordinance, which lays down that the battalion square against cavalry is formed in six ranks, has more drawbacks than advantages, because it reduces fire by two-thirds. But, in arranging the terrain as indicated above, six ranks could fire at once. What cavalry could advance under such terrible fire! The conjunction of these three means or works–manoeuvres guarantees the infantry effective protection.[19]

Clausewitz too was conscious of the advantage of 'using the terrain like an auxiliary force'. Projecting himself into the future, he added: 'If it is a question of large devices, this advantage could apply in most cases.'[20]

9

Fortified Positions and Entrenched Camps

In this chapter, Clausewitz envisages positions so fortified that the intention is to render them impervious to attack. Fortification was now intended not to improve the conditions of the battle, but to prevent it and protect a whole sector. Inseparable from a fortress, entrenched camps aimed to protect not a space but an armed force. Clausewitz only really advocated them for places situated along the sea. Elsewhere, too many troops were required to occupy them and they did more harm than good.[1]

In criticizing Lazare Carnot's semi-official work on defending posts,[2] Napoleon offered his opinion on entrenched camps. While he could see their relevance, they must not be conceived any old how:

> There is in this book a most extravagant idea: that of placing all of a stronghold's war and artillery munitions in an entrenched camp, so that once the post has been captured, the camp can still be defended for a long time. One can see that M. Carnot has no experience of war and that he has not thought deeply about it. An entrenched camp is a poor defence against our current armies. One would be blasted in it by criss-crossed batteries that would make all points in it uninhabitable; one would be on the receiving end of death without being able to deal it; and the shortest course would be to quit it with bayonets fixed to make a breach. Against current weapons, there is no fortification that is shielded from an improvised attack. If it does not have great command, if it is not preceded by a large, deep ditch and swept in its periphery by flanking fire, if this ditch does not have a scarp and counterscarp in stone or wood (or tuff, but of such a kind that it is impossible to climb it without a ladder), no earth work, unless covered by a large ditch of water or the earth makes it possible to support them from a very obtuse angle, and supported by countersinks or palisades, can be shielded from an improvised attack—that is, shielded from the attack of an army corps with field equipment and disposing of 60–80 gun muzzles (especially 12-pounders or howitzers). Carnot's ideas are not clear on the important issue of the line of demarcation

separating what can be attacked without 24-pounder guns and without
mortars, and what can be attacked without an all-out siege but simply with
12-pounders and howitzers. In the very instance that Carnot offers—
surrounding Antwerp with an outer wall of 10,000 toises [19 km], to shield
the worksites from bombardment—it is preferable to do it with separate forts,
because the capture of a fort does not entail that of the whole wall and a fort is
defended until the last moment, without being liable to experience panic
terror, because all the points of its outer wall are in view of the garrison.[3]

Napoleon proposed detached forts around Antwerp, which the Belgian
army was to construct in the 1850s. He therefore foresaw the breaking-up
of the fortifications characteristic of the nineteenth century. While he did
not want to shut an army up in a camp of the Roman variety, he strongly
desired an entrenched camp of greater proportions, adapted to the armies of
the time:

> The principles on which fieldworks are constructed have not changed since
> the Romans. So are they not open to improvement? Could the drawbacks that
> have led to their abandonment not be counteracted so that we can obtain the
> benefits which Roman armies derived from them? For, if they were consid-
> erable for armies of 25,000–30,000 men, they would be even greater for an
> army of 100,000–200,000 men. These large armies are compelled to separate
> into four, six, eight corps and to march through so many different openings,
> separated from one another by woods, ravines, and mountains—something
> that exposes them to being attacked in isolation by double or triple forces.
> They cannot shield themselves from this peril without the aid of a fortified
> camp and without the habit of fortifying their camps every day.[4]

Napoleon's position on entrenched camps allows of no ambiguity. In fact,
he did not want 'to shut himself up' in any system, whether that of Carnot
or Rogniat. As regards the latter, he rejected his idea of altering Vauban's
work on the northern border:

> Vauban organized whole areas in entrenched camps, covered by rivers, floods,
> posts, and forests. But he never claimed that fortresses alone could close the
> border. He wanted this border, thus fortified, to protect a smaller army against
> a larger army; to afford an advantageous field of operations for maintaining
> oneself and preventing the enemy army from advancing, opportunities for
> attacking with advantage, and finally the means of buying time for allowing
> help to arrive.
>   In the reverses of Louis XIV, this system of strongholds saved the capital.
> Prince Eugène of Savoy lost a campaign taking Lille.[5] The siege of Landrecies
> offered Villars the opportunity to turn his luck around.[6] One hundred years

later, in 1793, at the time of Dumouriez's treason, the posts of Flanders saved Paris; the coalition lost a campaign taking Condé, Valenciennes, le Quesnoy, and Landrecies. This line of fortresses was also useful in 1814. The allies violated Swiss territory and entered the passes of the Jura in order to avoid those posts; and, circumventing them thus, in order to block them and observe them, they had to reduce their forces by a greater number than the total of the garrisons.[7]

Having thus characterized the areas covered by Vauban's iron belt as entrenched camps, and having advocated fortified camps for armies of 100,000 and over, Napoleon warned against the idea of the entrenched camp as advocated by General Rogniat. Rather than Vauban's fifty posts, Rogniat imagined five or six large entrenched camps on the border, followed by other similar lines to the centre of the country. This would obviously occasion enormous expenditure, whereas Vauban's posts had the merit of already existing. Rogniat thought like an engineer, used to devising technical and geometrical solutions. Napoleon sometimes happened to do this, but he had an easy time of it here criticizing his former general's abstract reasoning. He then drew attention to the danger of getting trapped in an entrenched camp:

> But must a capital be defended by protecting it directly or by shutting yourself up in an entrenched camp in the rear? The first course is more reliable: it makes it possible to defend the crossing of rivers, defiles, and even to create field positions, to reinforce yourself with all your troops from the interior, to prepare a good battlefield in a good position, while the enemy bit by bit grows weaker. It would be a bad course of action to allow yourself to be shut up in an enclosed camp. You would run the risk of being broken through there or at least blocked and reduced to emerging, sword in hand, to procure bread and fodder. 4[00] or 500 carriages a day are needed to supply an army of 100,000 men: the invading army, being larger by a third, would prevent the convoys arriving; without hermetically sealing this vast entrenched camp, it would starve it; deliveries would be so difficult that famine would rapidly set in.
>
> There remains a third course—that of manœuvring without letting yourself be pinned against the capital you are seeking to defend, or shutting yourself up in an entrenched camp in the rear. For that, what are required are a good army, good generals, and a good leader. In general, the idea of exposing the capital that is to be defended involves the idea of a detachment and has the drawbacks connected with any dispersal in the face of an already superior army.[8]

More than the idea of entrenched camps, what the Emperor was combating
was dogmatism. He ended his note with a paragraph that we have cited in
connection with grand tactics, referring to 'undefined physico-mathemat-
ical problems, which cannot be solved with the formulas of basic
geometry'.[9]

A passage from the *Mémorial* might even suggest that Napoleon criticized
General Rogniat in the first instance for expressing an idea that he had had
before him. Las Cases reports this:

> He said that he done a lot for Antwerp, but that it was still little compared with
> what he reckoned on doing. By sea, he wanted to make it a point of attack
> lethal to the enemy; on land, he wanted to make it a certain resource in case of
> major disasters, a true point of national salvation.[10] He wanted to make it
> capable of accommodating an entire army in its defeat and to withstand a year
> of open trench warfare—a period in which a nation had time, he said, to come
> en masse to deliver it and resume the offensive. Five or six posts of this kind,
> he added, were the new defensive system he planned to introduce in the
> future.[11]

IO

Defensive Mountain Warfare

Defensive mountain warfare had always had a reputation for efficacy and power. Obviously, this was due both to the difficulty of marching in long attack columns and the extraordinary strength that a small position in mountains could possess, thanks to the steep slope covering one's front and gorges on its right and its left. In the mountains, a judicious choice of position could in fact confer enormous obstructive power on a simple post with a small garrison.[1]

Napoleon's reflections were limited to this aspect, which pertains to tactics in the first instance. In February 1817, he discussed operations in the Alps during the revolutionary wars:

> I posit as a major principle in mountain warfare that the art of the offensive is to get oneself attacked, because, nature having constructed some very strong positions, there is generally a disadvantage in attacking unless one has adopted a really vital position. Thus, he who wishes to take the offensive must necessarily seek to take up some position on the flanks or in the rear of the enemy bothering him and force the enemy to seek to dislodge him and consequently to attack him.[2]

Mountains were conducive to the defensive. The point was developed in September 1817:

> In mountain warfare, it is necessary to get oneself attacked and not to attack. That is the art. The enemy is occupying a strong position? Then you must take up such a position that he is compelled to come and attack you or, indeed, to move off and take up another behind. [...] Mountains are bigger obstacles than rivers. You can always cross a river, but a mountain, like the Vosges, often has only two or three passages. If there are posts, you cannot cross. A few hours are enough to build a bridge, but six months are needed to build a road. I would not have crossed the Alps for Marengo if the King of Sardinia had not build roads to the bottom. You can make a supreme effort for two or three leagues of bad passages, but not for fifteen. Had there been enough people to defend the town and fort of Bard, I would not have crossed.[3]

Napoleon admits here that he staked his all in 1800 and perhaps would not have done so a second time. Curiously, he draws no lesson from the beginning of his first Italian campaign, when he took the offensive in the mountains, in the midst of Austrian and Piedmontese forces that he managed to separate. Clausewitz did it for him. He reckoned that the defensive capacity of a mountainous position was self-evident but relative, restricted to the tactical level. Strategically, the revolutionary wars showed that attack could concentrate its masses on a single point and stave in an overly extended line of defence. For him, defensive mountain warfare led to defeat in the main.[4] What campaign did he have in mind, if not the most brilliant one—that of Montenotte in April 1796?[5]

II

Defence of Rivers and Streams

The good thing about rivers and streams was that they established a line of defence. This was particularly true of northern Italy:

> The lines an Italian or French army must take up in order to oppose an invasion from the German side are those followed by the right bank of the rivers that flow into the Adriatic north of the Po. They cover the whole Po valley and from there enclose the peninsular and cover upper, middle, and lower Italy. These are the best lines of defence. The ones that follow the rivers which flow into the Po cut off the Po valley and expose middle and lower Italy; they require two armies manœuvring on the two banks of the Po.
> The lines of defence that cover the Po valley are those of the Isonzo, Tagliamento, Livenza, Piave, Brenta, and Adige.[1]
> The crossing of rivers is defended with artillery in particular.[2]

In a long note from 1809, Napoleon stressed the time factor involved in defending a line traced by a river or a stream:

> Only the following advantages can be expected from a line: making the position of the enemy so difficult that he throws himself into mistaken operations and is beaten by inferior forces, or, if a prudent general of genius is in charge, compelling him methodically to surmount obstacles created at leisure, and thereby gain time; on the side of the French army, on the other hand, assisting the general's weakness, rendering his position so exactly right and easy that he cannot make big mistakes, and finally allowing him time to wait for help. In the art of war, as in mechanics, time is the major factor between weight and power.[3]

In addition to this new allusion to mechanics, indicative of a certain turn of mind, this passage implicitly acknowledges that defence affords a surcharge of force, which brings us fairly close to Clausewitz's assessment of the superiority of the defensive. Along similar lines, the latter noted that the determination of space fell to defence, whereas the determination of time pertained to attack.[4]

When a position was situated near a waterway, it was important for an army that was withdrawing to dispose of a certain space between it and the fortifications:

> This should be a lesson for engineers, not only for constructing strongholds, but also for creating bridgeheads: they must leave a space between the position and the river, so that without entering the position, which would jeopardize its safety, an army can line up and rally between the position and the bridge.[5]

After the French forces had withdrawn behind the Ebro during summer 1808, Napoleon wrote how defence of this line was to be conceived:

> In holding the line of the Ebro, the general must have clearly foreseen what the enemy might do in any eventuality. [...] In any event, a lot of time must not be wasted deliberating and one must be able to deploy from right to left or from left to right, without making any sacrifice, for, in combined manoeuvres, trial and error and lack of resolution, which are born out of contradictory news arriving in rapid succession, result in misfortunes. [...] An observation that is not out of place here is that the enemy, who has an interest in concealing his forces by hiding the real point of his attack, operates in such a way that the blow he wants to deliver is never positively indicated, and the general can only divine it by truly profound knowledge of the position and by the way in which he engages his offensive system to protect and ensure his defensive system.[6]

It is also necessary to encompass resumption of the offensive:

> Any river or line can only be defended by having offensive points. For, when you have only defended yourself, you have run risks without obtaining anything. But when you can combine defence with an offensive move, you make the enemy run more risks than he makes the corps under attack run.[7]

One must be under no illusions: the line of a river or a stream is always forced open. It is absolutely necessary to create bridgeheads on the other bank:

> Even when it is as wide as the Vistula, as rapid as the Danube at its mouth, a river is nothing if one [does not have] outlets on the other bank and a mind prompt to resume the offensive. As for the Ebro, it is less than nothing; one views it as but an outline.[8]
>
> Never has a river been regarded as an obstacle that held things up for more than a few days and its crossing can only be defended by placing troops in numbers in bridgeheads on the other bank, ready to resume the offensive as soon as the enemy starts his crossing. But, if one wishes to limit oneself to the

defensive, no other course of action is open but to dispose one's troops in such a way as to be able to concentrate them en masse and fall on the enemy before his crossing has been completed. But the locality must allow it and all dispositions have to be made in advance. [...]

Nothing is more dangerous that to try to seriously defend a river by skirting the opposite bank. For once the enemy has surprised the act of crossing—and he always does—he finds the army in a highly extended defensive order and prevents it from rallying.[9]

These considerations setting out advantages and disadvantages prefigure those of Clausewitz. Were the defence of a river to be broken at any point, any subsequent defence could not last, unlike in mountains. Examples of a river being defended effectively were fairly rare in history. It had been concluded from this that streams and rivers did not form sufficiently solid barriers. Yet their defensive advantages were indisputable.[10]

12

Defence of Swamps— Inundations

S tudying the terrain on which operations were soon going to resume against the Russians in spring 1807, Napoleon sent adjutant commander Guilleminot[1] on reconnaissance. He wanted to exploit the swamps of eastern Prussia:

> It is necessary to reconnoitre the whole of this position and the advantage that might be taken of the swamps and natural obstacles. In this instance every obstacle is good, since it tends to shield a smaller corps from a larger one and compels the enemy to make dispositions that provide us with time for acting.[2]

With the exception of its dikes, a swamp was completely unviable for infantry. It was more difficult to cross than any river. If the crossings were not numerous, marshes were among the strongest lines of defence there were. As for inundations, Holland was perhaps the only European country where they amounted to a phenomenon worthy of attention:[3]

> The British and the Russians are very good soldiers; the Dutch are mediocre and many of them desert. But in a country like Holland, where, at every step, one meets with advantageous or impregnable positions, because it is completely covered by non-fordable canals, swamps, or inundations, the defensive can be sustained to advantage with troops that are inferior in numbers.[4]

13
The Cordon

'By a cordon we mean any system of defence in which a series of interconnected posts is intended to give direct protection to an area.'[1] This line of posts could only protect against fairly weak attacks. The Great Wall of China was constructed with this aim in view, to prevent Tartar incursions. The cordon was a protection against Asiatic peoples for whom war was a quasi-constant but diffuse state. A defensive cordon was established in Europe against the Turks. The objective was to halt minor operations. During the revolutionary wars, the cordon was used outside of its context. The Austrian and Prussian general staff believed at certain points that a defensive cordon could serve as a cover against any attack whatsoever:[2]

> General Beaulieu wanted to defend Mincio with a cordon. This system is the worst thing in the defensive order. [...] In dispersing his army along that river, he weakened it. It would have been stronger had he occupied a good position on the hillocks between Lake Garda and the Adige, before the Rivoli plateau, and protecting himself with entrenchments.[3]

In March 1800, General Masséna was on the defensive in Italy, while the First Consul began to form an army of Reserve. Napoleon recommended to him to avoid a defensive cordon:

> Do not have a line, but keep all your troops concentrated and grouped around Genoa, keeping your depots in Savona. These are the true military principles: in acting thus, you will beat 50,000 men with 30,000 and you will cover yourself in undying glory.[4]

It is impossible to hold all points. Placed on the throne of Naples, Joseph had this repeated to him in 1806:

> You will not hold your points by placing troops everywhere; it is by letting them march.[5]

If you seek to guard all the points of your kingdom of Naples, there will not be enough French forces.[6]

Are you seeking to defend a whole border with a cordon? You are weak everywhere, for ultimately everything human is limited: artillery, money, good officers, good generals—all that is not infinite and, if you are obliged to disperse everywhere, you will not be strong anywhere.[7]

During summer 1808, the same Joseph had to evacuate his new capital: Madrid. Napoleon wrote a note for Marshal Berthier about the defensive system adopted by Joseph:

Has the system of cordons been adopted? Is it contraband or the enemy that one seeks to prevent crossing? [. . .] After ten years of war, must we revert to such idiocies?[8]

A month later, he repeated:

We have already made it known that the system of cordons is among the most harmful and that a line, like the Rhine and even the Vistula, can only be sustained by occupying bridges that make it possible to resume the offensive.[9]

The reference to contraband was taken up again in a letter written on behalf of the Emperor by General Bertrand to General Maison, who was defending Belgium in January 1814:

His Majesty definitely does not approve of the plan for a line of twenty leagues. That is good for contraband, but this system of war has never succeeded.[10]

14

The Key to the Country

In criticizing the operations conducted in Germany in 1796, Napoleon referred to the city of Ulm,

> an important position on the Danube, which is the key to this part of Germany and the Danube basin that prevails from the middle of Thuringia to that of the Tyrol.[1]

During the battle of 14 January 1797, the Rivoli plateau was 'the key to the whole position'.[2] The phrase was also used in connection with Ostend:

> The post of Ostend is of the utmost importance. We must seriously concern ourselves with perfecting its state: it is the key to Belgium. [...] It is the place that must be defended, because, were the enemy to control it, he could advance into Belgium or on to Antwerp. This is what leads me to desire above all two forts on the dunes, on the left and on the right, 400 or 500 toises from the position's ramparts.[3]

Napoleon probably adopted the notion of key from Lloyd, who used it a lot. Clausewitz criticized the phrase, which endowed certain theories with a semblance of scientific truth, but did not prevent confusion. In fact, it sometimes referred to the most exposed site, and sometimes the most solid site, in the country. For him, it was justified if used to refer to an area without control of which one would never commit the reckless act of entering a country. But it could not be a point that determined possession of everything. For then it would no longer be a question of good sense, but of 'magic'.[4]

15

Retreat to the Interior of the Country

Napoleon scarcely theorized what he had practised in 1814, during the French campaign. Nevertheless, he stressed the need for depth for rallying one's troops in a retreat:

> When you have been ejected from an initial position, you need to rally your columns sufficiently in the rear so that the enemy cannot forestall them. For the most untoward thing that can happen to you is for your columns to be attacked separately prior to their concentration.[1]

Napoleon doubtless had in mind his disastrous expedition to Russia when he made the following reflection:

> It is impossible to make a man fight despite himself if he agrees to abandon the country and allows everything to be captured.[2]

Equally marked by the 1812 campaign in Russia, Clausewitz analysed at length retreat into the interior of the country, where the defender withdraws voluntarily so that the enemy attack exhausts itself. He acknowledged that no European state possessed expanses comparable to Russia's and that rare were those where a line of retreat of 100 miles was conceivable. '[T]he conditions that produced an army like that of the French in 1812 are unlikely to recur—to say nothing of the disproportion that obtained between the two antagonists at the outset, when the French had more than twice the number of men, with immense prestige into the bargain.'[3] Clausewitz cites other examples, involving smaller distances: Dumouriez's retreat in the face of the Duke of Brunswick in 1792 and Wellington's before Masséna in Portugal in 1810. In these two cases as well, the opponent exhausted himself and a major battle was not required to cause him to fail.

If Chaptal is to be believed, Napoleon respected Wellington's retreat:

There is a man [. . .]; he was forced to flee before an army that he did not dare to confront, but he established a desert of 80 leagues between the enemy and himself; he delayed its march; he weakened it by privations of every kind; he knew how to destroy it without fighting it. In Europe, only Wellington and I are capable of implementing such measures. But there is a difference between him and me: it is that this France, which is called a nation, would blame me, whereas England will approve of him. I was only ever free in Egypt. So there I permitted myself similar measures. Much has been said of the burning of the Palatinate and our wretched historians still denigrate Louis XIV on this score. The glory of that deed does not pertain to the king. It is wholly his minister Louvois's and, in my view, it was the finest act of his life.[4]

In 1689, Louvois had systematically burnt the villages and 'wasted' the fields of the Palatinate to prevent the imperial troops from finding subsistence there and threatening France. This act elicited indignation in Europe. Once again, Napoleon regretted having had to wage war while respecting certain limits.

16

The People in Arms

Napoleon broached this issue in connection with the fruitless attempts by the court of Rome in February 1797 to arm the people of its states against the French:

> When a nation does not have cadres and a principle of military organization, it is very difficult for it to organize an army. If France organized good armies so promptly in 1790, it is because it had a good stock, which the emigration improved rather than impoverishing. Romagna and the mountains of the Apennines were turned fanatical; the influence of priests and monks was all-powerful; the methods of missions, preaching, and miracles were effective. The peoples of the Apennines are naturally brave; some sparks of the character of the ancient Romans are to be found there. Yet they were unable to put up any resistance to a handful of well-disciplined, well-led troops. Cardinal Busca appropriately cited the Vendée. The Vendée found itself in particular circumstances: the population was belligerent and contained a large number of officers and non-commissioned officers who had served in the army, whereas the troops dispatched against it had been raised in the streets of Paris and were commanded by men who were not soldiers, and who simply committed idiocies—which bit by bit hardened the Vendeans. Finally, the measures adopted by the Committee of Public Safety and the Jacobins left these people no *mezzo termine*: if you are going to die anyway, you might as well defend yourself. It is perfectly clear that if in this war against the Holy See, rather than employing calming measures and winning victories, we had first of all experienced defeats and resorted to extreme, sanguinary measures, a Vendée could have broken out in the Apennines: severity, blood, and death create enthusiasts, martyrs; they breed courageous, desperate resolutions.[1]

The incompetence of the Gauls in the face of Caesar could be compared with that of the Italian cities. In emphasizing the need to maintain a line army, Napoleon repeated that one should not count too much on the people in arms:

> At this time they [the Gauls] possessed no national spirit and not even a regional one; they were dominated by a town spirit. This is the same spirit

that has since forged the shackles of Italy. Nothing is more opposed to the
national spirit, to general ideas of liberty, than the particularistic family or
small town mentality. A further result of this division was that the Gauls had
no line army which was maintained and drilled and consequently no military
art or science. Thus, were Caesar's glory to be based solely on his conquest of
the Gauls, it would be problematic.

Any nation that lost sight of the importance of a standing line army, and
entrusted itself to national levies or armies, would experience the fate of the
Gauls, but without even having the glory of mounting the same resistance,
which resulted from the barbarism of the time and a terrain covered with
forests, swamps, and mires, without roads—something that made conquest
difficult and defence easy.[2]

At the time of the Amiens peace, the First Consul addressed the Conseil
d'État to similar effect, rejecting the plan for local auxiliary battalions, made
up of young conscripts before their enrolment and commanded by
reformed officers:

> The organization of auxiliary battalions does not meet the objective. On the
> contrary, it would imbue conscripts with the spirit of locality rather than of
> the army. Moreover, what do you want us to do with so many men in
> peacetime? We should only raise the number required to fill the army and
> leave all the rest free. All I need is to go and annoy, create discontent ... We
> must think of the arts, the sciences, the professions ... We are not Spartans.
> We can only organize a reserve for the eventuality of war.[3]

When France was about to be invaded, a decree of 4 January 1814 organized
a *levée en masse* in nine directly threatened departments. Free corps and
partisan bands were to be raised, but there were not many of them.[4] In 1815,
confronted with the European coalition, Napoleon still planned a *levée en
masse*, which he described to Marshal Davout, Minister of War:

> The *levée en masse* consists in the organization of the national guard, all forest
> guards, all the gendarmerie, and all good citizens and employees who would
> like to join it. It must be organized by department and under the orders of a
> *maréchal de camp*, or the person charged with organizing the national guards, or
> the person who commands the department. This *levée en masse* must assemble
> at the sounding of the tocsin and the generals who are commanders-in-chief
> of the armies must indicate the rallying points that it is to occupy en masse,
> such as the defiles of bridges or mountain gorges, or arrange rendezvous on
> days of action for it to come and support the army and fall on the enemy's
> flanks and rear. The supreme commanders will be in charge of entrusting a
> general from the general staff with all liaison and leadership of *levées en masse*.

In each department, the supreme commander will indicate the bridges, passes, and closed towns which the *levées en masse* must defend in particular. He must give orders for the inhabitants to work on the spot to put their town, their gates, their bridges in a defensive state with barriers, palisades, or bridgeheads, depending on the locality; so that enemy light cavalry, the officers bearing orders, convoys, and enemy foragers cannot spread out anywhere.[5]

The *levée en masse* was even less organized in 1815 than 1814, with Napoleon refusing to arm the *'fédérés'* who supported him.[6] Arming the people in a war that had become a national war was alluded to in a conversation of January 1818. Gourgaud said that conquests could now no longer be made as they had been in the past, by destroying nations. Napoleon added:

Firearms are contrary to conquests, because a mere peasant with such a weapon is worth a brave man. Thus, in a national war where the whole population takes up arms there is an adventure in this new state of affairs.[7]

For Napoleon, the war in Spain proved nothing. But he said this to exculpate himself, blaming circumstances, his generals, and, above all, his brother Joseph. Moreover, the works of Charles Esdaile have clearly established that the role of the guerrillas was unduly glorified by politicians and publicists and that the regular army played the essential role alongside Wellington's forces. The Spanish and Portuguese guerrillas played a certain role, but were more like collections of brigands than a 'people in arms'.[8]

However that may be, the tasks to be entrusted to the people in arms were not those of the regular army. Essentially, their role was to supply positions with garrisons:

The garrisons of strongholds must be drawn from the population and not from active armies. Regiments of the provincial militia used to serve this purpose: it is the most splendid prerogative of the national guard.[9]

For Clausewitz, 'a popular uprising should, in general, be considered as an outgrowth of the way in which the conventional barriers have been swept away in our lifetime by the elemental violence of war. It is, in fact, a broadening and intensification of the fermentation process known as war.'[10] According to Clausewitz, however, a war in which the people are involved must always be combined with the war conducted by the standing army, and conducted in accordance with a single overall plan. Popular levies must not be used against the enemy's main corps or even against any sizeable corps: 'They are not supposed to pulverize the core but to nibble at the shell and around the edges.'[11] These ideas coincide with those of the Emperor.

17
Defence of a Theatre of War

In April 1800, prior to placing himself at the head of the army of Reserve, the First Consul set out his conception of the defence of the Republic:

> The system of war adopted by the government is to keep all the troops massed at a few points that are conducive to both defence and attack. The border departments must therefore not worry if several points, which throughout the war have been well supplied with troops, are not so today. Let them look to their left and right, and they will see numerous armies, which are more formidable the more concentrated they are, not only threatening the enemy who would like to advance into French territory, but also being mobilized to atone with brilliant victories for the affront suffered by our arms in the last campaign.[1]

This text, where Napoleon presents a different system of defence from the cordon, can be related to an important idea introduced by Clausewitz in the chapter entitled 'centre of gravity'. According to him, an invader must direct his blows against that part of the country where the enemy's most numerous units were concentrated. Strategic judgement must identify this neuralgic point accurately. A theatre of war, whether large or small, with its armed force of whatever size, represented a unit that could be reduced to a single centre of gravity. 'That is the place where the decision should be reached; a victory at that point is in its fullest sense identical with the defence of the theater of operations.'[2]

Napoleon did not only envisage offensive wars. He was always concerned with the defence of France, remembering the invasion of 1792 and sensing that it might occur again via the Belgian plain:

> One cannot purport to hermetically seal a territory like that of France; [...] on the contrary, the opening via Cologne, which leads to Brussels and from there to Antwerp and Ostend, delivers possession of a fine country, severs Holland, combines the operation with our eternal enemies. It might be said that when you have arrived at Ostend or Antwerp, you have succeeded in helping the British to land.

Of all the campaign plans that the combined powers might attempt against us, this is the one that must be opposed most resolutely.[3]

At different points in his long military career, Napoleon ventured general considerations on the defence of certain theatres of war involving geographical peculiarities, which Clausewitz took into account when referring to the 'territory', but whose various aspects do not enter into his theoretical analysis. His campaigns in Egypt and Syria enabled Napoleon to appreciate the obstacle represented by a desert in defending a theatre of war. He had experience of crossing the Sinai desert:

> Of all the obstacles that can protect the borders of empires, a desert like that is unquestionably the greatest. Mountain chains, like the Alps, come second and rivers third. For if one has so much difficulty transporting an army's food supplies that one rarely succeeds completely in doing so, that difficulty become twenty times greater when you have to drag with you water, fodder, and wood, three very heavy things which are very difficult to transport and which armies usually find *in situ*.[4]

The defence of a remote theatre posed particular problems. The army to be found there was a secondary one and could not hope for immediate reinforcements if attacked. It was there as an expeditionary corps. Napoleon developed a whole line of thinking on this subject when his forces, entrusted to Marmont, had to occupy Dalmatia, ceded to the Kingdom of Italy by Austria at the peace of Pressburg:

> There is no way of preventing an army double or triple the strength of the army I would have in Dalmatia from landing on some point of 80 leagues of coastline and rapidly securing a decisive advantage over my army, if its constitution is proportionate to its size.
>
> It is also impossible for me to prevent a stronger army, which would come via the border of Austria or Turkey, securing advantages over my army in Dalmatia.
>
> But do 6,000, 8,000, or 12,000 men, whom general political events might lead me to keep in Dalmatia, have to be destroyed and rendered destitute after a few engagements? Do my munitions, hospitals, and depots, dispersed on a whim, have to succumb and become the prey of the enemy, as soon as he has acquired superiority in the field over my army of Dalmatia? No: that is what I have to anticipate and avoid. I can only do it by establishing a large position, a depot post that is something like the recess of the whole defence of Dalmatia, which contains all my hospitals, my depots, my installations, where my troops can come to regroup, rally, either to shut themselves up there or to resume the campaign, if such are the nature of events and the strength of the enemy army.

I shall call this position the central position. As long as it exists, my troops can lose engagements, but will only have suffered the usual casualties of war. As long as it exists, after taking a breather and having a rest, they can seize victory again, or at least offer me these two advantages: occupying a number three times theirs in the siege of this position and giving me three or four months to come to their aid. For, as long as the place is not captured, the fate of the province is not decided and the enormous materiel attached to defending such a large province is not lost. [...]

Once a central position is established, all my generals' plans of campaign must relate to it. If a larger army has landed at some point, the generals' concern must be to direct all operations in such a way that their retreat to the central position is always ensured.

If an army attacks the Turkish or Austrian border, the same concern must govern all the operations of French generals. Not being able to defend the whole province, they must regard the central position as the province.

All the army's depots will be concentrated there, all the means of defence will be dispensed there, and a constant objective will be given to the generals' operations. Everything becomes simple, easy, defined; nothing is vague when, at arm's length and by higher authority, a country's central point is established. People can feel how much security and simplicity is vouchsafed by this central point and how much contentment it creates in the minds of the individuals who make up the army. The importance of preserving it acts sufficiently on everyone for people to sense how exposed they are: on one side, the sea covered with enemy vessels; on the other, the mountains of Bosnia populated by barbarians; on a third, the grim mountains of Croatia, almost impassable in retreat, when this territory must above all be regarded as enemy territory. Too much anxiety stirs the army if, in this position, it does not have a simple, mapped out plan for all eventualities. This simple, mapped out plan is the ramparts of Zara.[5] When, after several months of campaign, you can always go and shut yourself up in a strong town with abundant provisions as a last resort, you have something more than security of life: security of honour.[6]

In terse fashion, Napoleon summarized for his brother Louis what he must do to counter a British landing on the coasts of his kingdom of Holland:

A reserve at a central point must guard the circumference; I have already explained this to you in my instructions.[7]

The defence of the Kingdom of Naples against a possible British landing posed similar problems:

Assuming the British had lots of forces in Calabria and seriously wanted to sustain such a disproportionate war, with an advance guard at Cassano, supported at a few marches' distance by two or three brigades, you will be reinforced in three days by 9,000 men; and if, ultimately, they did not think themselves sufficiently strong, they would withdraw by a march and would then be joined by 3,000 men. This is how you wage war when you have several points to guard and do not know via which one the enemy will attack you.[8]

To prepare for any eventuality, it must be possible to rapidly concentrate one's troops at any point:

I see that you have too many troops all over the place. [. . .] The art of war is to dispose one's troops so that they are everywhere at once. For example, you put more than 2,600 men in Apulia; three-quarters of them must be positioned in such a way that one section can head for Cassano in two days, another section in four days. Everything you had at Gaeta must be placed so as to be able to return to Gaeta in one day, if necessary, or proceed to Naples. I would like to have an army half the size of yours and have more people at Cassano, Gaeta, if it was necessary, in Abruzzo, and in Apulia. I beg you not to take this lightly. The art of troop placement is the great art of war. Always place your troops in such a way that, whatever the enemy does, you can bring them together in a few days.[9]

Joseph received a further letter of the same kind some days later:

These detachments must be linked up and you must take the utmost care to keep your corps unified. [. . .] If you do not start from the principle that, in the case of any point with no objective, the enemy will not attack it in force; if you guard every point, you will never achieve anything. Concentrate all your dragoons and form a reserve out of them. [. . .]

Your troops would no sooner be brought together to recuperate and adapt at the end of September than they would have a high opinion of their strength, which would spread throughout the kingdom—an opinion that would persist for longer than through the actual appearance of the forces.[10]

Similar ideas had been formulated as early as 1793—additional evidence that Napoleon had manifestly had the basics of his ideas on war in mind for twenty years:

Over the last twenty years, a great deal of money has been spent on fortifying various positions on Corsica; never has money been so ill used. No point of the island is in a state to resist the smallest squadron. The reason is simple: what has been attempted is the fortification of a large number of different points, without registering the fact that it is not possible to prevent landings on an island that has so many gulfs. What is needed is to stick to a single one, to choose it well, and fortify it with all the resources of the art; in case of attack, to concentrate one's defence there, make it the heart of liaison with the continent, like a centre of resistance defending the rocks of the interior step by step.

In essence, this point must be as close as possible to France, must possess an anchorage capable of holding a fleet, be amenable to a defence effort, and combine with the advantage of being on the coast the facility of setting out from there to defend the interior. It is Saint-Florent that combines all these advantages. Had the sums thrown away on Bastia, Ajaccio, and Bonifacio been employed fortifying it, we would have a position capable of repelling the efforts of any fleet. By contrast, Bastia and Ajaccio are in no condition to resist and, in defending themselves, would only hasten their ruin. As for Saint-Florent, it is in a truly criminal state of defencelessness.[11]

We find the same ideas in a reflection dating from 1794, when the young General Bonaparte recommended to the Committee of Public Safety that, rather than remain on the defensive in Piedmont, it should go on the offensive:

When two armies are on the defensive, the army that can unify different posts most promptly in order to eradicate the one opposed to it in the defensive order inevitably needs fewer troops and, when troop strength is equal, always secures advantages.[12]

Against a power that has control of the seas, it is impossible to defend islands:

[. . .] the British and the Russians, who are masters of the sea, will always seize islands when they wish to, by committing five times as many men there.[13]

In northern Italy, defence relied on a line of waterways and posts. This line could not be conceived as a cordon. It was necessary to know how to use it to mask troop concentrations at one point or another depending on the enemy attack:

Whatever the enemy does, the terrain is disposed in such a way that with half the strength and equal skill everything is easy for the French general; for him everything presages and indicates victory, while everything is difficult and tricky for the enemy. That is the only advantage fortifications can bring to war. Like cannons, places are but weapons that cannot accomplish their objective alone; they need to be used well and handled properly. For these operations, one senses that communications from Ronco, via the right bank of the Adige, to Anguillara and Venice, need to be taken care of. They must be reconnoitred and maintained in good condition, so as to be able to direct the army onto one of its extremities during the night in two or three marches. No experienced, prudent general will risk himself in front of this large curve of fortifications from Ronco to Malghera, where the French army, manœuvring behind the water, renders any espionage and communication impossible for the enemy, and at every sunrise is to be found three marches in his rear or on one of his flanks, with all its forces concentrated against his dispersed forces.[14]

Conclusion to Book VI

Defence affords additional strength to the weaker party: it is therefore the strongest form of war. Napoleon implicitly anticipated this famous assertion by Clausewitz in advocating the defensive when troops were not battle hardened. In most instances, however, he was a supporter of the offensive in tactics—an offensive based on firepower. In strategy, he assessed the advantages of attack or defence in accordance with circumstances and terrain. According to him, strategic defence must always plan echelons, intermediate positions, and placement of troops such that they could support one another.

Napoleon found fortresses useful in many respects, in defensive and offensive war alike. Depot posts were essential during his campaigns. Even smaller positions, with multiple defects, could render major services. He wanted to fortify Paris. For him, a place must be defended to the last. Its defence must be active and not allow the enemy to establish himself on any dominant point. The same men must defend the same sector all the time.

In the field, choice of a good position was crucial. It was bound up with the possibility of deploying maximum fire. On Saint Helena, Napoleon felt a kind of regret that he had not sufficiently used field fortifications and called for development of this art. He also advocated vast entrenched camps, covered by detached forts. But he warned against the danger of allowing oneself to be trapped inside them.

It could be useful to temporarily establish one's defence along a waterway, but it was necessary to envisage all possibilities, to remind oneself that the enemy would always succeed in crossing, and to be capable of resuming the offensive with a strategic reserve and bridgeheads on the opposite bank. Troops should never be dispersed in a cordon: that entailed lacking strength anywhere. In case of manifest inferiority, retreat in depth, into the interior of the country, could be a good method of defence if it made it possible to exhaust the attacker.

Napoleon was not in favour of arming the people. Even if he considered it in 1814 and 1815, he did not resort to it. He nevertheless envisaged the population supplying garrisons for defensive positions. In defending a theatre of war, he advocated concentration of forces. For the defence of a remote theatre, this could take the form of a major depot post, a 'central position' capable of serving as a recess to whose preservation all campaign plans must be subordinate. When not knowing where the enemy was going to attack, one should not be everywhere—hence nowhere—but position one's troops in echelons so that they could be rapidly concentrated anywhere.

BOOK VII

Attack

The concept of attack forms a 'true logical antithesis' to the concept of defence. Each of them is fundamentally implied by the other. Thus, Book VII of *Vom Kriege* is but a complement of the system of ideas examined in Book VI. Moreover, Clausewitz was only able to sketch out this book.

I

Attack in Relation to Defence

Napoleon practised attack much more than defence. Artillery was his preferred means:

> Artillery has given attack an extreme advantage over defence, whose methods have remained the same. Previously, a camp protected by a feeble ditch was impregnable. Today, artillery would devastate everything inside it and render it uninhabitable. Machiavelli wrote about war like a blind man reasons about colours. I shall dictate some notes on these points to you.[1]

He who attacked seized the initiative, giving himself enormous advantages, as is suggested by a bulletin reporting the victory at Jena:

> [...] there are moments in war when no consideration must counterbalance the advantage of anticipating the enemy and attacking him first.[2]

In this regard, Napoleon felt himself to be very close to Frederick II:

> To put it plainly, I think like Frederick; one should always attack.[3]

2

The Nature of Strategic Attack

Pressing his strategic attack as far as Vienna and into Moravia (1805), Napoleon shattered the third coalition in three months, whereas neutralizing the first two coalitions had taken five years and two years, respectively. This temporal contraction of operations was bound up with their spatial extension. In striking at the very heart of Austria, Napoleon not only beat its armies, but occupied its nerve centres of decision-making and force creation, thus arresting the whole country's will to fight. In 1806, against Prussia, the Emperor orchestrated a projection of strength of more than 700 kilometres, from the shores of the Rhine to the Baltic.[1]

Attack, Clausewitz stressed, is not a homogeneous whole. An attack could not be pursued without interruption. When an army invaded a country, it must protect the space it had left behind it. In strategy, attack therefore always contained defensive elements.[2] This emerges very clearly from the following text by Napoleon:

> An army that marches to the conquest of a country has its two wings supported by neutral countries or major natural obstacles, whether large rivers or mountain chains, or it has only one of them, or none at all. In the first eventuality, it has only to see to it that it is not broken through on its front line; in the second, it must rely upon the supported wing; in the third, it must keep its various corps well supported in its centre and never separate. For if having two flanks exposed is a problem to be overcome, the disadvantage is double if there are four of them, triple if there are six, and quadruple if there are eight—that is, if one is divided into two, three, or four different corps. In the first case, an army's line of operations can equally lean to the left or the right; in the second, it must lean to the supported wing; in the third, it must be perpendicular to the middle of the army's line of march. In all cases, it is necessary every five or six marches to have a stronghold or entrenched position on the line of operations, for bringing together food and munitions depots, organizing convoys, and making them a centre of motion, a reference point that shortens the line of operations.[3]

For Clausewitz, the 'diminishing force of the attack' was a key subject of strategy. In particular, it derived from the need to occupy territory and protect lines of communication, from losses due to combat and illness, and from the increasing distance from sources of supplies and reinforcements. Gradually, attack thus progressed towards its 'culminating point', beyond which 'the scale turns and the reaction follows', often more violent than the initial clash. It was important to instinctively grasp this culminating point. 'Often it is entirely a matter of the imagination.'[4]

Napoleon prefigured this notion somewhat in a reflection on his own career as a conqueror—which imparts to it a dimension that is even more political than strategic. He confided to Bertrand that it was in Russia that he had attained his 'culminating point'. He should have died in the Kremlin. He was at the height of his glory and reputation. After Waterloo, things were different. He had then carried out his retreat from Russia. People now knew his limits.[5]

3

The Offensive Battle

'The main feature of an offensive battle is the outflanking or bypassing of the defender—that is, taking the initiative.'[1] As Napoleon said in Book IV, Marengo, Austerlitz, and Jena were attributable to the plan of campaign itself. Clausewitz agreed when he said that these defensive battles, conducted with enveloping lines or reversed fronts, resulted from favourable lines of communication and also psychological and physical supremacy. For him, the flank attack—that is, a battle where the front has been reversed—was generally more effective than the outflanking form. The latter was a question of tactics. The former pertained to operational art: it was better to surprise the enemy by falling on his flank than to make tactical dispositions within eyeshot to outflank him. On this point, Clausewitz was in full agreement with Napoleon's comments on Frederick II's most famous victory.

Frederick II's battles and oblique order

Frederick II's victory over the Austrians at Leuthen on 5 December 1757 has remained a model offensive battle:

> The Battle of Leuthen was a masterpiece of moves, manoeuvres, and determination. It alone would be enough to immortalize Frederick and afford him a position among the greatest generals. He attacked an army that was stronger, in position, and victorious, with an army composed in part of troops who had just been beaten, and won a complete victory without purchasing it with large casualties disproportionate to the outcome.[2]

Numerous commentators have thought that at Leuthen Frederick had demonstrated the validity of a particular order of attack: 'oblique order'. Napoleon mocked those who regarded it as some sort of recipe for victory:

It was a wonderful battle, but it did not involve oblique order. At Kolin and elsewhere,[3] oblique order hardly succeeded for him. It is nothing but charlatanism. [...] The greatest of all errors is to march in front of your enemy presenting the flank to him. These oblique attacks are idiocies. Things turned out badly for Frederick at Kolin, as they did for his general at [Zorndorf][4] in front of the Russians. At Austerlitz, the Russians also wanted to make a flanking march. Things turned out badly for them. Either your move is made at a distance, without the enemy seeing you, or before his very eyes. If he does not see you, it is better to attack him in his rear than on his wing. These are my moves, which Jomini clearly developed and understood. If you make them in the enemy's presence, and he sees you, it is very dangerous. Had Frederick attempted that against me, he would have found himself in a position like the Russians at Austerlitz. But these inanities belong more to his commentators than to Frederick, who was a great general. [...]

Besides, I do not understand what this oblique order means. It is Prussian charlatanism. If an army occupies a front of 10,000 or 30,000 toises [19–59 kilometres], then we can understand that it protects itself by a flanking march on one wing; you reach a wing rapidly enough for it to be crushed before aid arrives. But when the army occupies 2,500 toises [4.9 kilometres], and there is a second line, and the reserve of the centre arrives with its artillery and cavalry at least after a quarter of an hour on the wing, it is not clear what might halt the enemy. It must therefore be assumed that you do not see him coming, that he does not occupy any height, and that he does not move. You can then march on the enemy's flank without exposing your line of operations, unless you change it—which is the great art of war as I have formulated and practised it. [...]

At Austerlitz, the Russians wanted to proceed in oblique order and, judging their movement from the day before, I said that I would not allow them to march calmly in front of me, presenting the flank to me.

At Waterloo, I had taken up position on the right flank of the enemy while supporting my left on Mont Saint-Jean, where I attacked. But if the British marched on my left, I would lose my line of communication. When I saw Blücher's movement, I was on the point of focusing on the enemy's right, changing my line of operations via the Nivelles road. That would have been more rational.[5]

When the aim is to turn a wing of the enemy in a battle, liaison with the forces carrying out this move must be maintained. The Austrian Field-Marshal Daun neglected to do so at the Battle of Liegnitz (15 August 1760) and the isolated corps of Loudon (or Laudon) was overwhelmed by the Prussians:

At Liegnitz, where he was at the head of such sizeable forces, he isolated Laudon without establishing communications with him via an intermediate corps, in such a way as to attack in concert and be informed hourly of what was happening on his right. The art of war stipulates that it is necessary to turn and outflank a wing without separating the army.[6]

Underlying all these reflections on Frederick II's battles, there is a basic distinction reiterated by Napoleon: a flanking march by some of the troops during a battle in view of the enemy is a terrible mistake, which costs very dear. By contrast, a manoeuvre by the bulk of the army out of view of the enemy to fall on one of its wings or, better still, on its rear, is the sign of a great general. That is what Frederick did at Leuthen. It did not involve a so-called oblique order of battle, for, by definition, an order of battle is seen by the enemy. The Emperor was playing on words somewhat, but this is because a clear distinction between battle and operations was still not made in his day. Frederick's attacking manoeuvre at Leuthen pertained to operational art— Napoleon might have said grand tactics—and not the tactics of battles.[7]

The form and the means of attack

As we saw in Book IV on the engagement, Napoleon hardly concerned himself with the detailed tactics of units on the day of battle, leaving that to his generals' initiative. However, he gave some general directives, such as the following:

> You will ensure that your two attacks are combined in such a way that they unmask themselves in time, so that one of the two does not bear the full brunt of the enemy's endeavours on its own. Your first concern will be to establish your communications between one attack and the other, so that, if one of them is repulsed, the other, taking the enemy from behind, might not only restore the advantage, but also render any early success he might have had extremely deadly.[8]

As we can sense, combined attacks involved risks. In Egypt, Bonaparte confessed to Desaix that he did not like them:

> You know that in general I do not like combined attacks. Arrive in the presence of Mourad Bey by whatever route you can and with all your forces. There, on the battlefield, if he resists, you will make your dispositions to do him the maximum damage possible.[9]

The combined attack involved dividing one's army into at least two columns, generally so that one of them turns the enemy. The combat dispositions were left to Desaix's wisdom. A reflection formulated on Saint Helena attests to Napoleon's great mental flexibility when it came to combat tactics:

> Today, ideas about the way to attack are not fixed. A general never knows if he must attack in lines or columns. If he attacks in lines, he is weak against a cavalry attack that takes him in the flank. Faced with the Russians, for example, who employ their cavalry marvellously, that is very dangerous. If you attack in columns, you have no fire and it seems that at Waterloo the Guard did not have time to deploy, did not open fire, leading to its rout.[10]

Shortly afterwards, the Emperor inclined to the column:

> In war, you have to be utterly simple. The attacking column suffices; it is formed promptly and by simple procedure. Nothing can be too simple in war.[11]

Field fortification can provide the column with the firepower it lacks:

> The infantry's order of battle is in three ranks; in columns by battalion, out of nine. The column's drawback is that in it firing is no good, because, with only the first two ranks being able to fire, seven-ninths of muskets cannot employ their firepower. With a little labour one can have the nine ranks fire simultaneously without disrupting their order and, after a powerful fusillade, march in columns, bayonet at the end of musket, on the enemy line, which is shaken, thinned out, weakened by such formidable fire.[12]

Napoleon always stressed the value of artillery. On 7 June 1809, he wrote to Eugène de Beauharnais, who commanded the army of Italy and was seeking to do battle with Archduke Jean's Austrians:

> Let me know how many artillery pieces and shots you have. From your advance guard to the tail of your park, there should not be more than three or four leagues. As for the artillery, this is what you must attend to: as soon as you have decided on your attack, support it by a battery of 30–36 guns. Nothing will withstand that, whereas the same number of guns dispersed over the line would not yield the same results.[13]

On 14 June, Eugène won the Battle of Raab. Napoleon congratulated his son-in-law, but added this remark:

> Since you attacked in echelons on the right, why did you not put 25 cannons at the head of your echelons? That would have strengthened your attack and intimidated the enemy. Like all other weapons, the cannon must be concentrated en masse if you want to achieve a significant result.[14]

Naval battles

At sea, the principle should be the same. The First Consul advocated firing
wherever results would be achieved most rapidly. The Navy Minister was
to make this known to all officers:

> You will get them to sense the disadvantage of always firing to dismast and just
> how true, in all circumstances, is the principle that one should start by doing
> the maximum possible damage to one's enemy.[15]

The French navy must rediscover the meaning of attack when it did battle
with the British. Starting out from that, Napoleon made interesting comparisons between two maritime traditions, with their respective strengths
and weaknesses:

> Often superior in strength to the British, we have not known how to attack
> them and have allowed their squadrons to escape, because we have wasted
> time on pointless manoeuvres. The first law of naval tactics must be that, as
> soon as the admiral has given the signal that he wishes to attack, each
> commander has to make the moves required to attack an enemy vessel, take
> part in the engagement, and support his neighbours.
>
> This principle is that of British tactics in recent times. Had it been adopted
> in France, Admiral Villeneuve at Aboukir would not have thought himself
> blameless for remaining inactive for 24 hours, with five or six vessels—i.e. half
> the squadron—while the enemy was destroying the other wing.[16]
>
> The French navy is called upon to acquire superiority over the British navy.
> The French understand construction better and French vessels, as the British
> themselves admit, are all better than theirs. The guns are superior in calibre by
> a quarter to British pieces. That represents two great advantages.
>
> The British have more discipline. The squadrons at Toulon and on the
> Scheldt had adopted the same practices and customs as the British and ended
> up with equally strict discipline, with the difference entailed by the character of
> the two nations. British discipline is a discipline of slaves: it is the master faced
> with the slave. It is maintained solely by the practice of the most appalling terror.
> A similar state of things would degrade and debase the French character, which
> needs paternal discipline, based more on honour and feelings.
>
> In most of the battles that we have lost against the British, either we were
> inferior; or we were conjoined with Spanish vessels which, poorly organized
> and in recent times degraded, weakened our line, rather than strengthening it;
> or, finally, the supreme commanders, who sought battle and were advancing
> towards the enemy until they were face to face, then hesitated, retreated on
> various pretexts, and thereby jeopardized the most brave.[17]

4

River Crossings

A major river blocking the line of an attack was always a major problem. It was rare for one to be able to tactically force the crossing of a river that was defended. For that what was required was very considerable psychological and physical superiority.[1]

On 7 May 1796, the advance guard of the French army of Italy crossed the Po at Piacenza in a dozen boats. A bridge then had to be built. That took two days.

> The river at Piacenza is very rapid; its width is 250 toises. Crossing rivers of this size is the most critical operation in war.[2]

It was necessary to try to cross as quickly as possible:

> Operations like crossing a river such as the Rhine are so delicate that the troops must not remain exposed so long without communications.[3]

However,

> When you have the volition to start a campaign, nothing can stop you; and, since history began recounting military operations for us, a river has never constituted a real obstacle.[4]
>
> With artillery, you can cross all rivers; around twelve hours are needed to make a good bridge of boats. You begin in the evening and cross in the morning.[5]

When General Moreau, at the head of the army of the Rhine, passed the Inn on 28 November 1800, he did so at six different points. For Clausewitz, it was a mistake to attempt a crossing at several points, unless they were very close and amenable to a single stroke.[6] Napoleon always had an excessive tendency to criticize Moreau, but he laid down this rule in connection with his crossing of the Inn:

> When the army opposed to you is protected by a river on which it has several bridgeheads, it should not be tackled head on. Such a disposition disperses

your army and exposes you to being cut off. The river you wish to cross must be approached in columns in echelons, so that there is only one column—the most advanced—which the enemy can attack without exposing his flank. During this time, your light troops will line the bank and, when you have decided on the point at which you wish to cross—a point that must always be remote from the head echelon, so as to thoroughly fool your enemy—you will head for it rapidly and lay your bridge.[7]

Well placed in a dominant position, artillery could facilitate the offensive crossing of rivers:

In siege warfare, as in field warfare, it is the cannon that plays the main role; it has caused a total revolution. High stone ramparts have had to be abandoned for grazing fire, and covered by mounds of earth. The custom of retrenching every day, when setting up camp, and finding safety behind terrible stakes planted alongside one another has also had to be abandoned.

As soon as you are master of a position that dominates the opposite bank, if it is sufficiently large for you to be able to position a fair number of artillery pieces there, you can easily acquire facilities for crossing the river. However, if the river is 200–300 toises in width [380–600 metres], the advantage is smaller, because, with your grapeshot no longer reaching the other bank and the distance enabling the enemy to slip away easily, the troops defending the crossing are able to bury themselves in tunnels that shield them from fire from the opposite bank. If the grenadiers ordered to cross to protect construction of the bridge manage to overcome this obstacle, they are destroyed by enemy grapeshot, which, placed 200 toises from the opening of the bridge, has the range to generate utterly murderous fire, and yet is 400–500 toises' distance from the batteries of the army seeking to cross, so that the artillery advantage is entirely with him. Thus, in this instance, crossing is possible only when you manage to take the enemy completely by surprise, and when you are favoured by an intermediate island or a very pronounced curve, which makes it possible to install batteries intersecting their fire on the gorge. This island or curve then forms a natural bridgehead and confers all the artillery advantage on the attacking army.

When a river is less than 60 toises, the troops who are thrown on the other shore, protected by marked superiority in artillery and the great command which the bank where it is positioned is bound to have, find themselves with so many advantages that, provided the river forms a curve, it is impossible to prevent the bridge being installed. In this case, the most skilful generals have been content, when they have been able to anticipate their enemy's plan and reach the crossing point with their army, to oppose the crossing of the bridge, which is a really narrow pass, by placing themselves in a semicircle around about, eluding the fire from the opposite bank, 300–400 toises from its heights. This is the manoeuvre carried out by Vendôme to prevent Eugène from using his bridge at Cassano.[8]

Napoleon explained this point when recalling the episode of the bridge of Lodi on 10 May 1796:

Crossing a river is nothing when it is only 60 or 100 toises [118–195 metres], when it is under cannon fire. When it is 300 toises, things are different: then the opposite bank is no longer within cannon range and the enemy can remain there. This is what explains the crossing of the bridge at Lodi, which was certainly a bold operation. The river must be only 60 toises. I first of all had batteries placed right and left that intersected in front of the bridge, kept the enemy at a distance, and counter-attacked his batteries. It was possible to have 30 muzzles on each side, but what was decisive was the battery placed near the bridge itself. We were under the reinforced fire of the enemy there. He made it hot. I dismounted and led the battery in situ myself; then the bridge found itself under the cover of my cannon. 'The bridge is ours', I said. Soon the enemy battery was compelled to withdraw.

Something that facilitated the crossing was that the village had an overhang on this side, so that the troops were sheltered from the enemy fire, which did them no harm, and by one on the right and one on the left they found themselves at their post and on the bridge without having suffered any harm. The column was half way from the bridge when the enemy spotted it. Some grapeshot threw some men to the ground. There was a moment's hesitation, but Lannes and a few brave men saved the head [of the column] and the tail crossed. It was the infernal column of the elite grenadiers. It was a sure bet. We only risked a battalion. We were soon in strength on the other bank. The enemy, I believe, had 10,000 men.[9] The artillery initially fired for around an hour and a half.

When I observed that the enemy artillery had withdrawn, 'This is the time!', I said to Masséna. We have the infernal column. Astonished at first, Masséna did not have to be told twice. The column, by one on the right, leapt onto the bridge and the crossing succeeded.

River crossings are a matter of cannons. They are infallible. [...] The enemy must be forced back. They are infallible. There is only one material obstacle capable of stopping them: a broken arch, a house directly facing the bridge. This is what is often responsible for the fact that one does not have time to burn the bridges. The enemy artillery warms up, a column dashes, you cannot stop it, you are no longer in time to stop it.

After Brienne's retreat, I ordered Ney to burn the bridge.[10] I had the sappers prepare faggots for setting it alight beforehand. Despite everything, these precautions would have been useless if, having had a house at the bridge opening crenelated, I had not left 50 men of my guard there, whose fusillade halted the enemy. Without these 50 brave men, despite the orders issued, despite the precautions taken to set fire, Ney tells me he would not have been able to burn the bridge. The enemy column was rushing forward and nothing could have stopped it.[11]

5

Attack on Defensive Positions

If the attacking army could pursue its objective ignoring the enemy's defensive positions, it would be a mistake to attack them. If it could not pursue its objective, it must seek to manoeuvre on the flank to expel the enemy from them. It was better to attack sideways, for it was always dangerous to tackle a determined opponent holding a good position.[1]

In his 1655 campaign on the banks of the Scheldt,

> Turenne was loyal to two maxims: 1 *Do not attack head on positions that you can obtain by bypassing them.* 2 *Do not do what the enemy wants, for the simple reason that he desires it; avoid the battlefield that he has reconnoitred, studied, and, even more carefully, the one he has fortified and where he is entrenched.*[2]

At Kolin on 18 June 1757 Frederick II suffered his most painful defeat, but he had been far too audacious in his attack on the Austrian position:

> At the Battle of Kolin, it is difficult to justify his attempt to circumvent Daun's right by making a flanking march of 3,000 toises [5.9 kilometres], 500 toises [990 metres] from the heights crowned by the enemy army. It was such a rash operation, one so contrary to the principles of war: *Do not perform a flanking march in front of an army in position, especially when it occupies the heights at whose foot you have to march!* Had he attacked the left of the Austrian army, he would have been perfectly placed for that. But to march under the cannon and musket fire of a whole army occupying the highest position around,[3] in order to outflank an opposite wing, is to assume that that army possesses neither cannons nor muskets. Prussian writers have said that the manoeuvre failed only as a result of the impatience of a battalion leader who, fatigued by the fire of the Austrian skirmishers, ordered right into battle and thus engaged the whole column. This is inaccurate. The king was present; all the generals were aware of his plans and, from head to tail, the column was no more than 3,000 toises. The move made by the Prussian army was required of it by the foremost concern—the need for safety and every man's instinct not to let himself be killed without defending himself.[4]

This excellent commentary coincides with recent analyses of the Battle of Kolin.[5] Before 1807, when pride had not yet seized hold of him, says General Bonnal, Napoleon was prudent and respected his enemies.[6] He recommended to his generals not to attack an enemy occupying a good position. In 1806, despite the numerical superiority of the Grande Armée, he was sensitive to the Prussians' reputation for excellence and cautioned prudence to Marshal Soult:

> If the enemy comes against you with forces inferior to 30,000 men, you can, in concert with Marshal Ney, concentrate your troops and attack him. But if he is in a position that he has long occupied, he will have taken care to reconnoitre it and dig in. In that case, proceed with caution.[7]

Even two days before Jena, when he felt that he had increasing ascendancy over the Prussian army, Napoleon wrote to Lannes:

> The art today is to attack anything one encounters, so as to beat the enemy comprehensively while he is concentrating. When I say that one should attack anything one encounters, I mean one should attack anything on the march and not what is in a position that renders it far superior.[8]

In 1807, he wrote to Murat, knowing that he often embarked on reckless attacks:

> You should not indulge in head-on attacks, but bypass the enemy's positions and march on Königsberg.[9]

In 1809, he cautioned Prince Eugène de Beauharnais:

> It is not impossible that Prince Jean has selected a good position and is waiting for you. In that case, I recommend that you reconnoitre well and firmly establish your system before attacking him. A forward movement, without strong combinations, may succeed when the enemy is retreating. But it never succeeds when the enemy is in position and determined to defend himself. Then, it is a system or scheme which cause a battle to be won.[10]

After his defeat by Wellington at Talavera (28 July 1809), Marshal Jourdan found himself criticized for having attacked a strong defensive position head on. Napoleon had the War Minister Clarke write to him

> [...] that it had taken the conjunction of all these mistakes for an army like my army of Spain to be thus bettered by 30,000 British; but that, as long as you seek to attack good troops like the British in good positions, without reconnoitring them and checking if you can take them, you will lead my men to death for absolutely no reason.[11]

Napoleon attacked Wellington head on at Waterloo. On Saint Helena, Gourgaud remarked to him that he also attacked the Russians at Borodino head on. Napoleon accepted this, but added that it was because he sought a battle at any price:[12]

> At the Battle of Borodino, I could have turned the Russian position on the right and forced them to abandon it. But I confess that I did not adjudge it to be as strong as it proved; and furthermore, I had need of a battle. I wanted to exploit the opportunity to prevent Kutuzov from dragging me further into the Russian interior in pursuit of him, without a major battle having determined peace.[13]

Napoleon was conscious of what he was doing and believed that a brush with the Russians, who had hitherto shied away from it, was the most important thing.[14] While he did not often acknowledge his mistakes, he made an exception in the case of Borodino:

> At Borodino, I mistakenly attacked the Russians' entrenched position; it was because I was thirsting for a major battle. For an army with a sizeable, good cavalry, which can manoeuvre behind a system of good redoubts, supported by an army, should not be attacked. It must be made to abandon its position by manoeuvres.[15]

During the 1813 campaign in Germany, he recommended to his generals that they use their artillery, particularly their howitzers, to destroy entrenched positions:

> You must not lose people in front of villages and entrenched positions, but must straight away have the 32 12-pounders brought up from your four reserve batteries, with 40 howitzers, by means of which you will destroy all the field fortifications in two hours.[16]
>
> If there is some entrenched village or some redoubt, they must be devastated with shells before attacking. You remember the experience we had in Vienna: neither the blockhouse, nor the exterior, nor the interior of the redoubt could withstand shells or superior artillery.[17]

6

Attack on a Mountainous Area

Mountain chains should not be attacked. When he commanded the army of Italy in March–April 1799, General Schérer was wrong to launch troops against them:

> It was a great mistake by Schérer to send divisions into the Graubünden and the Tyrol. The defence of Italy is confined to Corona and guarding Rocca d'Anfo. I never bothered with the Graubünden and knowing whether there was an enemy on that side. What does anything that might occur beyond the Graubünden, in a territory where there are no paths, matter! What does the Tyrol! There is only the Trento road; all the rest is nothing. Hold Corona and the Adige, and everything is covered. That is where all the troops should be concentrated. Making your troops manoeuvre by engaging in actions and attacks in mountains cannot yield any result: it is absurd. The example of the army of Reserve proves nothing. First of all, this movement was utterly unexpected—which contributed to its success; next it was feasible, because there was only a small mountain position to cross at this point; there was a road over most of it. Never did I think of the Saint Gothard.[1]

In 1800, the army of Reserve crossed through the Saint Bernard Pass. In the mountains, even when you wish to attack, you need to get yourself attacked:

> In the mountains are everywhere to be found a large number of positions that are extremely strong in themselves, which one must refrain from attacking. The genius of this war consists in occupying camps either on the flanks or in the rear of the enemy's, which leave him only with the alternative of evacuating his positions without fighting to take up others further back, or to emerge from them to attack you. In mountain warfare, he who attacks is at a disadvantage. Even in offensive warfare, the art consists in having nothing but defensive engagements and compelling the enemy to attack.[2]

Napoleon did the opposite at the Somosierra Pass on 30 November 1808, when he launched his Polish light horse in an attack on the Spanish position. But he then became impatient because he felt superior and his power blinded him, to the point that he felt capable of contravening his own principles:

> The Prince of Condé violated one of the principles of mountain warfare: *never attack troops who occupy good positions in the mountains, but flush them out by occupying camps on their flanks or in their rear.* [. . .] The French army succeeded on the first day by unheard-of endeavours of courage to force open the first positions. But it failed the following day, because in the mountains, after one position has been lost, you find another that is just as strong for stopping the enemy.[3]

It was in developing this point that Napoleon ventured his famous formula on the art of war, 'a simple art, in which everything is a matter of execution':

> Mountainous country depends on the plains that feed it and only has influence on them to the extent that they are within range of its cannon. The borders that protect empires are made up of plains, hilly country, and mountainous country. If an army wishes to cross them and is superior in cavalry, it will do well to adopt its line of operations through plains. If it is inferior in that arm, it will prefer hilly country. But for mountainous country, it will be happy in all cases to observe the mountains while bypassing them. In fact, a line of operations must not cross through mountainous country: 1 because you cannot live in it; 2 because in it you encounter at every step narrow passes that would have to be occupied by fortresses; 3 because marching there is difficult and slow; 4 because columns of brave men can be halted in it by ragged peasants straight from ploughing, vanquished, and defeated; 5 because the genius of mountain warfare is never to attack: even though you want to conquer, you must pave the way by positional manoeuvres that leave the army corps charged with defence with no alternative but to attack or fall back; 6 finally, because a line of operations must serve retreat and how can one think of withdrawing through gorges, narrow passes, and precipices? When they could not do otherwise, great armies have crossed mountainous country, to arrive in fine plains and beautiful country. Thus, it is necessary to cross the Alps to arrive in Italy. But to make supernatural efforts to cross inaccessible mountains and still find oneself in the midst of precipices, narrow passes, and rocks, without any prospect but the same obstacles to overcome and the same fatigue to endure for a long time; to be anxious at each new march knowing there are tight spots in the rear; to be in greater danger of dying of hunger every day—and this when one could do otherwise: this is to enjoy difficulties and to wrestle with giants; it is to act without good sense and consequently contrary to the spirit of the art of war. Your enemy has large cities, beautiful provinces, and capitals to protect—march on them via plains. The art of war is

simple, everything is a matter of execution. It has nothing vague about it; all of it is good sense and nothing is ideology.[4]

Taking the logic of these observations further, Clausewitz reckoned that a country like Spain had an interest in concentrating its forces behind the Ebro, rather than dispersing them among the fifteen defiles of the Pyrenees. A principal battle accepted on the plain did not exclude the preliminary defence of mountains by auxiliary forces. For Clausewitz, the mountain contained a basis of strength and could serve as a refuge for the weak on a secondary level. But were a whole army to seek to install itself there in a defensive system, it would always resemble a cordon. And the attacking army would always find a way of breaking through it.[5] However, it would avoid head-on attacks in favour of flanking operations that 'must be aimed more at actually cutting off the enemy forces than at tactical assaults in flank or rear, for even the rear of a mountain position can offer strong resistance if the forces are available. The fastest way of getting results is always to give the enemy reason to fear having his line of retreat cut. That fear is aroused more quickly and effectively in mountain warfare, for there it is not so easy to cut one's way out if the worst comes to the worst.'[6] Clausewitz's arguments complement those of Napoleon.

7

Manoeuvre

For Napoleon, manoeuvring signified attacking other than frontally. When your army advances against the enemy, and he leaves a strong rearguard in an advantageous defensive position, it is necessary to manoeuvre. Here we are at the tactical level. In an engagement like that of Arcola in November 1796, the manoeuvre consisted in bypassing the enemy or, rather, in making him believe that he had been bypassed:

> The enemy's left was supported by some swamps and by superiority of numbers it imposed on our right. I ordered citizen Hercule, an officer from my scouts, to choose 25 men from his company, to follow the Adige for half a league, to bypass all the swamps supporting the enemy's left, and then to fall at a great gallop on the enemy's back, sounding several trumpets. This manoeuvre succeeded perfectly: the enemy infantry was shaken.[1]

A manoeuvre involved more 'science'. After his great victory at Hohenlinden on 3 December 1800, General Moreau received this compliment from the First Consul:

> I cannot tell you how much interest I have taken in your wonderful, expert manoeuvres; you have once again surpassed yourself in this campaign.[2]

At the level of operations—what Napoleon called grand tactics—the manoeuvre was first of all synonymous with marches and major movements. It involved the idea of fooling the enemy as to the true point of attack. We can see it in this extract from a plan for the army of Italy:

> It has three kinds of movements to make: 1 movement to protect the attacking divisions; 2 movement of attacking divisions; 3 movement to bluff the enemy as to our true attack.[3]

In 1805, Napoleon wanted to adapt his plan to what the Austrians and Russians would do. However, he knew how he was going to attack them, as he confided to his plenipotentiary minister with the Elector of Bavaria:

[...] it is above all by manoeuvres and marches that I want to finish things with ease.[4]

The same day, he wrote to Fouché:

We have crossed the Rhine. We shall soon begin to manoeuvre.[5]

At the start of October, he wrote to Marshal Brune:

I am at Ludwigsburg; part of the army is at Stuttgart and we are engaged in major manoeuvres to bypass the enemy. I hope in a few days for some notable events. As yet, no blood has been shed.[6]

Manoeuvres did not exclude battle. They paved the way for it by making it as advantageous as possible:[7]

My intention is that, if the enemy continues to remains in his positions and prepares to accept battle, it will not occur tomorrow, but the day after, so that Marshal Soult and his 30,000 men are in it, so that he outflanks the enemy's right, attacks him while bypassing him—a manoeuvre that guarantees us certain, decisive success.[8]

In commenting on Caesar's victory over Pompey's lieutenants at Ilerda (Lerida) in Spain (49 BCE), Napoleon stressed its exceptional character, for there was no battle:

Caesar cut down an army equal in strength to his own solely by means of the superiority of his manoeuvres. Such results can only be obtained in civil wars.[9]

Caesar certainly possessed clear superiority over Pompey's forces. In this mountainous region, he managed to cut his opponents off from any source of provisions, prompting some of them to switch sides.[10]

Commentators, especially Jomini and Camon, have identified in Napoleon two typical manoeuvres, used at the operational (or strategic) and tactical levels alike: that in the rear and that in a central position. These manoeuvres have been performed since antiquity. Napoleon resorted to them as and when the situation lent itself to one or the other. On 6 May 1796, before executing it, he described to the Directory his manoeuvre at Piacenza in the rear of Beaulieu's Austrian army:

Yesterday, during the day, we cannonaded with the enemy positioned beyond the Po. This river is very wide and very difficult to cross. My intention is to go across it as close as possible to Milan, in order not to encounter any further obstacle to reaching that capital. With this measure, I shall bypass three lines of defence that Beaulieu has put in place along the Agogna, the Terdoppio, and the Ticino. I am marching on Piacenza today. Pavia has been bypassed and, if the enemy persists in defending that city, I shall find myself between it and its stores.[11]

Enlargement of the front, with the appearance of army corps, facilitated the opponent's outflanking and even envelopment, by making possible control over all possible lines of retreat. Manoeuvre in the rear in preparation for battle became more effective because the enemy, pushed into a corner, could no longer escape. Marshal Soult received a letter at the start of the Prussian campaign:

Give me your news more frequently. In a combined war like this, one can only achieve good results by very frequent communication. Put this at the top of your list of concerns. This is the most important moment of the campaign; they were not expecting what we want to do; woe betide them if they hesitate and lose a day![12]

The movements of these large units, which took different routes, but which were intended to contribute to the same objective, must be combined. The phrase 'combined war' was encountered above (Book I, Chapter 7). The difficulty consisted in coordinating the movements of different army corps properly, in three successive phases: the manoeuvre intended to fool the enemy with an auxiliary or imaginary corps;[13] interception of the line of retreat; dislocation or destruction of enemy forces. The battle was merely the outcome of the manoeuvre, the plan of campaign, as we saw in Book IV. This presupposed an opponent whose forces were not organized in army corps. In 1813, they had become so, rendering the Emperor's manoeuvres much less effective: the allies no longer allowed themselves to be caught out.[14]

Manoeuvre in a central position was explained on Saint Helena:

Waterloo was like the Battle of Millesimo on the entry into Italy: I separated the British and Prussian armies just as I had separated the Austrian and Piedmontese armies. [. . .]
    When I am forming a campaign plan, I have no respite until I have finished, until my ideas are established. I am like a woman giving birth. My Waterloo plan was made in the first fortnight of May. I had Jourdan, who knew the ground, summoned. I assessed how the Sambre would cover my manoeuvre and how I would manage to separate the two armies.[15]

Napoleon opted for manœuvring in a central position when he was in a position of numerical inferiority. He then aimed to disrupt the links between the enemy armies. The manoeuvre was more difficult to accomplish and more risky. It was at the beginning and end of his military career that he had to confront enemy armies that were numerically superior.[16] However, he laid down no particular rule for performing one manoeuvre or the other apart from simplicity:

> War being a profession of execution, all complicated ploys must be excluded from it. Simplicity is the precondition of all good manoeuvres; it is better to do three or four marches more, and concentrate one's columns further back and far from the enemy, than to effect their conjunction in his presence.[17]
>
> An operation of concentration should never be contrived near the enemy. The art of war does not require complicated manoeuvres; the most simple are preferable. Good sense is needed. After that, it is difficult to conceive how generals make so many mistakes. It is because they want to be clever. The most difficult thing is to divine the enemy's plans, to perceive what is true in all the reports one receives. The rest requires only good sense. It is like a fist fight; the more blows one delivers, the better. It is also necessary to know how to read a map properly. I am sure that Suvorov did not have maps. Henry IV was a true military man, but back then war only demanded courage and good sense. It was very different from war with great masses of men.[18]

Along with simplicity, rapidity is more important than anything else in manoeuvres:

> Prince Charles and the Austrians in general manoeuvred well. They have a good stock of staff [officers] and instructors, but they manoeuvred slowly— which meant that faced with me they lost their heads. I did not allow them time to carry out their movements. Faced with Moreau and others, they had all the time in the world to carry out their movements, to attract their wings or detachments to them. In front of me, they were dazed.[19]

Finally, there are cases where it is better not to seek to manoeuvre. The nature of the country intervenes, as in this order which Berthier had to relay in Egypt:

> You shall give General Desaix the order to attack Mourad Bey wherever he finds him, but while always keeping his forces unified. My intention is definitely not that he should divide his forces with a view to enveloping the enemy, such manoeuvres being too unpredictable in cut off countries like the one he finds himself in.[20]

Like Napoleon, Clausewitz was convinced that there were no rules of any kind in matters of strategic manoeuvre. Implicitly taking aim at Jomini, he characterized as 'misleading rules and maxims' the established contrast between outflanking the enemy and operating on interior lines, on the one hand, and concentrating one's forces and dispersing them over numerous posts, on the other. All this was a question of good sense and circumstances. Where the latter were favourable, 'superior application, precision, order, discipline, and fear will find the means to achieve palpable advantages'.[21] Desirous of differentiating himself from Jomini, Clausewitz wanted to stress psychological strength, rather than manoeuvres—something for which he was to be criticized.[22]

8

Attack on Fortresses

In campaigns that 'aim at a great decision', Clausewitz would not have any siege undertaken, unless it was completely unavoidable, as long as the decision was still in the balance. Fortresses would only be captured after the outcome was clear.[1] The spectre of the campaign of 1806 haunted him. Napoleon undertook the siege of Prussian positions only after Jena and Auerstedt. More so than Clausewitz, Napoleon offered some thoughts on attacking fortresses. In many respects, he was closer to the wars of the *ancien régime* than his illustrious commentator. When a position was besieged, it called for method. One could not do too many things at once:

> There are only two methods of ensuring the siege of a place. The first is to begin by beating the enemy army, removing it from the field of operations, casting what remains of it beyond some natural obstacle, such as mountains or a large river, and in this time opening the trench and capturing the place. But if you want to capture the place in the face of the relief army, without risking a battle, you need to be supplied with siege equipment, to have munitions and supplies for the likely duration of the siege, to form lines of contravallation and circum-vallation,[2] making use of the locality, whether heights, woods, swamps, or inundations. No longer needing to maintain any communication with the depot posts, it is now only a matter of containing the relief army and, in this case, you create an observation army that does not lose sight of it and which, blocking the path of the place to it, always has time to arrive on its flanks or in its rear, if it was stealing a march on it; or finally, taking advantage of the lines of contravallation, using part of the besieging corps to give battle to the relief army.
>
> But to do three things at once—1 besiege a fortress and contain its garrison without contravallation; 2 guard one's communications with depot posts situated six days' march away; and 3 contain the relief army without being aided by any natural obstacle or lines of contravallation—that is a mistaken combination which can lead only to catastrophes, unless one has double the forces of the enemy.[3]

Clausewitz did not trouble himself with such considerations. He quite simply reckoned that lines of circumvallation had 'gone completely out of fashion'.[4] As already noted above, the theoretician pushed his interpretation of the radical changes in the art of war further than the practitioner.

For Napoleon, sieges were costly, even if one followed Vauban's method, which was intended to spare lives. Perhaps it was better to cut to the quick and attempt a bloody but decisive assault:

> The Duke of Wellington is accused of being a butcher of men; above all, he is criticized for his assault on Badajoz.[5] However, it is a calculation that has to be made: perhaps it is better to attack thus and, ultimately, lose fewer men in a day's butchery than by daily, albeit gradual, losses in the ordinary course of a siege. And then, the time gained is immense! It would be intriguing to conduct research on this subject; Gourgaud, you should take it on. Somewhere in the camp libraries we must have detailed accounts of sieges in recent wars.[6]

The result of this inquiry has not come down to us, but Gourgaud alludes to his research on sieges in Spain in his journal, leading the Emperor to add this reflection:

> Breaches must be attacked with a few men at one time. Committing too many men causes only casualties and confusion.[7]

When sieges are especially bitter, street fighting sometimes ensues. This was the case at Jaffa in Palestine. Bonaparte wrote as follows to his commander-in-chief of the engineers:

> Recommend above all to the men of your service that they do not go into the streets; they should seize the entrances and advance with caution from house to house.[8]

When Murat was his lieutenant in Madrid in April 1808, before the rebellion broke out, the Emperor made the same recommendation to him:

> You must remember circumstances in which, under my orders, you have waged war in large towns and cities. One does not engage in the streets; one occupies the houses at the top of streets and sets up good batteries.[9]

Three months later, Spain was on fire:

> Saragossa has not been taken. It is encircled and a city of 40,000–50,000 souls defended by a popular movement is only taken with time and patience. The histories of wars are full of the greatest catastrophes as a result of armies rushing and diving into the narrow streets of towns. The example of Buenos Aires and the 12,000 elite British troops who perished there is proof of it.[10]

9

Invasion

War of invasion and routine war

Although he does not cite him, Clausewitz was referring to General Rogniat when he said that '[t]he French are always writing about *guerre d'invasion*', in contrast to routine war. The former refers to any attack deep into enemy territory, while the latter only nibbles away at a frontier.[1] Napoleon railed against this distinction:

> Any offensive war is a war of invasion; any war waged in accordance with the rules of the art is a routine war. Campaign plans are altered ad infinitum, depending on the circumstances, the genius of the leader, the nature of the troops, and the topography.[2]

Clausewitz was of the same opinion and perceived in Rogniat 'unscientific linguistic confusion. Whether an attack will halt at the frontier or penetrate into the heart of the enemy's territory, whether its main concern is to seize the enemy's fortresses or to seek out the core of enemy resistance and pursue it relentlessly, is not a matter that depends on form: it depends on circumstances. . . . In some cases it may be more methodical and even more prudent to penetrate some distance rather than stay close to the frontier.'[3]

When Napoleon invaded an enemy country, he sought battle with the main opposing army. Leading the Prussians in 1792, the Duke of Brunswick had not had this determination:

> Brunswick behaved very foolishly in the Champagne campaign. When you want to invade a country, you should not fear giving battle and looking everywhere for your enemy in order to fight him.[4]

An army of invasion must look to its flanks, which by definition were vulnerable, and not multiply them:

The flanks are the weak parts of an invading army. You must strive to support them, if not both, then at least one, on a neutral country or a large natural obstacle. Ignoring this principle of war, the French army [in 1796 in Germany], dividing into three separate corps, created six flanks for itself, whereas had it manoeuvred well it would have been easy for it to march unified supporting both wings.[5]

An army that disembarked at some point on a coast had advantages which it must hasten to exploit:

> The advantage of any army that arrives by sea is that it arrives without warning, like the wind; you know its strength only when there is no longer time to prevent it from landing and doing what it plans to do. Nothing can justify the British government in having violated this principle and renounced this advantage—the greatest of all—of any army that carries war to a country by sea. One must disembark straight away. The art consists in the admiral allowing himself to be seen from land only when there is no time, when he is already one league from it. He must calculate his navigation so as to be one or two leagues from the beach where he wishes to land at 3 o'clock in the morning in summer and 6 o'clock in the morning in winter. He must reach it with all sails out and, half an hour later, without even having moored, having taken control of the beach, and bring onto land during the day and the following night the whole army, all the field artillery, and food for fifteen days. Everything must be on land and ready to receive the army if it presents itself 24 hours after reports of an enemy squadron, convoy, or army have been made from land. Operating with this speed, it does not matter if the sea is as calm as a lake, provided that there is no storm and that a strong wind does not put one on the coast [sic]—which happens rarely. Moreover, one should disembark in a bay where the prevailing wind for the season does not dash you against the coast. It is because they ignore these principles that the British always lose so many men in their landings; they have often succeeded in them—which should not have been the case. When bad weather forces mooring, and not disembarking for 5–6 days, there is but one course of action for the admiral to adopt—weighing anchor and landing elsewhere, where he is not expected.[6]

Conciliating invaded peoples

In a famous passage, Clausewitz notes that the conqueror is always a friend of peace, as Napoleon himself said, and always wants to enter the state he is invading without encountering any opposition.[7]

It is true that Napoleon often recommended conciliating invaded peoples. To General Joubert, who had entered Trento and was soon to march on the Tyrol, the commander-in-chief of the army of Italy wrote in late 1797:

> Neglect nothing to please the inhabitants of the regions you conquer. You should not look to wretched inhabitants of mountains for money or resources; you should seek only to win their approval, so that they are more content with us than with the Austrians.[8]

On the eve of landing in Egypt, a proclamation was addressed to French soldiers and warned them:

> The peoples with whom we are going to live are Muhammadans; their first article of faith is this: 'There is no God but God and Muhammad is his prophet.'
>
> Do not contradict them; behave with them as we behaved with the Jews and the Italians; have consideration for their muftis and their imams, just as you had for the rabbis and bishops.
>
> For the ceremonies prescribed by the Koran, for mosques, have the same tolerance you had for convents and synagogues, for the religion of Moses and Jesus Christ.
>
> The Roman legions protected all religions. Here you will encounter customs that are different from those of Europe; you must become accustomed to them.[9]

A few days later, General Kleber, who was ill, remained in Alexandria, where he commanded the garrison. He received precise instructions so that relations with the local population and authorities were as good as they could be:

> To the extent that it is possible, maintain good intelligence with the Arabs; have the greatest consideration for the muftis and the country's principal sheiks. [. . .] We must accustom these people gradually to our manners and our way of seeing things and, meanwhile, allow them great latitude in their domestic affairs, above all not interfering in their justice which, being wholly based on divine laws, stems entirely from the Koran.[10]

General Menou,[11] charged with the command of Rachid, received even more precise instructions. He was recommended:

> 14—To protect the muftis, the imams, and religious worship, and to use the country's principal sheiks to command the mass of the population, taking particular care, however, to disarm as far as possible and study the men of

whom we must assure ourselves if ever an unfortunate event forced us to take measures to contain the population.

15—To discover who are the youth of 16–20 whom we could take with us, half willingly, half forcibly, on the pretext of teaching them the profession of arms and to serve us as hostages when circumstances require it.

16—I think that at Rachid, as here, there will be three parties: that of the men who are attached to the government; that of pure Muhammadans or holy, virtuous men, which is the party that has a great mass of public opinion for it; and finally that of the men who have been agents of the government and who are currently in disgrace.

17—You must pander to the opinion of the second party and inspire great hopes in it for an order in which the just are protected.

18—You will flatter the third about restoring them to their position. But as far as possible, you will use those who are currently in the government, after having made them take the pledge of obedience and do nothing contrary to the interests of the army.[12]

The commander-in-chief of an army of invasion had an interest in behaving in a way that respected local morals and customs. Napoleon tried to get his brother Jérôme, who was leading a life of pleasure in his kingdom of Westphalia in 1809, to understand this:

Adopt manners and habits in conformity with those of the country you govern. That is how you will win over the inhabitants by esteem, which only ever goes with the opinion of customs and simplicity.[13]

On the difficulties of occupation and the placement of troops

If we are to believe him, Napoleon's Corsican childhood very early on made him understand the difficulties involved in occupying a conquered country:

When the French occupied Corsica in my youth, the government took a lot of trouble to find out what was happening, who had committed what crime, and could not find out, whereas all the inhabitants of Ajaccio knew everything. However great the effort made by the government to win them over, it found it very difficult.

What I saw then has served me in conquered countries: I was never astonished by the hatred that prompts the worst follies and the difficulty we have in securing the submission of some fanatics.

Corsica owes a lot to France. It is not really grateful, people say. France wanted to extinguish hatreds, stop the assassinations that had been encouraged by the weak government in Genoa in order to rule Corsica and dominate it more surely. But the Corsicans could not find that reasonable. It is not enough to render peoples a service; it is further necessary to serve them as they understand it and adapt to their prejudices.[14]

On Saint Helena, Napoleon offered a reflection on fanatics that crosses the ages:

Of all assassins, he continued, fanatics are the most dangerous: only with great difficulty can one protect oneself against the ferocity of such men. A man who has the intention, the desire, to sacrifice himself is always the master of another man's life and, when he is a fanatic, especially a religious fanatic, delivers his blows with all the more assurance. History is teeming with such actions: Caesar, Henri III, Henri IV, Gustavus, Kléber, etc., were among their victims. Religious fanatics, political fanatics, all are to be feared. The accomplices of these tigers, if there are any—for these great criminals often have no accomplices but themselves—are always wrapped in an impenetrable veil that shields them from the most active, most precise searches.[15]

General Bonaparte was confronted with an occupied people during his first Italian campaign. Not all Italians greeted the French as liberators. There were several revolts. In Egypt, he faced similar difficulties. To ensure peace and quiet in Cairo, he wrote as follows to the person he had appointed governor:

Great vigilance is more necessary for the peace of the place than a major dispersal of troops. Some duty officers who do the rounds of the city, some orderly sergeants who cross the city on donkeys, some adjutant majors who visit the key places, a few Franks who wend their way in the markets and the various districts, and some reserve companies who can be sent into places where there is a problem—these are more useful and less tiresome than fixed guards in squares and at intersections.[16]

One should not fire blanks, but real bullets and live fire directly at the populace. Otherwise, the crowd would become emboldened, and confrontations last longer and involve more casualties:

For, with a rabble, everything depends upon the first impressions made upon them. If they receive a discharge of firearms, and perceive the killed and wounded falling amongst them, a panic seizes them, they take to their heels instantly, and vanish in a moment. Therefore, when it is necessary to fire at all, it ought to be done with ball at first. It is a mistaken piece of humanity, to use powder only at that moment, and instead of saving the lives of men, ultimately causes an unnecessary waste of human blood.[17]

Occupation forces should not be dispersed on a national scale, but maintained at a short marching distance, so that they could be concentrated rapidly. Napoleon wrote this several times to his brother Joseph, who had gone to occupy the throne of Naples in 1806:

> Be aware that there should not, as it were, be a village in your kingdom that has not seen your troops and yet the inhabitants must not have anything to complain about. It is appropriate not to disperse your forces. It is in fact better to have 600 men who make six journeys to different points, or to send patrols everywhere, but in such a way that the bulk of this corps remains concentrated, than to have the 600 men distributed, with 100 men in each place, to six different points. Endeavour to keep the battalions united.[18]

Rather than proliferating small garrisons, what were required were mobile camps:

> [...] above all proscribe small garrisons, for otherwise you will have a lot of casualties. The true system is that of mobile camps: 1,800 men under the orders of a major general, positioned around Cosenza, and constantly supplying columns, 500–600 men doing the rounds of the country, are the best means. [...] let there nowhere be fewer than 400 men. Place small detachments only in fortresses and well-fortified posts. [...] If Colonel Laffon[19] had attacked the rebels boldly, he would have brought them to their senses with 400 men. Any troop that is not organized is destroyed when one marches on it.[20]

The idea that an organized military force always triumphs over a crowd, even one superior in numbers, is taken up again in this advice given to Joseph, who had become King of Spain:

> Whatever the numbers of the Spanish, you must march straight at them with firm determination. They are incapable of resisting. You must not sidestep them or manoeuvre them, but run over them.[21]

The strength of organized, concentrated troops could be counted on. Not only did they possess greater military effectiveness, but they gave themselves confidence and impressed populations more:

> In general, I desire that my troops should be concentrated as much as possible, for the simple reason that the people, who are used to seeing them thus, rebel as soon as they learn that the troops have headed elsewhere.[22]

The psychological aspect was extremely important here. The occupation of the Iberian Peninsula must also consist in holding the key points:

> In any country whatsoever, by holding the main cities or posts, one easily contains it, having under one's thumb the bishops, magistrates, main property-owners, who have an interest in order being maintained under their responsibility.[23]
>
> Write again to General Drouet[24] that I attach the utmost importance to having news of the army of Portugal; that the principle of occupying all points is impossible to implement; that one must make do with occupying the points where the depots and hospitals are, and having one's troops in hand in order to direct them where they are needed [...].[25]

The situation in Spain was compared to a civil war:

> In civil wars, it is the important points that must be held; it is not necessary to go everywhere.[26]

Counter-insurgency methods

Mobile columns were an initial method. Napoleon advised his brother Joseph to organize them in Calabria, with Italian-speaking Corsican units:

> Organize four mobile columns commanded by intelligent, upright, and firm officers, each of them consisting of 700–800 men, some cavalry, and a lot of infantry, assigned to the different parts of this province, and sending detachments everywhere. These columns will not have been established a month before they will know all the localities, blend in with the inhabitants, and will have conducted a good hunt for brigands. The latter must be shot on the spot as soon as they have been arrested. [...] but stop the generals from stealing. If they behave arbitrarily, if they vex and dispossess citizens, they will arouse the provinces. The first one to steal must be hit hard, removed in disgrace, and handed over to a military commission.[27]

Mobile columns went hand in hand with keeping the troops concentrated. Napoleon explained this clearly to Murat on the eve of the Madrid uprising of 2 May 1808:

> I definitely do not approve of you dispersing your troops. I am informed that you have sent a regiment into a village from the Escurial. You can detach a regiment to make an example of it; but it must return on the spot. If, every time a riot occurs, you send a regiment or battalion, I will not have an army. If

you get the villages used to having garrisons, they will rebel as soon as you withdraw them. You must send mobile columns, which must not be away for more than eight days, and must return as soon as their mission is over.[28]

When Spain went up in flames, orders were always given to the same effect:

My cousin, write to Generals Dorsenne, Caffarelli, and Thouvenot[29] that in the country where they are an odious system is being followed; that enormous forces are assembled in villages against gangs of brigands who are active, so that we are continually exposed to disagreeable events, whereas the opposite should be done; that principal points should be occupied and that mobile columns should set out from there in pursuit of the brigands; that, were things to be conducted in this way, many particular woes would be avoided; that they should hasten to implement this plan and wage an active war on the brigands; that the experience of the Vendée has proved that the best thing is to have mobile columns, dispersed and multiplied everywhere, and not stationary corps.[30]

'Making examples' was a second method. In Egypt, methods were adapted to oriental customs. Bonaparte wrote to the governor general of the province of Menouf:

You must treat the Turks with the utmost severity. Every day I have three heads cut off here and carried through Cairo: that is the only way of overcoming these people.[31]

In the province of Mansourah, where villages had rebelled, Bonaparte had the main culprit burnt, took hostages, and threatened the other villages with similar reprisals.[32] But these repressive methods alternated with attempts at appeasement, as is attested by this sentence from a letter to General Vial:[33]

Try to get the mass of inhabitants of El-Choa'rah and Lesbe to change their minds, by granting them a general pardon.[34]

Blowing hot and cold: such was also the essence of the instructions to General Brune, commander of the army of the West, on the way to behave towards the Chouans at the start of the Consulate:

Any individual who will submit, receive him. But do not permit any further meetings of the leaders; hold no kind of diplomatic negotiation.

Show great tolerance for priests; severe acts towards the large communes to compel them to guard themselves and protect the small ones. Do not spare communes that behave badly. Burn some smallholdings and large villages in the Morbihan and start to make some examples.

Let your troops lack neither bread, nor meat, nor pay. In the culpable *départements*, there is enough to maintain your troops. It is only by making the war terrible for them that the inhabitants will unite against the brigands and finally realize that their apathy is fatal to them.[35]

Junot, who had set off to repress an insurrection in the states of Parma and Piacenza, was reminded in several dispatches of the value of severe examples, justified for their efficacy but also, paradoxically, their humanity:

> Tranquillity is not maintained in Italy with fine phrases. Do as I did in Binasco:[36] have a large village burned down; have a dozen rebels shot; and form mobile columns to seize brigands everywhere and give an example to the people of these regions.[37]
>
> Remember Binasco: it earned me the tranquillity that Italy has enjoyed ever since and spared the blood of many thousands of men. Nothing is more salutary than terrible examples aptly given.[38]
>
> Without a harsh example, the peoples of Italy will always be ready to rebel.[39]
>
> I am pleased to see that the village of Mezzano, which was the first to take up arms, will be burnt. Give great pomp to this: have a big description given of it in all the papers. There will be much humanity and mercy in this act of severity, because it will prevent other revolts.[40]

While Napoleon always enjoyed a certain popularity in Italy, the severity he displayed towards many rebellions, and the harshness of his comments on the inhabitants, reveal a dark side of his personality, in which the conceptions of the age must be given their due. In Naples, Joseph assumed the crown. His brother warned him that he must expect an insurrection and that it should be disarmed preventively:

> Properly include in your calculations that, fifteen days sooner or later, you will have an insurrection. It is an event that constantly occurs in conquered countries. [...] Whatever you do, you will never sustain yourself in a city like Naples by opinion. [...] I imagine that you have cannons in your palace and have taken all measures required for your safety. You will not be able to watch over all your people too much. The French are of unparalleled assurance and levity. [...] Disarm, disarm! Put this huge city in order. Keep your artillery parks in positions where the rabble cannot capture your cannons. Reckon that you will have a riot or a small insurrection. I would very much like to be able to help you with my experience in similar affairs.[41]

Napoleon, let us not forget, began his career as a general putting down the royalist insurrection of 13 vendémiaire. He ended up desiring a rebellion in order to be relieved, rather as one is after a child has overcome an infantile illness:

> I would very much like the rabble of Naples to rebel. As long as you have not made an example of them, you will not be their master. Any conquered people requires a rebellion and I would view a rebellion in Naples in the same way as a father regards smallpox in his children, provided it does not weaken the patient unduly. It is a salutary crisis.[42]

Joseph tried to win the sympathy of the Neapolitans by promising not to impose war taxes and by forbidding French soldiers from demanding food from their hosts. He found himself reprimanded:

> You do not win people over by cajoling them [...]. If you do not make yourself feared from the outset, misfortunes will come to you. The imposition of a tax will not have the effect you imagine; everyone expects it and will find it natural. This is what happened in Vienna, where there was not a sou and where it was hoped that I would not introduce a tax. A few days after my arrival, I introduced a tax of 100 million francs: it was found very reasonable. Your proclamations to the people of Naples savour too little of the master. You will not gain anything by overly caressing. If they do not perceive a master, the peoples of Italy, and peoples in general, are disposed to rebellion and mutiny.[43]
>
> In a conquered country, generosity is not humanity. Several Frenchmen have already been killed. In general, it is a matter of political principle to project a positive image of one's generosity only after having demonstrated severity towards bad elements.[44]

This reflection emerged from a reading of Machiavelli's *The Prince*. Reflecting on Saint Helena on conquered Italy in 1796 and, in particular, on the Pavia rebellion, Napoleon formulated the following thoughts:

> The conduct of a general in a conquered country is surrounded by pitfalls. If he is harsh, he vexes people and increases the number of his enemies; if he is soft, he creates hopes that give greater prominence to the abuses and vexations inevitably bound up with the state of war. Whatever, if in these circumstances an act of sedition is pacified in time, and the conqueror knows how to employ a mixture of severity, justice, and softness, it will have had but one good effect: it will have been advantageous and will be a new guarantee for the future.[45]

While Napoleon was mistaken about the strength of Spanish national sentiment prior to risking his pawns in the Peninsula, he clearly understood the kind of war entailed by the insurrection. This rapidly went beyond a mere uprising. The Spanish army and people were supported by the British, who landed in Portugal. In particular, towns held by the Spanish—in the first instance, Saragossa—had to be reduced to obedience. Orders were given as to the particular methods to be employed:

> I require a large quantity of shells because of the large number of projectiles one is compelled to use in an insurrectionary war.[46]
> The Spanish war is like that of Syria; as much will be done by mines as by cannon.[47]

The siege of Saragossa against a fanatical population was indeed to witness bombardments by howitzers and patient work blowing up houses by miners from the engineers.[48]

Means of pacification

To succeed in pacification, non-military methods were also required:

> Conquered provinces must be kept obedient to the victor by psychological methods, the responsibility of the communes, the mode of organization of the administration. Hostages are one of the most powerful means, when peoples are convinced that the death of these hostages is the immediate effect of their breaking their word.[49]

The cooperation of local elites was indispensable for ensuring pacification. In Egypt, Bonaparte did not approve of the way that they were treated by the general commanding the province of Menouf and at the same time, understanding the difficulties of such a situation, he spared him and assured him of his confidence:

> I do not approve of the fact that you had the divan arrested without having gone into whether he was guilty and then had him released twelve hours later: that is not the way to conciliate a party. Study the peoples you are among; distinguish those who are most amenable to being employed; sometimes make just and severe examples, but never do anything that approximates to caprice and levity. I sense that your position is often difficult and I am full of confidence in your good will and your knowledge of the human heart. Know that I do you the justice due to you.[50]

In leaving Egypt, Bonaparte left the command and a series of instructions to Kléber, including the following:

> You know, Citizen General, my outlook on the internal politics of Egypt; whatever you do, the Christians will always be our friends. They must be

prevented from being too disrespectful, so that the Turks do not have the same fanaticism against us as against the Christians, which would make them irreconcilable. While waiting to uproot it, fanaticism must be lulled. In winning the good opinion of the major sheiks of Cairo, one has the good opinion of the whole of Egypt and all the leaders this people might have. There is no one less dangerous for us than sheiks who are fearful, do not know how to fight, and who, like all priests, inspire fanaticism without being fanatics.[51]

A report was dictated on the way in which Egypt had been administered:

It is impossible for us to pretend to an immediate influence over peoples for whom we are so foreign. In order to rule them, we need intermediaries; we must give them leaders, for otherwise they will choose them for themselves. I preferred ulamas and doctors of law: 1 because they were natural leaders; 2 because they are the interpreters of the Koran and the biggest obstacles we have experienced still derive from religious ideas; 3 because these ulamas have gentle customs, love justice, are wealthy and inspired by good ethical principles. Indisputably, they are the most honest people in the country. They do not know how to mount a horse, are not accustomed to any military manoeuvres, are unqualified to figure at the head of an armed movement. I have interested them in my administration. I have used them to speak to the people, and have composed the councils of justice of them; they have been the channel I have used to govern the country. I have increased their wealth; in all circumstances, I have given them the greatest marks of respect. I have had them awarded the first military honours. In flattering their vanity, I have satisfied that of the whole people. But taking all this care of them would be in vain, if you did not show yourself to be permeated by the profoundest respect for the religion of Islam and if you allowed the Christian, Greek, and Latin Copts emancipations that changed their habitual relations. I wanted them to be more submissive, more respectful towards the things and persons attached to Islam than in the past. [...]

Great care must be taken to persuade the Muslims that we love the Koran and venerate the Prophet. A single word, a single ill-judged initiative can destroy the work of several years. I have never allowed the administration to interfere directly with the persons or income of the mosques; I have always dealt with the ulamas and allowed them to act. In any contentious issue, the French authority must be favourable to the mosques and the religious foundations. It is better to lose a few rights and not give rise to disparagement of the administration's secret arrangements in these highly delicate matters. This method has been the most powerful of all and the one that has contributed most to making my government popular. [...]

It is necessary to conform to the manners of the Orientals, abolish hats and tight trousers, and impart to the clothing of our troops something of the dress of the Maghrébins and Arnauts. Dressed thus, they would appear to the inhabitants like a national army; this would be in keeping with the country's circumstances.[52]

The French left Egypt without having improved their relations with the inhabitants—quite the reverse. The latter continued to regard them as the 'Christian enemy', the successors of the crusaders. Bonaparte's conceptions anticipated colonialism but, as Steven Englund remarks, he also innovated in this respect: he was the first to hit upon answers that would only become 'old' much later. He may legitimately be regarded as the origin of a French school of counter-insurgency that developed subsequently in North Africa and Indochina. The French were expelled from Egypt by a British exped-ition, not an insurrection by the inhabitants. If the French were not able to win real sympathy, and if their religious policy carried little conviction, their methods were increasingly effective. Kléber and Menou proved good colonial governors.[53]

Napoleon dispensed the lessons drawn from his Egyptian experience to Spain, as indicated by this note which he had written for General Savary:

> The Emperor considers that it is not enough to render the authorities respon-sible; they must of course be made responsible, but they must be provided with the resources for it. For that, it is necessary to disarm and form four companies of national guards from the country's most commendable elem-ents, to support the *alcaldes* and maintain tranquillity; they will be responsible if they do not maintain it. You will join to the responsibility of the most considerable elements in each town the responsibility of bishops, of convents. This is how public tranquillity has been able to be maintained in France. Without them, France would have succumbed since 1789 to the most terrible anarchy.[54]

To Joseph, who had transferred from the throne of Naples to that of Madrid, Napoleon spelt out how the local elites were to serve him:

> But, for a country to be properly submissive, the intendants, *corregidors*, and higher magistrates whom the people are accustomed to obeying must be appointed by you and go into these provinces, issue proclamations, pardon rebels who surrender and bring in their weapons, and above all issue circulars to the *alcaldes* and priests, whereby the latter understand that they are under your government. This measure will have the advantage of reorganizing the police and finances, and give leadership to these peoples. The intendants and *corregidors* must also communicate with your ministers and make known to them the various items of intelligence that come to their knowledge.[55]

If the conqueror wanted not only to occupy a country, but to raise troops there, certain conditions had to be fulfilled for the public spirit to be favourable. Above all, the occupying army must not live off the land. This was the case in Italy in 1796:

> This circumstance—being obliged to live off local resources—will significantly set back the public spirit of Italy. On the contrary, had the French army been able to be maintained by funds from France, from day one we would have been able to raise numerous corps of Italians. But seeking to call a nation to liberty and independence, wanting a public spirit to be formed in its midst so that it raises troops, and at the same time carrying off its main resources— these are two contradictory ideas and the art consists in reconciling them.[56]

Troops could be raised if a certain enthusiasm manifested itself. If it was missing, it was better not to raise any:

> Do not follow the system of provincial guards; nothing would be more dangerous. Those people will boast and believe they have not been conquered. Any foreign people that has this idea has not submitted.[57]

In Spain, confronted with what he called a 'people's war', Napoleon was even more concerned with opinion:

> Retrograde measures are dangerous in war; they must never be adopted in people's wars. Opinion does more than reality; knowledge of a retrograde move, which the ringleaders attribute to whatever they wish, creates new armies for the enemy.[58]
>
> Conquered peoples only become subjects of the conqueror out of a mixture of policy and severity and through their amalgamation with the army. These things have been wanting in Spain.[59]

... and in the Vendée:

> Only political and moral means can keep peoples conquered; the elite of the armies of France was not able to contain the Vendée, which only has a population of 500,000–600,000.[60]

For a proclaimed champion of the military approach, this confession is powerful and revealing of Napoleon's human depth. It also appears in this famous sentence, addressed to the president of the Institute after the first Italian campaign:

> Genuine conquests, the only ones inspiring no regret, are those made over ignorance.[61]

This must not lead us to forget the number of repressive orders, inciting generals to the utmost severity, in Napoleon's correspondence. Unquestionably, he only believed in strong-arm tactics and shootings for the sake of examples.[62] The balance sheet of his occupation policy in Egypt is terrible on a human level.[63] But most subsequent colonial expeditions were at least as bloody. As we said above, his methods proved effective overall. In the Iberian Peninsula, too, had the Russian expedition not intervened to transfer forces, the guerrilla was on the point of being defeated at the end of 1811.[64]

Conclusion to Book VII

Napoleon was basically inclined to attack first. In strategy, in his view, any attack must be based on a line of operations equipped with defensive sites or positions. In an offensive battle, the attack generally sought to outflank or envelop the enemy. Above all, this should not be done within view of the enemy, but should be aimed at the enemy's wing, or rather rear, without him knowing it. This is what Frederick II did at Leuthen. Combined attacks contained the risk of not being sufficiently masked. Napoleon had no preference as regards the details of the attack, but saw the advantages of the simplicity of formation in columns and obviously counted on a concentration of artillery. At sea too, he wanted the commanders of vessels to attack more and exploit their guns more. Crossing a river was always tricky, but could slow down the attack. Here again, judicious use of artillery, and rapid concentration of forces at a particular point, must make it possible to lay a bridge. A defensive position should not be attacked head on, but bypassed. Napoleon recommended this on several occasions, but contravened his own rule when he had an absolute desire for battle, as at Borodino or Waterloo. He confessed his mistake on Saint Helena, in a conversation reported by Gourgaud.

In mountains, even in an offensive war, Napoleon prescribed not attacking an enemy who was in position, but flushing him out by approaching his flank or his rear. Good sense indicated avoiding mountains. Clausewitz's arguments complement those of Napoleon on this score very well. This is less the case when it comes to manoeuvres. Napoleon had more to say about them, stressing two essential characteristics: simplicity and rapidity. He also enlarged more on siege warfare. Older than Clausewitz, an officer in a cultured service of the French army, he was manifestly more interested in the issue of attacking fortified places. There is doubtless an important observation to be made here: the theoretician had a tendency to accentuate what was new in the practitioner.

For Napoleon, as for Clausewitz, the more or less in-depth invasion of enemy territory was a question not of method, but of circumstances. A conqueror always proclaims himself a friend of peace, wrote Clausewitz. On several occasions, Napoleon did indeed seek the favours of conquered peoples. He took particular precautions with the Muslim population in Egypt. From his youth in Corsica he appreciated the difficulties of an occupation. His first campaigns, in Italy and Egypt, confronted him with these problems and led him to formulate some solutions as regards troop placement: not fixed, dispersed guards, but concentrated, mobile forces; not small garrisons, but mobile camps; holding only the main towns where important figures, interested in the maintenance of order, resided. To counter an insurrection, mobile columns were required, which pursued the 'brigands', but did not pillage. It was necessary to make severe examples and then issue pardons in order to appease. To succeed in pacification, hostages could be taken, but the imperative thing was to ensure the cooperation of local elites.

The methods are those of all occupiers since. But Napoleon was a real innovator in this area. He may even be regarded as the founder of the French school of counter-insurgency. In the wake of the Enlightenment, he regarded himself as bringing progress to populations dominated by obscurantism and fanaticism. In this sense, he could not comprehend national motives for insurrections. However, alongside an implacable severity, we find a particular desire to understand the human heart in order to manage an occupation. This did not preclude the exactions and horrors inseparable from this type of situation and his responsibility is well-nigh complete, at any rate in the invasions of Egypt, the Kingdom of Naples, Portugal, and Spain. Were the orders issued followed by results? Even 'exemplary' pacifications, like that of Suchet in Aragon, did not meet with a success that could last, but the withdrawal of men for the Russian expedition adversely affected a situation that had turned in favour of the French.[1] Napoleon's writings on these perennial issues merit attention. In them he applies his long experience, his knowledge of history and men. While they do not provide a recipe for victory, they can always be meditated on and at least suggest what should never be done.

BOOK VIII

War Plans

In this last book, Clausewitz reverts to war as a whole and consequently takes up various ideas from Book I. For him, the violence of the Napoleonic Wars disclosed the pure concept of war, its 'absolute' character. 'No one would have believed possible what has now been experienced by all.'[1] It was in these pages that Clausewitz dubbed Napoleon 'the God of War'.[2] The unfinished character of the book, where some chapters do not even have a proper title, has led us to collect Napoleon's reflections under the title that best approximates to the ideas developed by Clausewitz.

I

The Plan of Campaign

On 26 October 1799, a few days before the coup d'état of Brumaire, General Bonaparte confided to Pierre Louis Roederer:

> No man is more pusillanimous than I am when I make a military plan; I inflate all the possible dangers and woes attendant upon the circumstances; I am in an appalling state of agitation. This does not prevent me from appearing very serene before the people around me; I am like a girl in childbirth. And when my decision is taken, everything not conducive to a successful outcome is forgotten.[1]

Between the siege of Toulon and his first Italian campaign, Napoleon was essentially a maker of plans for two years (1794–6). On the project of invading northern Italy alone, he wrote fourteen notes or reports.[2] He was therefore well placed to criticize the plans of others, like that of Generals Moreau and Jourdan against Archduke Charles in Germany in 1796. He criticized them in particular for their lack of an overall view:

> In the first instance, you need to know what you want to do and have a plan.[3]
> Actions are meditated at length and, to achieve success, you need to think for several months about what might happen.[4]
>   I am in the habit of thinking for three or four months in advance about what I must do and I reckon on the worst.[5]
>   A plan of campaign must anticipate anything the enemy might do and contain the means of foiling him.[6]

In imagining anything the enemy might do, Napoleon conceived plans with several strands. He had been heavily influenced in his youth by the ideas of the engineer Pierre de Bourcet, which were the first to be developed in connection with plans of campaign.[7] He must have read him in Valencia in 1791.[8] For Bourcet, a campaign plan must have several strands and it was necessary 'to base examination of operations on the greater or lesser obstacles that [would] have to be overcome, the disadvantages or

advantages that would result from success in each strand, and, having raised
the most plausible objections, to plump for the course that [might be]
conducive to the greatest advantages [...]'.[9]

Before leaving to take things in hand in Spain, Napoleon criticized his
marshals *in situ*—in particular Jourdan—for their lack of a 'fixed' plan:

> If the enemy is coming to Burgos, should we attack him or wait for him? In
> the latter case, why has Bessières not camped behind Burgos? You succeed in
> war with certain, powerfully conceived plans. Should we abandon Burgos in
> the same way that we abandoned Tudela? It would not be a way of imparting
> morale to an army to expose it to a retrograde step resembling a failure, if
> 10,000 men entered Burgos and captured the garrison in the citadel. This is
> what you expose yourself to when you do not have a fixed plan.[10]

Napoleon never undertook operations without a plan with several strands;
only rarely was it revealed to his subordinates and, above all, it was adapted
to circumstances. This was the case, for example, with the move towards the
Tyrol and on Bassano at the start of September 1796 in pursuit of the
Austrian General Wurmser:

> An operation of this kind can be meditated in advance and conceived in its
> entirety. But its execution is gradual and only authorized by the events that
> occur daily.[11]

On 20 September 1806, the Emperor revealed to his brother Louis the
general idea of his plan of campaign in the event of war with Prussia.
Starting from his kingdom of Holland, Louis must create a diversion and
draw the Prussians northwards:

> You must have put in your news sheets that a sizeable number of troops are
> arriving from all points of France, that at Wesel there will be 80,000 men
> commanded by the King of Holland. I want these troops to be on the march in
> the first days of October, because it is a counter-attack that you will make to
> attract the enemy's attention while I manoeuvre to bypass him. All your
> troops must head for the territory of the Confederation [of the Rhine] and
> spread out to its confines without going beyond them or committing any
> hostile act. [...] On 30 September I shall be at Mainz. All this is for your
> information only; everything must be kept secret and mysterious.[12]

As we said above, Napoleon did not always use the right words. It is clear
that 'counter-attack' means 'diversion' here.[13] On 30 September, a long
letter, from which we here only cited some extracts of general significance,
clearly refers to this diversion and indicates both the existence of a general

idea of manoeuvre—the 'plan of operations'—and the share left for the enemy's reaction:

> My intention is to concentrate all my strength on my extreme right, leaving all the space between the Rhine and Bamberg completely empty, so as to have nearly 200,000 men concentrated on one and the same battlefield. If the enemy presses some encounters between Mainz and Bamberg, I shall scarcely be concerned, because my line of communication will be established at Forchheim, which is a small stronghold, and from there Würzburg. [...] What might occur is incalculable, because the enemy, which supposes my left to be on the Rhine and my right in Bohemia, and which believes my line of operations to be parallel to my battlefront, has a great interest in outflanking my left, and in that case I can throw him into the Rhine. [...] My initial marches threatened the heart of the Prussian monarchy and the deployment of my forces will be so imposing and so rapid that it is likely that the whole Prussian army of Westphalia will deploy on Magdeburg and everyone will start forced marches to defend the capital. [...] I am counting on your corps solely as a means of diversion and to keep the enemy entertained until 12 October, which is when my operations will be unveiled [...]. In all this, I am going as far as human foresight allows.[14]

Even if a plan is required, circumstances and accident must always be given their due:

> Campaign plans are altered ad infinitum, depending on circumstances, the genius of the leader, the nature of the troops, and the topography. There are two kinds of campaign plan: good ones and bad ones. Sometimes the good ones fail as a result of chance circumstances; sometimes bad ones succeed by a whim of fortune.[15]

Wurmser's plan before the Battle of Rivoli made the great mistake of not taking account of the movements that his opponent would carry out:

> That scheme would have been wonderful, if men, like mountains, were stationary. But he had forgotten the popular saying that *while mountains are stationary, men walk and meet.* Austrian tacticians have always been full of this false system. The Aulic Council,[16] which had written Wurmser's plan, assumed that the French army was stationary, fixed to the spot of Mantua. This gratuitous assumption led to the loss of the House of Austria's finest army.[17]

Napoleon addressed a similar criticism to the officer commanding the engineers in Dalmatia in late summer 1806:

> The engineers' leader, rather than trying to answers the questions he was asked, threw himself into campaign plans that are manifestly ridiculous, since

they depend on the strength and make-up of the enemy army and the strength
and make-up of the French army.[18]

In the vast Russian theatre, with enormous armies, it was more difficult to
get all the components to contribute to an overall plan. However, as
Napoleon wrote to his brother Jérôme, it was indispensable:

> In this profession, and in such a large theatre, you can only succeed with a
> well-formed plan and elements that are properly coordinated. You must
> therefore study your orders closely and do neither more nor less than you
> have been told, especially when it comes to combined movements.[19]

2

The Military Objective and the Enemy's Centre of Gravity

To assess the scale of the military objective and the efforts required, the character of the hostile government and its people had to be considered and a number of different circumstances and relationships determined: 'Bonaparte was quite right when he said that Newton himself would quail before the algebraic problems it could pose.'[1] However, '[o]ut of these characteristics a certain centre of gravity develops, the hub of all power and movement, on which everything depends. That is the point against which all our energies should be directed.'[2] Continuing his implicit critique of Rogniat in favour of Napoleon, Clausewitz wanted operations to be directed at this hub of enemy power to destroy it, wanted victory to be pursued without interruption or restriction, rather than adopting a slow, so-called methodical system.[3] The centre of gravity did not necessarily refer to the bulk of forces: it is what kept the enemy forces together. Were it to be struck, the enemy would be thrown off balance. It was therefore necessary to consider the enemy as a whole, analyse the connections between his different elements, and accurately determine his centre of gravity, which might be his main arm, or his most important ally, or his capital, for example.[4]

The 'scale of the military objective' and the 'efforts required' are implicit in the following reflection, where Napoleon recognizes that he has 'surfed' on the wave of the French Revolution:

A man is only a man; he needs elements in order to act. You need to know about fire, which makes the kettle boil. You would need to know the situation of Arabia twenty years before Muhammad. [...] How could Muhammad and his successors have made so many and such astonishing conquests with a weak people? He would not make them today with the Arabs as they are. It is always in the aftermath of revolutions that men are ready for great things. The great successes of the Empire, and had France succeeded in dominating Europe—she would have owed it all to the Revolution. No doubt it is I who led her; but I had found some elements and made the most of them.[5]

Here Napoleon senses the importance of the passions for war. He often referred to peoples. Clausewitz was to include these factors in his description of war as 'a paradoxical trinity—composed of primordial violence, hatred, and enmity, which are to be regarded as a blind natural force; of the play of chance and probability within which the creative spirit is free to roam; and of its element of subordination, as an instrument of policy, which makes it subject to reason alone. The first of these three aspects mainly concerns the people; the second the commander and his army; the third the government.'[6]

On Saint Helena, Napoleon enjoyed criticizing the military operations of the Revolution that he had not led. We need to put things in perspective and realize that he was engaged in self-promotion. But this does not prevent his observations from often being pertinent and bringing out mistakes in the general leadership of the war. The notion of the military objective and even that of centre of gravity are implicit in these observations on the plans of campaign in Germany and Italy at the start of 1799:

> The French government ordered its armies to take the offensive in Germany and Italy alike. It should have remained on the defensive in Germany, because there we were not able to concentrate forces superior to those of the enemy. At all events, the three armies of the Danube, Helvetia, and the Bas–Rhin should have formed but a single army. [...]
>
> We had to take the offensive in Italy, because the French forces in the peninsular were much larger than the enemy's and because it was essential to expel the Austrians from their position on the Adige before the arrival of the Russians. But we should have taken the offensive with all our forces concentrated. Schérer attacked with 60,000 men;[7] a few days earlier, he had been weakened by 14,000 men, whom he had detached in Valtellina and Tuscany. The outcome of the battle on the Adige would impact on the fate of Valtellina and Tuscany, whereas actions in those provinces could not have any influence on the success of the battle. We should also have recalled 30,000 men of the army of Naples. The French army, then highly superior, would have beaten the Austrian army, would have pushed it beyond Piava, would have caused it to suffer major reverses, and would have seized Legnago, which would have entailed the loss of Suvurov's corps—an event all the more important in that it would have given the Tsar pause for thought. Anything that is mere fantasy

and not based on a genuine interest does not withstand a reverse. [. . .] Had
we pursued this plan, the second coalition would soon have been dissolved.[8]

Clausewitz complements Napoleon: 'The first task, then, in planning for a
war is to identify the enemy's centers of gravity, and if possible trace them
back to a single one. The second task is to ensure that the forces to be used
against that point are concentrated for a main offensive.'[9]

It was sometimes necessary to distinguish between a main theatre and a
secondary theatre, in order to focus one's energies appropriately. Napoleon
had been highly conscious of this in 1800 at the start of the Consulate:

> In this campaign, the German border was the predominant one; the border of
> the river at Genoa was the secondary border. In fact, events that occurred in
> Italy would have no direct, immediate, and necessary impact on the affairs of
> the Rhine, whereas events that occurred in Germany would have a necessary
> and immediate impact on Italy. Consequently, the First Consul concentrated
> all the Republic's forces on the predominant border—that is, the army of
> Germany, which he reinforced, the army of Holland and the Bas-Rhin, the
> army of Reserve, which he concentrated on the Saône, in range to enter
> Germany should it prove necessary.
>
> The Aulic Council concentrated its main army on the secondary border, in
> Italy. This error, this violation of a major principle, was the true cause of the
> Austrians' catastrophe in this campaign.[10]

These considerations were formulated on Saint Helena, but they expressed
long-held convictions. In July 1794, when he only commanded the artillery
of the army of Italy, General Bonaparte already advocated focusing energies
on the principal front, in terms that anticipated Clausewitz's notion of
centre of gravity:[11]

> The type of war to be waged by each army is therefore to be decided:
>   1 by considerations deduced from the general spirit of our war;
>   2 by political considerations that are a development of them;
>   3 by military considerations.
> Considerations deduced from the general spirit of our war.
> The general spirit of our war is to defend our borders[.] Austria is our most
> bitter enemy; the type of war of the different armies must therefore deliver the
> maximum number of direct or indirect blows against that power.
> Were the armies on the borders of Spain to embrace the offensive system,
> they would undertake a war that would be a separate war for them alone.
> Austria and the powers of Germany would experience none of it. It would
> therefore definitely not be part of the general spirit of our war.

Were the armies on the borders of Piedmont to embrace the offensive system, they would force the House of Austria to guard its states in Italy and consequently this system would form part of the general spirit of our war.

What is true of sieges of places also applies to systems of war: concentrate your fire on one point and, when the breach is made, the balance is broken and everything else becomes useless and the position is taken.[12]

It is Germany that must be crushed. That done, Spain and Italy will fall of their own accord.

Our attacks must therefore not be dispersed, but concentrated. [...]

Political considerations.

The political considerations that should determine each army's type of war furnish two points of view.

1 Create a diversion that forces the enemy to reduce his strength on one of the borders where he would be too strong.

Were our armies in Spain to embrace the offensive system, we would not obtain this advantage; this absolutely isolated war would not force the coalition into any diversion.

The offensive system embraced by the armies in Piedmont necessarily creates a diversion on the border of the Rhine and the North.

2 The second viewpoint of political considerations is to offer us the prospect, in one or two campaigns, of overturning a throne and changing a government.

The offensive system of our armies in Spain cannot reasonably supply us with this result.

Spain is a great state. The weakness and ineptitude of the court of Madrid, and the degradation of the people, make it far from formidable in its attacks. But the nation's patient character, the pride and superstition prevalent in it, and the resources afforded by a large mass will render it formidable when it is pressurized at home.[13]

Spain is virtually an island; it will have major resources in the coalition's superiority at sea.

Portugal, which is null and void in our current war, would then powerfully aid Spain.[14]

It therefore cannot enter a cool head to take Madrid; such a project would definitely not be in accordance with our current position.

Piedmont is a small state; its people are well disposed; it has few resources against fortunate events, no mass, no real national spirit. It is reasonable to anticipate that, by the next campaign at the latest, this king will be on the wander like his cousins.

Military considerations.

The topography of the border of Spain is such that, with equal forces, the advantage of the defensive is completely with us.

The Spanish army that would be opposed to ours would have to be stronger in order not to suffer any setback and hold us in mutual respect. [...]

We must therefore adopt the defensive system for the Spanish border and the offensive system for the Piedmontese border.

The considerations derived from the general spirit of our war, the political considerations, and the military considerations come together to lay down the law for us.

To strike Germany—never Spain or Italy. Should we obtain major successes, we must never allow ourselves to become embroiled by pushing into Italy as long as Germany offers a formidable front and has not been weakened.[15]

While national pride and vengeance summoned us to the imminent campaigns in Rome, politics and interest should always direct us towards Vienna.[16]

Bonaparte construed 'the general spirit of our war' in the sense of Montesquieu's *The Spirit of the Laws*.[17] The latter was also a model for Clausewitz.[18] We must return to the most famous sentence in this long note, used as an epigraph by Generals Camon and Colin, where an analogy is made with making the breach when laying siege to places. It expresses the idea encountered above of focusing energies—the principle which, according to Colin, had pride of place in Napoleon, who formulated it very early on. Colin stresses that this focusing of energies became possible and necessary in combat only with a field artillery capable of making a breach on the battlefield: 'To discover it, it was necessary to be of the school of Gribeauval.'[19]

The principle of sticking to essentials was expressed in a different way to the Austrian general of Belgian origin, Chasteler, who was a plenipotentiary commissioner for the recovery of the Venetian state and the determination of the new borders after the Peace of Campoformio in 1797. In the course of a conversation, General Bonaparte said to him:

Nothing is simpler than war. Most generals achieve nothing great because they are attached to too much at once or they are especially attached to the most inconsistent, secondary things. The key to the position—that is the objective towards which I gear all my forces. As a result, they find themselves directly engaged before my very eyes. The day of battle having arrived, I concern myself neither with eventualities nor with my line of operations. Such concerns certainly do not fall to a commander. The state entrusts him with 100,000 men; if he commits only 60,000, he must answer with his head for the 40,000 who have not fought.[20]

In his *Mémoires*, General Berthezène offered a simplified version of this text, which (according to him) was uttered before the Austrian generals at Leoben in 1797:

> There are lots of good generals in Europe, but they see too many things at once. Me, I only see one—the masses. I try to destroy them, certain that the accessories will then succumb of their own accord.[21]

Berthezène had been made a prisoner by the coalition after the fall of Dresden in the autumn of 1813 and had several conversations with Chasteler during his captivity.[22] When the objective of a war was conquest, it could only be ensured by politics:

> There is no doubt that conquest is a combination of war and politics. That is what makes Alexander admirable. [. . .] What is admirable about Alexander is that he was idolized by the peoples whom he had conquered; it is that following a reign of twelve years his successors divided up his empire; it is that the conquered peoples were more attached to him than his own soldiers; it is that he was forced into acts of severity to compel his closest generals to conduct themselves politically.
>
> Alexander conquered Egypt by going to the temple of Jupiter Hammon. This initiative ensured him his kingdom. Had I gone to the mosque [in Cairo] with my generals, who knows what effect it would have had? It would have given me 300,000 men and the empire of the Orient.[23]

3

War and Politics

In Book VIII, Clausewitz repeats the famous formula from Book I: 'war is simply a continuation of political intercourse, with the addition of other means.'[1] War is merely an instrument of politics. The latter does not enter into every detail, but its influence is decisive at the level of war as a whole, a campaign, and often even a battle. The military standpoint must therefore be subordinate to political authority.

Influence of the political objective on the military objective

It was the political authority that decided the military objective—what each army was to do. Bonaparte affirmed this in his note of July 1794 to the Committee of Public Safety:

> The kind of war that each arm is to fight can only be determined by the higher authority.
> It is by such considerations above all that one imbibes the absolute necessity, in an immense struggle like ours, of a revolutionary government and central authority which has a stable system, which imparts to each spring its operation, and which by profound views directs courage and renders our successes solid, decisive, and less bloody.[2]

In a reproach to O'Meara, his Irish doctor on Saint Helena, Napoleon put his finger on the way in which the British were able to employ their naval forces to achieve their political designs:

> You have the great advantage of declaring war when you like, and of carrying it on at a distance from your home. By means of your fleets you can menace an attack upon the coasts of those powers who disagree with you, and interrupt their commerce without their being able materially to retaliate.[3]

The army, rampart of the state

> With the events that have just occurred before your very eyes, you
> have learnt that, without force and good military organization, states
> are nothing.[4]

France in particular, surrounded as she was by enemies, must have an army.
And to that end, she must not be ruled by an assembly:

> The government of a Chamber is doubtless better than that of a single man.
> But I doubt whether it could ever come to pass in France. When a large army
> is necessary, it is very problematic if the government is not to be the master.
> The British today are quite right to want to destroy the army, if they wish to
> preserve their liberty. But the army is France's primary need; the primary need
> is not to be conquered by foreigners. France surrounded by power cannot do
> without an army. An army is necessarily obedient. The insurrection of an
> army is a rare event and akin to the reversal and overthrow of the state.[5]

General de Gaulle would not have disavowed these statements. They attest
to Napoleon's conviction that, from Valmy to Waterloo, despite thrusts
towards Venice, Cairo, Vienna, Madrid, and Moscow, France was simply
defending itself. The invasions of 1814 and 1815 portended others and gave
rise to a preoccupation with defence that was to dominate French strategy.
Although the *ancien régime* had witnessed foreign incursions, the scales had
changed and the army was more clearly becoming the indispensable
rampart of the nation than in the past. Charles de Gaulle's reflections, in
Vers l'armée de métier (1934) and *La France et son armée* (1938), would
continue in the same vein. His distrust of parliamentary regimes was
attributable to the same reasons as Napoleon's and the events of 1940
served only to confirm him in that opinion. That said, by his stubborn
determination to maintain a dominant position, Napoleon had himself
contributed to increasing the hostility of the other powers and, ultimately,
to weakening France.

Primacy of the civil over the military

An officer by training and a general at the time of the Brumaire coup d'état,
Napoleon nevertheless did not install a military regime.[6] Having become

First Consul, he took care to dress for ceremonies, receptions, and official portraits in a coat of red velvet attached to his political office. During a session of the Conseil d'État, on 14 floréal an X (4 May 1802), he unequivocally established the superiority of the civil power over the military power. For him, every general must possess essentially civil qualities. This was the case in the disciplined, organized armies of the Greeks and Romans. Physical strength and skill in using arms were the primary qualities of war leaders in the Middle Ages. With the advent of gunpowder, the military system had revived the organization and discipline of the phalanx and legion. This was a key element in the 'military revolution of modern times':[7]

Since this revolution, what has the strength of a general consisted in? His civil qualities, the *coup d'œil*, calculation, intellect, administrative knowledge, eloquence—not that of the jurisconsult, but that appropriate to the head of armies—and, finally, knowledge of men. All this is civil. Now it is not a man of 5 feet 10 inches who will do great things. If strength and courage sufficed to be a general, every soldier could lay claim to command. The general who does great things is the one who possesses these civil qualities. It is because he passes for having more intellect that the soldier obeys and respects him. You need to hear the soldier reasoning in his bivouac; he has more esteem for the general who knows how to calculate than the one who has more courage. It is not that the soldier does not esteem courage, for he would scorn the general who lacked it. Mourad Bey was the strongest, most skilful man among the mamluks; without that, he would not have been Bey. When he saw me, he could not conceive how I could command my troops; he only understood it when he knew our system of war. The mamluks fought like knights, hand to hand, haphazardly; that is what led to us defeating them. If we had destroyed the mamluks, emancipated Egypt, and formed battalions in the nation, the military spirit would not have been crushed; on the contrary, its power would have been greater. In all countries, force cedes to civil qualities. Bayonets are lowered before the priest who speaks in the name of heaven and before the man who impresses with his science. I have predicted to military men, who had misgivings, that military government would never catch on in France unless the nation was rendered senseless by fifty years of ignorance. All attempts will fail and their authors will be the victims. It is not as a general that I govern, but because the nation believes that I have the civil qualities suitable for government. If it did not have this opinion, the government would not be supported. I knew full well what I was doing when, as an army general, I assumed the status of *member of the Institute*; I was certain of being understood even by the last drummer.

We do not need to argue from centuries of barbarism to the present day. We are 30 million human beings united by the Enlightenment, property, and trade. 300,000–400,000 soldiers are nothing among this mass. In addition to the fact that the general commands only through civil qualities, as soon as he is no longer serving, he returns to the civil sphere. Soldiers themselves are but the children of citizens. The army is the nation. Were we to consider the military man ignoring all these relations, we would be persuaded that he knows no law but force, that he relates everything to it, that he sees nothing but it. The civilian, on the other hand, sees only the general good. The peculiarity of the military man is to want everything despotically; that of the civilian is to subject everything to discussion, truth, reason. They have their different prisms; they are often misleading. Yet the discussion generates light. When it comes to the issue of primacy, I therefore have no hesitation in thinking that it unquestionably belongs to the civil. Were honours to be differentiated into military and civilian, two orders would be established, whereas there is but one nation. If we only awarded honours to soldiers, that preference would be even worse, for then the nation would no longer be anything.[8]

The Eighteenth Brumaire was not a pronunciamiento. Bonaparte did not seize power in the name of the army. As is well known, the army was not fully behind him—far from it.[9] When it came to conscription, he wanted the civil authority to have primacy over the military authority, because recruits were citizens:

As for the designation of the men who will have to set off for the army, I would leave it to the civil authorities. It is a municipal affair. The military must receive them from civil society and simply examine if they are suitable for service. The civil authorities are less susceptible of injustice and less liable to corruption than military men, who are just passing through and who are unconcerned about what will be said of them after their departure.[10]

In 1806, the Emperor reproved Junot, who did not get on well with the Prefect of Parma:

Military authority is useless and out of place in the civil order; one should therefore not act like a corporal.[11]

Nor did the Emperor appreciate the disrespectful behaviour of the pupils at the school at Metz towards the inhabitants and he wrote to the Police Minister:

The first duty of these young men is respect for civil authority. Do not let them think themselves authorized to commit incivilities and imitate the

insolent petulance that young officers formerly permitted themselves; let them
know that the citizens are their fathers and that they are merely the children of
the family.[12]

This did not prevent the Napoleonic regime from having a 'strong military
impregnation'. For example, officers demanded places of honour in ceremonies. Incidents with the civil authorities were numerous in this
connection.[13]

4

The Supreme Commander
and the Government

The supreme commander must have a free hand in the theatre where he is in command, if only in order to confront the unexpected promptly. Even before taking charge of the army of Italy, General Bonaparte demanded such autonomy:

> The government must have complete confidence in its general, allow him great latitude, and simply present him with the objective it wishes to be accomplished. An answer for a dispatch from Savona takes a month and, in this time, everything can change.[1]

Following his initial victories, the commander-in-chief of the army of Italy was even clearer:

> If the Directory takes prompt measures, transfers part of the army of the Alps to me, transfers cavalry to me, and above all covers its plans and what it wants done in secrecy, I shall shortly be master of all Italy and will go to Rome to restore the Capitol. But if you do not explain to me what is wanted, and I am constantly inhibited by the fear of not carrying out your intentions and being accused of seeking to involve myself in diplomacy, it will be impossible for me to do great things. At this point, the war in Italy is half military and half diplomatic. To lull one party by suspending arms in order to have time to crush the other, or to secure a crossing and subsistence, is now the great art of the Italian war.[2]
>
> If you handicap me with obstacles of all kinds; if I have to refer every one of my steps to government commissioners; if they have the right to alter my movements, to deprive me of troops or to send me them, do not expect any good to come of it. [. . .] In the state of affairs of the Republic in Italy, it is indispensable that you have a general who has your complete confidence.[3]

To the unity of military and diplomatic action that he demanded in Italy, Bonaparte added a financial dimension:

Everything has to be confronted with a mediocre army: containing the German armies, laying siege to strongholds, guarding our rear, levying taxes on Genoa, Venice, Tuscany, Rome, Naples; we must be in strength everywhere. We therefore need unified military, diplomatic, and financial thinking. Here we must burn, have people shot, institute terror, and make a sensational example. Elsewhere, there are things we must affect not to see and hence not say, because the time has not yet come. At this point in time, diplomacy is therefore completely military in Italy.[4]

Napoleon admired Marlborough, because he had displayed qualities as a soldier and a diplomat:

He was not a man narrowly limited to his battlefield; he negotiated, he fought; he was at once commander and diplomat.[5]

On Saint Helena, Napoleon commented at length on the relations between the supreme commander and his government:

A supreme commander is not covered by an order from a minister or a prince, distant from the field of operations, and with little or no knowledge of the latest state of things. 1 Any supreme commander who takes responsibility for executing a plan that he considers bad and disastrous is criminal; he must make representations, insist that it be changed, and, ultimately, offer his resignation rather than be the instrument of the ruin of his own people; 2 any supreme commander who, as a result of higher orders, fights a battle in the certain knowledge that he will lose it, is likewise criminal; 3 a supreme commander is the first officer of the military hierarchy. The minister, the prince, issue instructions with which he must comply in his soul and conscience; but these instructions are never military orders and do not dictate passive obedience; 4 a military order itself requires passive obedience only when given by a superior who, present when he gives it, has a knowledge of how things stand, can attend to objections, and furnish explanations to the person who must execute the order.

Tourville attacked 80 British vessels with 40; the French fleet was destroyed.[6] Louis XIV's order certainly does not justify him; it was not a military order that required passive obedience; it was an instruction. The implicit clause was, if there is at least an even chance of success. In that case, the admiral's responsibility was covered by the monarch's order. But when as a result of the state of things the battle was certain to be lost, to execute this order literally was to misunderstand its spirit. If, approaching Louis XIV, the admiral had said to him: 'Sire, had I attacked the English, your whole squadron would have been destroyed. I have brought it back into some port', the king would have thanked him and the royal command would in fact have been executed.

[. . .] In his *Mémoires*,[7] General Jourdan says that the government had hinted to him that he should fight the Battle of Stockach. He thus seeks to justify himself for the bad outcome of that action. But this justification could not be accepted even had he received a positive, formal order, as we have demonstrated. When he decided to give battle, he believed that he had a good chance of winning; he was mistaken.

But might it not be the case that a minister or a prince explained his intentions so clearly that no clause could be implied? That he said to a supreme commander: *Give battle. Given the number and calibre of his troops, and the positions he occupies, the enemy will beat you; no matter, it is my will.* Should such an order be passively carried out? No. Were the general to understand the utility and, consequently, the morality of such a strange order, he would have to carry it out. But if he did not understand it, he should not obey it.

However, something similar often happens in war: a battalion is left in a difficult position to save the army; but the commander of this battalion receives positive orders from his leader, who is present when he gives them, who answers all objections, should there be reasonable ones to make. It is a military order given by a leader on the spot and to whom passive obedience is owed. But what if the minister or prince was in the army? Then, if they take command, they are supreme commanders; the supreme commander is now only a subordinate major general.

It does not follow from this that a supreme commander should not obey the minister who orders him to fight a battle; on the contrary, he must do so whenever, in his opinion, the odds are even and there are as many probably in favour as against. For the observation we have made applies solely to cases where the odds seem to be completely against.[8]

Recent interpretations of Clausewitz take account of the nuances formulated by Napoleon here. Clausewitz has long been reduced to the formula 'war is simply the continuation of politics by other means'. Reassuring during the Cold War and the 'balance of terror', this sentence is now regarded as but one definition of war among others and is today regarded as indicating the need to harmonize the military and the political, rather than subordinating the former to the latter. Political leaders must have sufficient knowledge of the true nature of war to avoid asking of it what it cannot accomplish. The politics understood by Clausewitz in Book VIII does not necessarily intervene in war at all levels. It fashions the war plan, but not tactics. It can supply war with its logic, but the latter still possesses its own grammar. In reality, politics adapts as much to war, its developments and circumstances, as war adapts to politics.[9]

In connection with certain generals in Spain, Napoleon reminded Berthier of the limits of military authority:

My Cousin, I am sending you some dispatches from Spain; in transmitting them to the Duke of Istria, tell him that it is ridiculous to observe General Dorsenne making decrees; that generals should not use formulas which pertain exclusively to sovereignty; that they should act only on the orders of the day and that they can neither *settle* nor *decree* anything.[10]

5

The War Plan when the Objective is the Destruction of the Enemy

At Vitebsk on 13 August 1812, Napoleon justified his invasion plan for Russia in a lengthy monologue reported by Baron Fain. The text takes up several ideas referred to in previous books: the drive and impatience of French troops, the political dimension of strategy, the ineluctability of risk-taking in war, seeking battle so as to avoid stalemate:

> The Russians, it is said, are retreating deliberately; they would like to draw us towards Moscow!—No, they are not retreating deliberately. If they have left Vilnius, it is because they could no longer rally there; if they have abandoned the line of the Danube, it is because they lost their hopes of being joined there by Bagration. If, recently, we have witnessed them surrendering us the fields of Vitebsk, to withdraw to Smolensk, it is in order to make a link-up that has been deferred so many times. The moment of battles is approaching. You will not have Smolensk without a battle; you will not have Moscow without a battle. An active campaign might have unfavourable odds. But a long-drawn-out war would have much more inauspicious odds and our distance from France would only multiply them!
>
> Can I think of taking up quarters in the month of July? [...] Can an expedition like that be divided into several campaigns? Believe me, the question is a serious one and I am preoccupied by it.
>
> Our troops are happily heading forward. They enjoy wars of invasion. But a stationary, prolonged defensive campaign is not in the French genius. To halt behind rivers, to remain there stationed in shacks, to manoeuvre every day in order only to be in the same place after eight months of privation and woes, is that how we are used to waging war?
>
> The lines of defence presented to you today by the Dnieper and Danube are merely illusory. Once winter arrives, you will see them filled with ice and effaced under snow.

Winter does not only threaten us with its cold weather; it also threatens us with diplomatic intrigues that might brew up behind our backs. Shall we allow these allies whom we have just won over, and who are still utterly astonished to no longer be fighting us and glory in following us, time to reflect on the oddity of their new position?

And why halt here for eight months, when twenty days would suffice for us to achieve our objective? Let us anticipate winter and reflections! We must strike promptly, on pain of jeopardizing everything. We must be in Moscow in a month, on pain of never entering it!

In war, fortune counts for half in everything. If one always awaited a complete conjunction of favourable circumstances, one would never finish anything.

In short, my plan of campaign is a battle and my whole policy is success.[1]

All of Napoleon is in this quotation, which is like a conclusion. In the final chapter of *Vom Kriege*, Clausewitz devotes several paragraphs to the 1812 campaign in Russia. For Napoleon, it was the first to fail. The excessive character of the undertaking, in terms of the forces committed and the theatre of operations itself, represented the culminating point of the Napoleonic Wars. For Clausewitz, the Emperor's failure did not stem from the fact that he advanced too rapidly or too far, as is generally believed. In essence, he could not have done differently. His campaign failed because the government of Russia remained firm and its people unshakeable. Napoleon was mistaken in his calculations. He did not assess his opponent correctly.[2]

Russian firmness and Napoleon's disarray emerge from this letter addressed to the Tsar from Moscow:

The beautiful, superb city of Moscow no longer exists. Rostopchin has had it burnt down. 400 arsonists have been arrested in the act; they have all stated that they were setting fires on the orders of the governor and the police chief: they have been shot. The fire seems finally to have stopped. Three-quarters of the houses have been burned down; a quarter remains. Such behaviour is atrocious and pointless. Is its object to deprive us of resources? But these resources were in cellars that the fire could not reach. Furthermore, why destroy one of the most beautiful cities in the world to achieve such a meagre objective? This is the conduct that has been followed since Smolensk, which has consigned 600,000 families to begging. The pumps of the city of Moscow had been broken or carried off, some of the weapons from the arsenal given to malefactors who have made it necessary for us to fire cannon shot on the Kremlin to expel them. Humanity and the interests of Your Majesty and this great city would have had it entrusted to me on loan, since the Russian army left it exposed; administration, magistrates, and civil guards would have been left there. This is what was done in Vienna, twice, Berlin, and Madrid.[3]

Napoleon had had a proclamation written for the emancipation of the serfs, but kept it confidential as a form of possible intimidation. He did not want to cross certain boundaries and confided his desire to negotiate to Caulaincourt:

> Hitherto, apart from the fact that Alexander burns his towns and cities so that we cannot live in them, we have fought a fairly good war. No disagreeable publications, no insults. He is wrong not to come to an agreement now that we have had a scrap. We would soon be in agreement and remain good friends.[4]

To fight a 'fairly good war': Napoleon was still sufficiently marked by the age of Enlightenment not to desire a full unleashing of force. As we saw in Book I, he quipped that he regretted not having resorted to it, but he did not open Pandora's Box. Such restraint was not due solely to the still relatively rudimentary state of technologies of destruction.[5]

On several occasions, Napoleon sketched an analysis of Russian power, which frightened him:

> Russia is a frightening power that seems set to conquer Europe. With its Cossacks, Tartars, and Poles, it can put thousands of cavalry on horse everywhere. There would not be enough horses in Europe to withstand it. In the past, three powers opposed its expansion: Sweden, but it has not been able to do anything since the loss of Finland; Poland, but it now forms part of the Russian Empire; and the Turks, who are null.[6]

Las Cases reports further considerations on Russia and, beyond it, on the future of Europe:

> The Emperor passed on to what he called the admirable situation of Russia against the rest of Europe, to the immensity of its mass in the event of invasion. He depicted this power situated under the pole, sustained by eternal ice which, as and when necessary, rendered it inaccessible. It could be attacked, he said, only for three or four months, or one-quarter of the year, whereas it had the whole year—twelve months—against us. It offered attackers nothing but the harsh conditions, suffering, and privations of a desert land, a dead or inert nature, whereas its peoples launched themselves with zest towards the delights of our south.
>
> In addition to these physical features, said the Emperor, joined to its sizeable, sedentary, brave, tough, devoted, passive population were enormous

tribes, whose normal state was deprivation and vagabondage. 'One cannot but tremble at the idea of such a mass, which can be attacked neither by the coasts nor in the rear; which floods over us with impunity, inundating everything if it triumphs or withdrawing into the ice, the heart of desolation, death, which have become its reserves if it is defeated; and all this with the ability to re-emerge immediately if required. Is this not the hydra's head, the Antaeus of the fable, who could be finished off only by grasping him bodily and suffo-cating him in one's arms? But where is Hercules to be found? It only fell to us to dare to pretend to the role and, it must be admitted, our attempts were clumsy.'

The Emperor said that in the new political set-up of Europe, the destiny of this part of the world was no longer bound up with the capacity, the dispositions of a single man. 'Let there be found, he said, an emperor of Russia who is valiant, impetuous, and capable—in a word, a tsar with a beard on his chin (something he expressed, moreover, much more energetically)—and Europe is his. He can begin his operations on German soil, 100 leagues from the two capitals of Berlin and Vienna, whose sovereigns are the sole obstacles. He can wrest the alliance of one of them by force and with his aid knock out the other at a stroke; and thenceforth he is at the heart of Germany, amid second-rank princes, most of whom are his relatives or depend on him for everything. If needs be, if required, in passing over the Alps he throws some firebrands onto Italian soil, ready for the explosion, and marches in triumph towards France, whose liberator he once again proclaims himself. Certainly, in such a situation, I would reach Calais at a set time and by stages, and find myself the master and arbiter of Europe...' And after a few moments of silence, he added: 'Perhaps, my dear chap, you are tempted to tell me, as Pyrrhus' minister did his master: *After all, what's the point?* I answer: to found a new society and protect against great misfortunes. Europe is awaiting and soliciting this good deed; the old system is on its last legs and the new system is not yet established and will not be without further prolonged and furious convulsions.'[7]

This realization of the break created by the wars of 1792–1815, and the presentiment that they would be followed by other conflagrations, partially anticipates Clausewitz's observation at the end of *Vom Kriege*: 'But the reader will agree with us when we say that once barriers—which in a sense consist only in man's ignorance of what is possible—are torn down, they are not so easily set up again. At least when the major interests are at stake, mutual hostility will express itself in the same manner as it has in our own day.'[8] Napoleon had already alluded to a 'new society'; he had already spoken of 'system' during his time on the island of Elba. In his *Mémorial de l'île d'Elbe*, General Vincent, at the time director of fortifications in

Florence, relates that in May 1814, during a conversation, the deposed emperor told him not to speak to him any more about war:

> You see, I have thought about it a lot; we have waged war all our life; the future will possibly force us to again, and yet war is going to become an anachronism. If we fought battles over the whole continent, it is because two societies were face to face: the one dating from 1789 and the *ancien régime*; they could not subsist together; the younger one devoured the other. I know full well that ultimately war overthrew me—me, the representative of the French Revolution and instrument of its principles. But no matter: it is a battle lost for civilization and civilization, believe you me, will take its revenge. There are two systems: the past and the future. The present is merely a painful transition. Who is to triumph, according to you? The future, no? So! The future is intelligence, industry, and peace; the past was brute force, privileges, and ignorance. Every one of our victories was a triumph for the ideas of the Revolution, rather than of its eagles. One day, victories will be won without cannons and bayonets! . . . Do not speak to me of war.[9]

Conclusion to Book VIII

War must be subject to reflection before the action begins. Napoleon advocated reflecting well in advance on all eventualities and anticipating all possible enemy behaviour and reckoning on the worst. Execution was gradual, as everyday events permitted. A plan could be altered ad infinitum and chance always played its part. The worst plans were those that assumed a stationary enemy. The scale of the military objective was bound up with a number of circumstances, but it always involved what Clausewitz was to call a centre of gravity, where efforts should be concentrated. It made it possible to clear a main theatre of operations. It was a crucial error not to identify this. If efforts were concentrated there, and a breach was made, the equilibrium was broken and the other theatres succumbed of their own accord. Things always came down to the concentration of forces.

If the military objective was conquest, political gestures were required to ensure it, like those made by Alexander the Great in the East. The political objective determined the military objective. Napoleon anticipated Clausewitz's famous formula. He reckoned that a general must above all possess civil qualities. This is what he possessed over and above courage and valour, which he shared with his men. According to Napoleon, a military government would never catch on in France. The civil had primacy over the military. This did not stop the supreme commander having to possess freedom of action in the theatre where he was in command. He must even be able to interfere in diplomacy and finances. Nor must he carry out the orders of a distant minister or prince if they were unreasonable. A supreme commander must resign, rather than execute a plan which he deemed bad.

Finally, if the Napoleonic Wars disclosed to Clausewitz the idea of absolute war, to such an extent had the unleashing of violence increased by comparison with the previous century, Napoleon himself sensed the risks of escalation and said that he sought to control them. Despite the violence of his temper, his unchecked desire to make the most of anything to expand French power, and his stubbornness in conceding nothing, he remained within the intellectual framework of Reason. He would not have wanted to push his opponents to the limit because he knew that war was always followed by negotiations. He thought of the balance of power in Europe and feared the development of Russian power. To make the transition from the old system to a new one, Europe would experience further 'prolonged and furious convulsions'. Between these words uttered on Saint Helena for posterity and the objective examination of diplomatic and military phenomena, there will always be food for thought and study.

Conclusion

Historical science is essentially critical and it is right that it should set about myths and reassess idols. Napoleon unquestionably pertained to these two categories and still does. Charles Esdaile has legitimately 'deconstructed' the wars of the Revolution and Empire and highlighted the serious strategic errors made by Napoleon: the conquest of the Iberian Peninsula from 1807 onwards and the invasion of Russia in 1812, for example, occasioned major human and political disasters.[1] Napoleon lost the conquests of the Revolution. He rendered France smaller than it was when he inherited it. Thierry Lentz does not regard Napoleon's legacy as his 'European system', but his fashioning of modern France.[2] In other words, domestic reforms, in the main realized under the Consulate, were enduring, unlike foreign policy, which was taken to extremes under the Empire: this is to recognize the ultimate failure of a strategy. John Lynn has ended up frankly posing the question: does Napoleon really have much to teach us militarily? Comparing him with Louis XIV, with whom he is more familiar, Lynn stresses the contrast between the transient conquests of the Eagle and the enduring conquests of the Sun King.[3] Taking up Paul Schroeder's thesis, he recalls that Napoleon had no long-term vision for Europe, no exit strategy, as we would say today. He refused to operate within the international norms of his time: his repeated acts of aggression are sufficient evidence of this.[4] His reputation as a master of strategy derives from a confusion with operations. Lynn is not wrong on this point, but unduly reduces Napoleon to the status of beneficiary of the Revolution's military gains. Napoleon's operational art, it is true, benefited from exceptional circumstances in his glory days. But Lynn does not name them. Napoleon, it is true, taught people how to win battles and not how to win wars. While it errs on numerous points of detail, Lynn's critique is salutary in the round, if only for the questions it raises. Overall, Napoleon failed in his foreign policy and therefore in his strategy or, rather, 'grand strategy'. For General Riley, this stems in part from the fact that

Napoleonic France only had a military view of strategy and did not genuinely combine with it naval and economic dimensions. Better equipped on these two levels, Great Britain was the only country capable of doing so at the time.[5] These issues will continue to be much debated, but it can be accepted that Napoleon's reputation is justified more in terms of operational art than strategy, inasmuch as the latter is the military translation of a foreign policy.

When the brains of the French army prepared for revenge on Germany by studying Napoleon, they focused on the operational level. Lieutenant-Colonel de Lort de Sérignan was the most critical, and also one of the few to envisage the political aspects of strategy. For him, the hatreds generated in Europe by twenty years of conquest and occupation, by countless vexations, had helped to plunge France into a political isolation from which she was still suffering in 1914. He also reckoned that Napoleon's influence in the French army had not been what one might think and that it had above all resulted in killing individual initiative within it.[6] Alongside these not unfounded criticisms, Lort de Sérignan agreed with the professors of the École supérieure de guerre in saying that there was no such thing as a precise, definitive Napoleonic doctrine. 'The procedures of war implemented', said General Bonnal, 'were scientific in the sense that they always derived from a calculation of moral and material strength applied to a definite objective, in clearly defined temporal and spatial conditions. But if Napoleonic warfare was scientific in its applications, it was empirical in its conceptions. Napoleon's calculations did not derive from adapting a principle to a concrete case. Principle existed in him in a latent state—in other words, in the depth of the unconscious—and it was unconsciously that he applied it when the clash of circumstances, objective and methods in his brain caused the spark to burst forth, and showed him the artistic solution par excellence—the one that reached the limits of perfection.'[7]

Napoleon's 'case' at bottom consisted in the exceptional adaptation of a personality to the vicissitudes of war. Rarely has such an assemblage of the qualities of war leader been encountered in history, brought together at the precise moment when circumstances were to permit of their manifestation and development. He instinctively discerned laws that were now going to impose themselves, for better or worse: 'The influence of number, the bringing into play of the masses, the principle of the economy of strength, that of the destruction of the bulk of the opponent as an immediate, principal, and overriding objective.'[8] The uniqueness, even strangeness, of Napoleon as a war leader is expressed in his mastery of paradoxes:[9] no one was more conscious of the influence of chance on the outcome of a battle and no one left less room for chance; no one was more daring in his conceptions or more rapid in execution and no one gave as much attention to the most minute details of army organization or proved more prudent when the situation demanded it.[10] For all these reasons, Napoleon remains unavoidable for those who wish to prepare to use their armed forces effectively. Thus, a centre of analysis carried out a study of 'military advantage' in history for the US Defense Department in 2002. Napoleon featured in it, following the Macedonians, Romans, and Mongols, and prior to implications being drawn from it for the USA.[11] Napoleon alone had a chapter named after him and the phenomenon is not an isolated one. There are few individuals who have so marked a period or series of wars that they have given their name to it. For the authors of the report, Napoleon offered generals 'a model by which they could engage in and win wars of manoeuver. His strategic vision was unmatched by any other general of his time and few since his death.' Napoleon created a synthesis of the tactics of arms that stressed combined action. On an operational level, he had the advantage thanks to a better system of information and he himself represented a 'key variable'. In strategy, he did not realize the importance of the navy, reducing it to an adjunct to the army. His enduring legacy consists in the operational conduct of a war: seeking to break up the enemy, confuse him, and bring him to battle at the time, and in the place, most advantageous to the French.[12]

Study of Napoleon's writings and sayings reveals even more. What he said about war is more important than has hitherto been thought. One cannot but be struck by the abundance of his reading, the importance of his reflection, by what Jean Guitton called 'the antecedent power of thought'. Guitton was convinced that there is a secret relationship between the method of the warrior and the method of the thinker and that they can illuminate and reinforce one another.[13] The father of French philosophy, Descartes, was also a soldier. As has been noted by the American David Bell, whose study of total war is not marked by particular sympathy for Napoleon, people who know any history at all instinctively do not place him in the same category as Stalin, Hitler, or Mao. One reason for this is that Napoleon always manifested a profound human quality in everything he said. Another is that, for better or worse, he continues to embody what a human being can or cannot attain through war, including the idea of a certain human greatness.[14] In Stalin, Hitler, and Mao, there is no longer any humanity. Like Clausewitz, Napoleon put man at the centre of his thought; he knew everything he depicted and often displayed a concern for permanent foundations, as opposed to transient forms.[15] His texts also represent a moment in the history of human intelligence whose richness we hope to have demonstrated. Study of Napoleon's campaigns has always been privileged over that of his thoughts on war. Yet the latter have retained all their relevance. More than maxims to be applied, they are reflections to be meditated on, in connection with any form of war, including the conflicts of the twenty-first century. If Clausewitz has just been rediscovered in this connection,[16] Napoleon likewise warrants being so. Let us reiterate some key ideas.

War in itself is not waged 'in namby-pamby fashion'. If the law is obeyed, it makes it possible to limit its violence. A supreme commander must calculate a good deal, in the knowledge that some things will always escape him and remain subject to chance. His strength of character is even more important than the faculties of his mind, which are nevertheless indispensable. He must take decisions in an uncertain context and for that he needs the moral courage 'of the early hours'. He must have as much intelligence as possible, but his 'tact' alone will enable him to discern the useful information that will guide his decisions. There are always unforeseen events in war. The skilful leader will exploit them rather than suffering them. There are always things to take from theoreticians, but the 'high parts' of war are learnt solely through experience and the study of history. Principles supply 'axes to which a curve is related'— basic truths supplied to the capacity for judgement of the leader, who will take what he deems useful depending on circumstances. A critical analysis of past campaigns will have formed his judgement beforehand. Strategy is less a question of moves to be conceived than of moral strength for executing them. 'Everything is opinion in war.' A sense of honour, emulation, esprit de corps, and a feeling of equality represent an army's virtues. Dominated by fire, combat aims at a significant destruction of enemy forces. It is a 'moral spark' that tips the scales. One does not engage lightly in a battle, for it is difficult to extricate oneself. Favourable odds are required. If they obtain, it is necessary to assemble all available force and be resolute.

In an army, unified command is essential. Supplies are an important task and local resources are there to contribute. If troops are not battle-hardened, it is better to entrust them with a defensive mission, which is accomplished better by men used to staying in the same sector, where they will have acquired habits. They must not be arranged in a cordon: that is equivalent to being strong absolutely nowhere. When you do not know where the enemy will attack, it is better for troops to be positioned in close echelons, so that they can be concentrated rapidly at the point under attack. If you attack, you will seek to outflank or envelop the enemy, but out of his view. It is always difficult to combine different forms of attack simultaneously. The simplest moves are the best. They are accompanied by concentrated fire. In mountainous terrain above all, you must not attack an enemy position head on. The enemy must be flushed out by making for his flank or rear. If you attack a town, it is better not to commit your infantrymen in the streets. An occupation is always difficult. In any event, forces must remain concentrated and be mobile. The main towns, where the important people whose collaboration is essential reside, must be held. Mobile columns will pursue insurgents, but without committing blunders. It is necessary to make examples, but also to issue pardons. Finally, all possibilities must be envisaged when one engages in war. It is necessary to reckon on the worst. Plans must be carefully prepared, but remain open to adaptation depending on circumstances. You must always ask yourself how the enemy will react. When it comes to the enemy, it is necessary to target the essentials—the main front—and direct your efforts there. If a breach is made, the secondary fronts will fall of their own accord. A conquest must always be accompanied by political gestures. Any military objective is subordinate to a political objective and every general must possess civil qualities. This does not preclude the supreme commander having to enjoy a certain freedom of action in the theatre where he is in charge. These observations can be applied to all wars, including those of the early twenty-first century.

Already detected by Jay Luvaas, the intellectual similarity between Napoleon and Clausewitz is no cause for astonishment.[17] They are practically in agreement on everything, differing only on points of detail, such as the division of the major phases of a battle into two or three acts. Clausewitz did not have Napoleon's correspondence at his disposal and had only read his *Mémoires*. But his thinking is entirely tributary to the Napoleonic campaigns. 'In the beginning was Napoleon', writes Andreas Herberg-Rothe at the start of his study of Clausewitz's puzzle. The 1806 campaign in Prussia led Clausewitz to develop a theory of operational success. Napoleon's defeats in Russia (1812), at Leipzig (1813), and then at Waterloo (1815) enabled him to elaborate a political approach to war. In other words, the successes, limits, and ultimate failure of Napoleonic war prompted Clausewitz to carry his reflection beyond the purely military aspects to envisage a general political theory of war.[18] Napoleon had not fundamentally changed his way of waging war between 1806 and 1815. He himself said that at the age of 20 he already possessed the ideas that he would retain for the rest of his career. Hew Strachan has stressed that *Vom Kriege* was conceived as a dialogue. Unlike most theoreticians of the time—Jomini, for example—Clausewitz did not present conclusions, but engaged in arguments and debates. He used this dialectical method in Books I and VIII in particular. Books II–V resorted to it less, insofar as they described the Napoleonic art of war.[19] Yet Napoleon essentially expressed himself on war in dialogues. This helps us understand why we have been able to regroup Napoleon's writings in accordance with the plan of *Vom Kriege*. The book had a pedagogical purpose: it sought to get Prussian officers to think. Here too there is something in common with Napoleon's texts. In his correspondence, he explained his way of waging war to a much greater extent than has been appreciated. Obviously, these nuggets have to be extracted from a mass of letters and the extracts assembled. Often what are involved are one-off explanations, adapted to a particular situation. But they sometimes assume a more general scope. We have noted it above all in the letters to Eugène de Beauharnais and to his brothers Joseph, Louis, and Jérôme. Napoleon turned pedagogue with the members of his family to whom he entrusted armies, and who had no military knowledge. Sometimes—rarely, it is true—he also sketched his plan of campaign for certain marshals, like Soult in 1806 and Marmont in 1809. He confided some important reflections to General Dejean. All this remains fairly limited, but qualifies what has generally been said of a lack of communication between Napoleon and his generals as regards the art of war. It remains the case that, unlike Frederick II, he never put on paper 'instructions' about to how to wage war and he did not create a military higher education system.

Napoleon and Clausewitz were on the hinge of the age of Enlightenment and the age of Romanticism. For Clausewitz, Napoleon embodied the *Geist* of the art of war. In the round, the Emperor was closer to the eighteenth century because he was older. One sign of this is that he frequently evoked the wars of antiquity—something Clausewitz practically never did.[20] For René Girard, Napoleon is behind all Clausewitz's ideas; he was a source of obsession for him and functioned as a 'model-obstacle', at once attracting and repelling him. Clausewitz thought against Napoleon, in both senses of the preposition, and his resentment was fertile: it enabled him to theorize. If he perceived reciprocal action in war so clearly, it is because he himself was gnawed away at by mimetism. He attempted, says Girard, to appropriate the being of his model.[21] There is therefore some legitimacy in using, as we have done, the plan of *Vom Kriege* to structure the ideas of Napoleon.

Notes

ABBREVIATIONS

AN   Archives nationales, Paris

AP   Archives privées

Corresp.   *Correspondance de Napoléon Ier publiée par ordre de l'empereur Napoléon III*, 32 vols, Paris, Plon et Dumaine, 1858–70. Vols 29–32 contain the *Œuvres de Napoléon Ier à Sainte-Hélène*

Corresp. gén.   Napoléon Bonaparte, *Correspondance générale*, 101 vols, published in November 2014, Paris, Fayard, 2004–

MS   manuscript

n.p.n.d.   no place no date

SHD/DAT   Archives du Service historique de la Défense, Département de l'armée de terre, Vincennes

INTRODUCTION

1. John R. Elting, *The Superstrategists: Great Captains, Theorists, and Fighting Men Who Have Shaped the History of Warfare*, London, W. H. Allen, 1987, p. 112.
2. David Gates, *Warfare in the Nineteenth Century*, Basingstoke and New York, Palgrave, 2001, p. 54.
3. Jeremy Black, *Rethinking Military History*, London, Routledge, 2004, p. 184.
4. [Raimondo Montecucoli], *Mémoires de Montecuculi, généralissime des troupes de l'Empereur...*, new edn, Paris, d'Espilly, 1746; [Maurice de Saxe], *Les Rêveries, ou Mémoires sur l'art de la guerre de Maurice, comte de Saxe...*, by M. de Bonneville, The Hague, Pierre Gosse Junior, 1756.
5. *Maximes de guerre de Napoléon*, Brussels, Petit, 1838.
6. *Maximes de guerre et pensées de Napoléon Ier*, 5th edn, revised and expanded, Paris, Dumaine, 1863.
7. M. Damas Hinard (ed.), *Dictionnaire Napoléon ou Recueil alphabétique des opinions et jugements de l'empereur Napoléon Ier*, 2nd edn, Paris, Plon, 1854. The term 'guerre' includes several entries, from 'Alpes' to 'vin'. The references for citations are given in summary fashion.

8. Antoine Grouard, *Stratégie napoléonienne. Maximes de guerre de Napoléon Ier*, Paris, Baudouin, 1898.

9. Honoré de Balzac, *Napoléon et son époque*, texts collected and annotated by Léon Gédéon, introduced by J. Héritier, Paris, Éditions Colbert, [1943], p. 295 (525 maxims by Balzac, pp. 295–360); Thierry Bodin, 'Les Batailles napoléoniennes de Balzac', *Napoléon, de l'histoire à la légende. Actes du colloque des 30 novembre et 1er décembre 1999 à l'auditorium Austerlitz du musée de l'Armée, hôtel national des Invalides*, Paris, Musée de l'Armée-Éditions In Forma-Maisonneuve et Larose, 2000, p. 107.

10. Frédéric Masson, *Petites Histoires*, 2 vols, Paris, Ollendorff, 1910–12, I, pp. 40–51.

11. [Napoleon], *Préceptes et jugements de Napoléon*, collected and categorized by E. Picard, Paris and Nancy, Berger-Levrault, 1913.

12. *Napoléon par Napoléon*, 3 vols, Paris, Club de l'honnête homme, 1964–5.

13. Paul Adolphe Grisot, *Maximes napoléoniennes*, extracted from the *Journal des sciences militaires* (May 1897), Paris, Librairie militaire Baudoin, 1897, 32 pp.

14. André Palluel, *Dictionnaire de l'Empereur*, Paris, Plon, 1969.

15. Jean Delmas and Pierre Lesouef, *Napoléon, chef de guerre*, 3 vols, Paris, Club français du livre, 1970.

16. The following chronological list is far from exhaustive: Napoleon I, *Manuel du chef*, Napoleonic maxims selected by Jules Bertaut, Paris, Payot, 1919; *Les Pages immortelles de Napoléon*, selected and explained by Octave Aubry, Paris, Corrêa, 1941; Napoleon, *Pensées pour l'action*, collected and presented by Édouard Driault, Paris, Presses Universitaires de France, 1943; Napoleon, *Comment faire la guerre*, texts assembled by Yann Cloarec, Paris, Champ libre, 1973; Lucian Regenbogen, *Napoléon a dit. Aphorismes, citations et opinions*, Paris, Les Belles Lettres, 1996; Charles Napoléon, *Napoléon par Napoléon. Pensées, maximes et citations*, Paris, Le Cherche Midi, 2009.

17. *A Manuscript Found in the Portfolio of Las Cases Containing Maxims and Observations of Napoleon, Collected During the Last Two Years of his Residence at St. Helena*, trans. from the French, London, Alexander Black, 1820.

18. David Chandler, 'General Introduction', *The Military Maxims of Napoleon*, trans. G. C. D'Aguillar, London, Greenhill Books, 2002, p. 25.

19. Chandler, 'General Introduction', pp. 22–3, 30–1.

20. Chandler, 'General Introduction', pp. 23, 27.

21. G. F. R. Henderson, *Stonewall Jackson and the American Civil War*, 2 vols, London, Longmans, Green and Co., 1900, II, pp. 394–5.

22. *Napoleon and Modern War: His Military Maxims Annotated*, trans. with notes by Conrad H. Lanza, Harrisburg, Pa., Military Service Publishing Co., 1943.

23. Jay Luvaas, 'Napoleon on the Art of Command', *Parameters: Journal of the US Army War College*, 15, 1985–2, pp. 30–6.

24. *Napoleon on the Art of War*, texts selected, translated, and published by Jay Luvaas, New York, The Free Press, 1999.

25. We have not adopted the division into three parts given that they do not possess titles, and references to Clausewitz's treatise are usually given exclusively to the books and chapters.

26. Luvaas, 'Napoleon on the Art of Command', p. 32.

27. Robert Van Roosbroeck, 'Der Einfluss Napoleons und seines Militärsystems auf die preussische Kriegführung und Kriegstheorie', in Wolfgang von Groote and Klaus Jürgen Müller (eds), *Napoleon I. und das Militärwesen seiner Zeit*, Freiburg im Breisgau, Rombach, 1968, p. 202.

28. Carl von Clausewitz, *La Campagne de 1796 en Italie*, Paris, 1901; Pocket, 1999, p. 8.

29. F. G. Healey, 'La Bibliothèque de Napoléon à Sainte-Hélène. Documents inédits trouvés parmi les *Hudson Lowe Papers*', *Revue de l'Institut Napoléon*, no. 75, 1960, p. 209; Frédéric Schoell, *Recueil de pièces officielles destinées à détromper les Français sur les événements qui se sont passés depuis quelques années*, 9 vols, Paris, Librairie grecque–latine–allemande, 1814–16. See also Jacques Jourquin, 'La Bibliothèque de Sainte-Hélène', in Bernard Chevallier, Michel Dancoisne-Martineau, and Thierry Lentz (eds), *Sainte-Hélène, île de mémoire*, Paris, Fayard, 2005, pp. 121–5.

30. Emmanuel de Las Cases, *Le Mémorial de Sainte-Hélène*, 1st complete critical edn, established and annotated by Marcel Dunan, 2 vols, Paris, Flammarion, 1983, II, p. 568.

31. Schoell, *Recueil de pièces officielles*, I, pp. xxix–xxx and II, pp. 289–342.

32. Jean-Jacques Langendorf, 'Clausewitz, Paris 1814: la première traduction; l'unique étude signée', *Revue suisse d'histoire*, no. 34, 1984, pp. 498–508.

33. Émile Mayer, 'Clausewitz', *Revue des études napoléoniennes*, no. 24, 1925, p. 158.

34. Vincent Desportes, *Comprendre la guerre*, Paris, Économica, 2000, p. 382.

35. Hew Strachan, *Carl von Clausewitz's On War: A Biography*, London, Atlantic Books, 2007, pp. 25–6, 105.

36. Antulio J. Echevarria II, *Clausewitz and Contemporary War*, Oxford, Oxford University Press, 2007, p. 26.

37. Gustave Davois, *Bibliographie napoléonienne française jusqu'en 1908*, 3 vols, Paris, L'Édition bibliographique, 1909–11, III, pp. 5–42. Despite its age, this classic text remains useful for a precise overview of Napoleon's published texts. It can be complemented by the *Correspondance générale* published by Fayard and the excellent clarification by Chantal Lheureux-Prévot, 'L'Exil de Napoléon Ier à Sainte-Hélène. Bibliographie thématique. Du *Bellérophon à La Belle Poule*', in Chevallier, Dancoisne-Martineau, and Lentz (eds), *Sainte-Hélène*, pp. 361–94.

38. Nada Tomiche, *Napoléon écrivain*, Paris, Armand Colin, 1952.

39. Interesting observations can be found on this subject in Tomiche, *Napoléon écrivain*, pp. 143–55, 241, and 244.

40. Napoleon I, *Commentaires*, 6 vols, Paris, Imprimerie impériale, 1867; *Correspondance de Napoléon Ier publiée par ordre de l'empereur Napoléon III*, 32 vols, Paris, Plon and Dumaine, 1858–70, XXIX, p. iii.

41. Hubert Camon, *La Bataille napoléonienne*, Paris, Chapelot, 1899, p. 10. This book is the first of a long series on Napoleonic warfare.

42. Hubert Camon, *La Guerre napoléonienne. Précis des campagnes*, 2 vols, Paris, Chapelot, 1903, I, p. 4.

43. Service historique de la Défense, Département de l'armée de Terre (SHD/DAT), series C, under the classification number 17 C 2.

44. SHD/DAT, 17 C 2, item no. 27; SHD/DAT, 17 C 3, items 50 and 53. The last item, entitled 'Instruction for the representatives of the people and the supreme commander of the army of Italy', is a minute in Junot's hand, corrected by Bonaparte, undated, but according to the analysis by the archivists of the Dépôt de la guerre probably from 5 Fructidor Year III (22 August 1795). The three items have been published under these numbers in Volume I of the Second Empire *Correspondance*, which nevertheless placed the last in July 1795.

45. Edmond Bonnal de Ganges, *Les Représentants du peuple en mission près les armées 1791–1797, d'après le Dépôt de la guerre, les séances de la Convention, les Archives nationales*, 4 vols, Paris, Arthur Savaète, 1898–9, II, pp. 164–5.

46. Jean Colin, *L'Éducation militaire de Napoléon*, Paris, Chapelot, 1900; Teissèdre, 2001, p. 295.

47. Napoleon, *Œuvres littéraires et écrits militaires*, edn prefaced and established by Jean Tulard, 3 vols, Paris, Société encyclopédique française, 1967; Claude Tchou, 2001, II, pp. 309–14. The attribution of the note to Bonaparte has hardly been followed by the historiography on Robespierre's fall. But the latter involves studies that pay little attention to military questions, as recognized by Luigi Mascilli Migliorini, *Napoléon*, trans. from the Italian, Paris, Perrin, 2004, p. 493.

48. *Mémoires pour servir à l'histoire de France, sous Napoléon, écrits à Sainte-Hélène, par les généraux qui ont partagé sa captivité, et publiés sur les manuscrits entièrement corrigés de la main de Napoléon*, 8 vols, Paris, Didot and Bossange, 1823–5 (2 vols written by Gourgaud, 6 vols written by Montholon). New edn: *Mémoires de Napoléon*, presented by Thierry Lentz, 3 vols, Paris, Tallandier, 2010–11.

49. According to the papers of the Governor of Saint Helena, Hudson Lowe, in the manuscript department of the British Museum and subsequently the British Library, Additional Manuscripts, 20.149, p. 50 (Philippe Gonnard, *Les Origines de la légende napoléonienne. L'Œuvre historique de Napoléon à Sainte-Hélène*, Paris, Calmann-Lévy, 1906, p. 35).

50. Jean Tulard and Louis Garros, *Itinéraire de Napoléon au jour le jour 1769–1821*, Paris, Tallandier, 1992, p. 510: Thursday, 15 April 1819 is mentioned as the day of the dictation of these notes; Yves de Cagny, *Archives provenant du général comte Bertrand 1773–1844. Autographes de l'empereur Napoléon Ier, du général Bertrand, des proches de l'Empereur à Sainte-Hélène... Deuxième vente*, Paris, Hôtel Drouot, Wednesday, 8 June 1983, no. 95: Napoléon Ier, Ouvrages militaires. Dictées de Sainte-Hélène, 1. Dix-huit notes sur l'ouvrage intitulé *Considérations sur l'art de la guerre*, par le général Rogniat: text of thirty-seven

pages wholly in the hand of Ali, with a page of notes and corrections made by Napoleon and twenty written corrections in the text. Ali evokes these dictations and Rogniat's book, but without giving a precise date (Louis-Étienne Saint-Denis, called Ali, *Souvenirs sur l'empereur Napoléon*, presented and annotated by Christophe Bourrachot, Paris, Arléa, 2000 [1st edn, 1926], pp. 240, 243; Gonnard, *Les Origines*, p. 75).

51. Bruno Colson, *Le Général Rogniat, ingénieur et critique de Napoléon*, Paris, ISC-Économica, 2006, pp. 539–644.

52. Gonnard, *Les Origines*, pp. 46–63; Tomiche, *Napoléon écrivain*, pp. 237–50.

53. On the imbalance between the structures of the dictations and the publication in the *Correspondance*, see Tomiche, *Napoléon écrivain*, pp. 251–8.

54. Carl von Clausewitz, *On War*, ed., trans., and introd. Michael Howard and Peter Paret, London, Everyman, 1993, VIII, 3 B.

55. Gonnard, *Les Origines*, pp. 63, 97, and 113–14.

56. Didier Le Gall, *Napoléon et le* Mémorial de Sainte-Hélène. *Analyse d'un discours*, Paris, Kimé, 2003, pp. 14, 16, 42, 46, 47.

57. E. De la Cases, *Mémorial de Sainte-Hélène ou Journal où se trouve consigné, jour par jour, ce qu'a dit et fait Napoléon durant dix-huit mois*, new edn, revised and expanded, 20 books in 10 vols, Paris, Barbezat, 1830.

58. Las Cases, *Le Mémorial . . .* , ed. Dunan.

59. Barry E. O'Meara, *Napoleon in Exile, or, A Voice from St. Helena: The Opinions and Reflections of Napoleon on the Most Important Events of his Life and Government in his Own Words*, 2 vols, Philadelphia, Jehu Burton, 1822; Gonnard, *Les Origines*, p. 148.

60. François Antommarchi, *Derniers Momens de Napoléon, ou Complément du Mémorial de Sainte-Hélène*, 2 vols, Brussels, Tarlier, 1825; Jean Tulard, Jacques Garnier, Alfred Fierro, and Charles d'Huart, *Nouvelle Bibliographie critique des mémoires sur l'époque napoléonienne écrits ou traduits en français*, new revised and expanded edn, Geneva, Droz, 1991, p. 23.

61. Gonnard, *Les Origines*, pp. 286, 292–6, 307, 308.

62. Gaspard Gourgaud, *Sainte-Hélène. Journal inédit de 1815 à 1818*, with a preface and notes by Vicomte de Grouchy and A. Guillois, 3rd edn, 2 vols, Paris, Flammarion, 1899; Gaspard Gourgaud, *Journal de Sainte-Hélène 1815–1818*, edn expanded in accordance with the original text, introduction, and notes by Octave Aubry, 2 vols, Paris, Flammarion, 1944.

63. Archives nationales (AN), Paris, Archives privées, 314, Fonds Gourgaud, carton 30 (314 AP 30), Journal de Sainte-Hélène.

64. Tulard, Garnier, Fierro, and d'Huart, *Nouvelle Bibliographie*, pp. 138–9.

65. Henri-Gatien Bertrand, *Cahiers de Sainte-Hélène*, manuscript deciphered and annotated by Paul Fleuriot de Langle, 3 vols, Paris, Sulliver and Albin Michel, 1949, 1951, and 1959.

66. AN, 390 AP 25, Cahiers de Sainte-Hélène.

67. Tulard, Garnier, Fierro, and d'Huart, *Nouvelle Bibliographie*, p. 215.

68. Gonnard, *Les Origines*, p. 306.
69. Jacques Garnier, 'Complément et supplément à la *Nouvelle Bibliographie critique des mémoires sur l'époque napoléonienne écrits ou traduits en français* de Jean Tulard', *Revue de l'Institut Napoléon*, nos. 172–3, 1996–3/4, pp. 7–80.
70. Chandler, 'General Introduction', p. 14.
71. Laurence Montroussier, *L'Éthique du chef militaire dans le* Mémorial de Sainte-Hélène, Montpellier, Université Paul-Valéry, 1998, pp. 155–6; Harold T. Parker, 'Towards Understanding Napoleon, or Did Napoleon Have a Conscience?', *Consortium on Revolutionary Europe 1750–1850, Selected Papers*, 1997, pp. 201–8.
72. Antoine Casanova, *Napoléon et la pensée de son temps: une histoire intellectuelle singulière*, Paris, La Boutique de l'Histoire, 2000, p. 221.
73. Gonnard, *Les Origines*, p. 7.
74. Napoleon I, *Manuel du chef*, p. 77.
75. Palluel, *Dictionnaire*, p. 16.
76. Steven Englund, *Napoleon: A Political Life*, New York, Scribner, 2004, chapter 3, n. 13.
77. Palluel, *Dictionnaire*, p. 17.
78. Jules Lewal, *Introduction à la partie positive de la stratégie*, notes by A. Bernède, Paris, CFHM-ISC-Économica, 2002 (1st edn, 1892), p. 121.
79. Andy Martin, *Napoleon the Novelist*, Cambridge, Polity, 2000.
80. For a more advanced study of the style, readers are referred to Tomiche, *Napoléon écrivain*, pp. 156–60, 170–84 (correspondence), 183–96 (proclamations, orders of the day, addresses), 198–207 (reports, memoirs and bulletins), and 259–74 (writings on Saint Helena).
81. Annotation will be limited to the information required to understand the text. For more detail on people, battles, and the theory of war, readers are referred to the following dictionaries: Owen Connelly, *Historical Dictionary of Napoleonic France 1799–1815*, Westport, Conn., Greenwood, 1985; Richard Holmes (ed.), *The Oxford Companion to Military History*, Oxford, Oxford University Press, 2001; Gordon Martel (ed.), *The Encyclopedia of War*, 5 vols, London, Wiley-Blackwell, 2012; George F. Nafziger, *Historical Dictionary of the Napoleonic Era*, Lanham, Md., Scarecrow Press, 2002; Alan Palmer, *An Encyclopedia of Napoleon's Europe*, New York, St Martin's Press, 1984.

BOOK I

1. Latin historian, author of *De viris illustribus*.
2. Guillaume Raynal, historian and philosopher, author of a book criticizing colonization and the clergy (1770).
3. Henri Bonnal, 'La Psychologie militaire de Napoléon', *Revue hebdomadaire des cours et conférences*, 22 February 1908, p. 423.

4. Carl von Clausewitz, *On War*, ed. and trans. with an introduction by Michael Howard and Peter Paret, London, Everyman, 1993, I, 1, p. 83. Only references to Clausewitz's *On War* will refer to the number of the book in Roman numerals and that of the chapter in Arabic numerals, followed by the page number. For all other references, the Roman numeral will refer to the number of the volume.

CHAPTER I

1. AN, 390 AP 25, Fonds Bertrand, Cahiers de Sainte-Hélène, MS of May 1817 (11 May), p. 3 (Henri-Gatien Bertrand, *Cahiers de Sainte-Hélène, janvier 1821– mai 1821*, manuscript deciphered and annotated by Paul Fleuriot de Langle, Paris, Sulliver, 1949, I, p. 223).

2. Jean Colin, *L'Éducation militaire de Napoléon*, Paris, Chapelot, 1900; Teissèdre, 2001, pp. 366–7.

3. In particular, Peter Paret, *The Cognitive Challenge of War: Prussia 1806*, Princeton, Princeton University Press, 2009, pp. 154–5.

4. Michael, Baron von Colli da Vigevano, an Austrian general 'lent' to the Kingdom of Sardinia to command its army.

5. *Corresp.*, I, no. 127, p. 128, to General Colli, Albenga, 19 germinal an IV—8 April 1796.

6. *Corresp.*, I, no. 738, p. 479, to General Beaulieu, commander-in-chief of the Austrian army of Italy, end of June 1796; no. 1484, p. 894, to Archduke Charles, Klagenfurt, 11 germinal an V—31 March 1797.

7. *Corresp.*, I, no. 955, p. 606, to General Berthier, 10 vendémiaire an V—1 October 1796.

8. *Corresp.*, II, no. 1092, p. 56, to General Wurmser, Modena, 25 vendémiaire an V—16 October 1796.

9. AN, 314 AP 30, Fonds Gourgaud, Journal de Sainte-Hélène, MS 39 (Gaspard Gourgaud, *Journal de Sainte-Hélène 1815–1818*, edn expanded in accordance with the original text, introduction and notes by Octave Aubry, 2 vols, Paris, Flammarion, 1944, II, p. 185).

10. Gourgaud, *Journal de Sainte-Hélène*, II, pp. 191–2.

11. AN, 390 AP 25, MS of 1818, p. 52 (Bertrand, *Cahiers*, II, p. 156). Henri de la Tour d'Auvergne, Vicomte de Turenne, Marshal of France. Having won several victories during the Thirty Years War, he was Louis XIV's master in the art of war. His Alsatian campaign in 1674–5 was his masterpiece.

12. A serious defeat inflicted on the Romans by Hannibal in 216 BCE.

13. AN, 314 AP 30, MS 40 (Gourgaud, *Journal*, II, p. 222).

14. The suffering endured by Masséna and 15,000 French in besieged Genoa from 19 April to 5 June 1800 was dreadful. In going to the limits of human resistance, Masséna made a major contribution to Bonaparte's success at Marengo on 14 June.

15. AN, 314 AP 30, MS 50 (Gourgaud, *Journal*, II, pp. 330 and 340).

16. *Corresp.*, XIII, no. 11281, p. 533, message to the Senate, Berlin, 19 November 1806.

17. Jonathon Riley, *Napoleon as a General*, London, Hambledon Continuum, 2007, p. 28.

18. Clausewitz, *On War*, I, 1, pp. 83–4.

19. Clausewitz, *On War*, I, 1, p. 98.

20. Claus von Clausewitz, 'Übersicht des Krieges in der Vendée 1793', *Hinterlassene Werke des Generals Carl von Clausewitz über Krieg und Kriegführung*, 10 vols, Berlin, Dümmler, 1832–37, X. We now possess a French translation: *Œuvres posthumes du général Carl von Clausewitz. Sur la guerre et la conduite de la guerre. Tome IX et tome X. Éclairage stratégique de plusieurs campagnes*, trans. G. Reber, Paris, La Maison du Dictionnaire, 2008, pp. 633–58.

21. *Mémoires pour servir à l'histoire de France, sous Napoléon, écrits à Sainte-Hélène, par les généraux qui ont partagé sa captivité, et publiés sur les manuscrits entièrement corrigés de la main de Napoléon*, 8 vols, Paris, Didot and Bossange, 1823–5 (2 vols written by Gourgaud and 6 vols written by Montholon), VI, pp. 207, 231, 246, and 255. This work will henceforth be cited as follows: *Mémoires*, Montholon (or Gourgaud), volume number, pagination.

22. *Corresp.*, VI, no. 4478, p. 5, to General Hédouville, Paris, 8 nivôse an VIII—29 December 1799.

23. 'Précis des guerres de Jules César', *Corresp.*, XXXII, p. 47.

24. 'Quatre notes sur l'ouvrage intitulé *Mémoires pour servir à l'histoire de la révolution de Saint-Domingue*', *Corresp.*, XXX, p. 526.

25. *Corresp.*, II, no. 1086, p. 47, to the executive Directory, Milan, 20 vendémiaire an V—11 October 1796.

26. [Antoine-Clair Thibaudeau], *Mémoires sur le Consulat. 1799 à 1804*, par un ancien conseiller d'État, Paris, Ponthieu et Cie, 1827, p. 396.

27. Bonaparte was appointed First Consul for ten years.

28. Note by Thibaudeau.

29. *Sic.* [Thibaudeau], *Mémoires*, pp. 390–5; italicized in the text.

30. Raymond Aron, *Paix et guerre entre les nations*, Paris, Calmann-Lévy, 1962, pp. 108–11.

31. *Corresp.*, X, no. 8282, p. 121, to General Pino, Paris, 2 pluviôse an XIII—22 January 1805.

32. *Corresp.*, XV, no. 12474, p. 151, to Talleyrand, Finkenstein, 26 April 1807, 22.00 hrs.

33. *Corresp.*, XI, no. 9561, p. 472, to Prince Joseph, Schönbrunn, 22 frimaire an XIV—13 December 1805.

34. Jean-Paul Bertaud, *Quand les enfants parlaient de gloire. L'armée au cœur de la France de Napoléon*, Paris, Aubier, 2006, pp. 42–4.

35. *Corresp.*, XI, no. 9575, p. 480, to Prince Joseph, Schönbrunn, 24 frimaire an XIV—15 December 1805.

36. *Corresp.*, XV, no. 12408, p. 91, to the King of Naples, Finkenstein, 18 April 1807.

37. AN, 390 AP 25, MS of August 1817 (26 August), p. 11 (Bertrand, *Cahiers*, I, p. 264).

38. Jacques-Olivier Boudon, *Histoire du Consulat et de l'Empire 1799–1815*, Paris, Perrin, 2003, pp. 138–40, 149, and 278–82; Thierry Lentz, *Napoléon et la conquête de l'Europe 1804–1810*, Paris, Fayard, 2002, pp. 205, 226, and 230–5; Jean Tulard, *Le Grand Empire 1804–1815*, Paris, Albin Michel, 1982, pp. 66 and 318–21.

39. Philip G. Dwyer, 'Napoleon and the Drive for Glory: Reflections on the Making of French Foreign Policy', in P. G. Dwyer (ed.), *Napoleon and Europe*, London, Pearson Education, 2001, pp. 118–35; Englund, *Napoleon*, pp. 254, 261–2; Charles Esdaile, *Napoleon's Wars: An International History, 1803–1815*, London, Allen Lane, 2007, pp. 91–2, 112–13, and 407.

40. Esdaile, *Napoleon's Wars*, pp. 130–1.

41. Englund, *Napoleon*, pp. 30, 469–70, 474.

42. *Lettres inédites de Napoléon Ier (an VIII—1815)*, published by Léon Lecestre, 2 vols, Paris, Plon, 1897, II, no. 1020, p. 248, to Prince Cambacérès, Archchancellor of the Empire, Dresden, 18 June 1813.

43. *Lettres inédites de Napoléon Ier*, II, no. 1029, p. 254, to Cambacérès, Dresden, 30 June 1813.

CHAPTER 2

1. Gunther Rothenberg, 'The Age of Napoleon', in Michael Howard, George J. Andreopoulos, and Mark R. Shulman (eds), *The Laws of War: Constraints on Warfare in the Western World*, New Haven and London, Yale University Press, 1994, pp. 86–97.

2. The accusation of violation of the territory of Parma has been challenged by François Furet and Denis Richet, *La Révolution française*, Paris, Fayard, 1973, p. 380. Similarly, the passage through the territory of Anspach in 1806 probably did not contravene the Treaty of Basle (Englund, *Napoleon*, p. 274).

3. Neutrality was not always better respected in the wars of the *ancien régime*. Prince Eugène of Savoy did not hesitate to violate Venetian territory in 1701 (Archer Jones, *The Art of War in the Western World*, Oxford, Oxford University Press, 1987, p. 283).

4. Emer de Vattel, *Le Droit des gens, ou principes de la loi naturelle, appliquée à la conduite et aux affaires des nations et des souverains*, new edn, 2 vols, Paris, Aillaud, 1830, II, p. 176 (1st edn in 1758).

5. Georges-Frédéric de Martens, *Précis du droit des gens moderne de l'Europe fondé sur les traités et l'usage. Pour servir d'introduction à un cours politique et diplomatique*, 2nd revised edn, Göttingen, Dieterich, 1801, pp. 425–6.

6. On 24 March 1797, during the pursuit of Archduke Charles's Austrians at the end of the first Italian campaign, General Samuel Köblös von Nagy-Varád defended the entrenched post of Chiusa di Pletz in Friuli. The post was stormed and Köblös taken prisoner.

7. *Corresp.*, II, no. 1632, p. 417, to the executive Directory, Goritz, 5 germinal an V—25 March 1797.

8. *Corresp.*, V, no. 3983, p. 329, before El-Arich, 2 ventôse VII—20 February 1799.

9. Djezzar-Pacha, troop commander in the province of Damietta.

10. *Corresp.*, V, no. 4035, p. 361, to the executive Directory, Jaffa, 23 ventôse an VII—13 March 1799.

11. Philip Dwyer, *Napoleon: The Path to Power, 1769–1799*, London, Bloomsbury, 2007, pp. 421–2.

12. Englund, *Napoleon*, p. 130.

13. *Corresp.*, XII, no. 10131, p. 304, to the King of Naples (Joseph Bonaparte), Saint-Cloud, 22 April 1806.

14. *Lettres inédites*, I, no. 333, p. 227, to Joseph Napoleon, King of Spain, Bordeaux, 31 July 1808, 23.00 hrs.

15. Claude-Louis-Hector, Duke of Villars, Marshal of France, the victor of Denain (1712).

16. François I at Pavia (1525) and Jean le Bon at Poitiers (1356).

17. The father of the three Horatii brothers who confronted the Curiatii brothers in Roman legend.

18. *Mémoires*, Montholon, V, pp. 275–81.

19. Dupont at Bailén (Spain) on 22 July 1808 and Sérurier at Verderio (Italy) on 28 April 1799.

20. AN, 390 AP 25, MS of 1818, p. 51 (Bertrand, *Cahiers*, II, p. 155).

21. Roger Dufraisse and Michel Kerautret, *La France napoléonienne. Aspects extérieurs 1799–1815*, Paris, Seuil, 'Nouvelle Histoire de la France contemporaine 5', 1999, p. 265.

22. *Corresp.*, VI, no. 5277, pp. 565–6, to Talleyrand, Paris, 21 nivôse an IX—11 January 1801.

23. *Corresp.*, VII, no. 5524, p. 116, appendix to a letter to Talleyrand, Paris, 22 germinal an IX—12 April 1801. The 'monsters' are those who detonated an 'infernal machine' on the rue Saint-Nicaise on the evening of 24 December 1800, shortly after the First Consul's carriage had passed. There were numerous casualties.

24. *Corresp.*, VI, no. 5216, p. 528, decision, Paris, 23 frimaire an IX—14 December 1800.

25. Emmanuel de Las Cases, *Mémorial de Sainte-Hélène où se trouve consigné, jour par jour, ce qu'a dit et fait Napoléon durant dix-huit mois*, new, revised and expanded edn, 20 vols in 10 bks, Paris, Barbezat, 1830, XVII, p. 107. References to this edition will be followed for purposes of information by the corresponding

pages in the best available edition—that of Marcel Dunan (Dunan edn): Emmanuel de Las Cases, *Le Mémorial de Sainte-Hélène*, 1st complete and critical edn, established and annotated by Marcel Dunan, 2 vols, Paris, Flammarion, 1983 (here, Dunan edn, II, p. 487). In a note, Marcel Dunan elaborates: 'The excuse of the British, incarcerating the mass of their prisoners in two or three bridges of old demasted vessels moored in the principal ports, was that it would have been difficult to monitor them effectively on land, their number having multiplied by the superiority of the British fleet out of all proportion to the captives taken by the French.'

26. Bertaud, *Quand les enfants parlaient de gloire*, p. 207.
27. Sophie Wahnich and Marc Belissa, 'Les Crimes des Anglais: trahir le droit', *Annales historiques de la Révolution française*, no. 300, April–June 1995, pp. 233–48.
28. Rothenberg, 'The Age of Napoleon', pp. 91–2.
29. 'Précis des guerres de Jules César', *Corresp.*, XXXII, p. 14.
30. *Corresp.*, III, no. 1971, pp. 157–8, notes on the events in Venice, assumed to be 12 messidor an V—30 June 1797.
31. *Corresp.*, XI, no. 9038, p. 57, to Talleyrand, camp of Boulogne, 15 thermidor an XIII—3 August 1805.
32. *Corresp.*, XIII, no. 10893, 10895, 10900, pp. 270, 271, 276, to Berthier, Murat, and Soult, Mayence, 29 September 1806.
33. *Corresp.*, XXVIII, no. 21760, p. 66, to General Caulaincourt, Paris, 3 April 1815.
34. *Corresp.*, V, no. 4198, pp. 468–9, to General Berthier, Cairo, 3 messidor an VII—21 June 1799.
35. *Inédits napoléoniens*, published by Arthur Chuquet, 2 vols, Paris, Fontemoing and de Bocard, 1913–19, I, no. 1014, p. 278, to Berthier, Dresden, 19 June 1813. The Prussian Major von Lützow had raised a free corps kitted out in black at the start of 1813. Composed of volunteers, the majority of them from German universities, this 'black corps' operated in the rear of the French army, harassing poorly defended convoys and positions. Major Enno von Colomb was Blücher's brother-in-law and likewise commanded cavalrymen who operated as partisans behind French lines in 1813.
36. *Mémoires*, Gourgaud, II, pp. 93–6.
37. The Italian Alberico Gentili (Latinized as 'Gentilis'), having taken refuge in Oxford in the late sixteenth century, was the first to free the law of war from theology. The Dutchman Hugo de Groot or Grotius lived for a long time in Paris. He published his *De jure belli ac pacis* in 1625. The Swiss Emer or Emmerich de Vattel distinguished in his book *Du droit des gens* (1758) between private wars and public wars. Even more than his predecessors, he established a body of norms intended to mitigate the woes caused by war.
38. Thus, for Napoleon's last victory at Ligny, on 16 June 1815, much damage was done to harvests, livestock, and the buildings of villages, but no civilian

casualties were signalled. The municipal archives of Sombreffe contain traces of the indemnities demanded for the damage suffered. These were paid after a report by the authorities—in the event, those of the Kingdom of the Low Countries.

39. David A. Bell, *The First Total War: Napoleon's Europe and the Birth of Warfare as We Know It*, Boston and New York, Houghton Mifflin, 2007; Jean-Yves Guiomar, *L'Invention de la guerre totale, XVIIIe–XXe siècle*, Paris, Éditions du Félin, 2004. The first of these books, while highly stimulating, exaggerates by only dealing with extreme cases.

40. In other words, all war munitions could be accused of being contraband and subject to seizure.

41. *Mémoires*, Gourgaud, II, pp. 96–100.

42. *Mémoires*, Gourgaud, II, p. 100.

CHAPTER 3

1. *Mémoires*, Montholon, V, p. 76.

2. Clausewitz, *On War*, I, 1, pp. 85–6.

3. Clausewitz, *On War*, I, 1, p. 115.

4. [Roederer, Pierre-Louis], *Œuvres du comte P.-L. Roederer . . .*, published by his son, 8 vols, Paris, Firmin Didot, 1856–9, III, p. 536.

5. Gustavus Adolphus II, King of Sweden at the start of the seventeenth century, had revolutionized tactics by slimming down the lines of infantry so as to enhance the musketeers' fire power. His victory at Breitenfeld (1631) over the troops of the German Holy Empire, during the Thirty Years War, earned him a great reputation.

6. Las Cases, *Mémorial*, vol. 18, pp. 113–14 (Dunan edn, II, p. 575).

7. Louis II of Bourbon, called the 'Great Condé', the victor at Rocroi (1643).

8. One of the greatest Austrian generals, he beat the Turks at Zenta (1697), the French at Turin (1706), and the Turks again at Peterwardein (1716) and Belgrade (1717).

9. [Rémusat], *Mémoires de madame de Rémusat 1802–1808*, published by her grandson, 7th edn, 3 vols, Paris, Calmann-Lévy, 1880, I, p. 333.

10. [Rémusat], *Mémoires de madame de Rémusat*, I, pp. 267–8. Napoleon was the author of several novels and novellas during his youth: among them were *Clisson et Eugénie* and *Le Comte d'Essex* (Napoléon Bonaparte, *Œuvres littéraires*, edn established, annotated, and presented by Alain Coelho, Nantes, Le Temps Singulier, 1979).

11. Pierre-Simon Laplace, *Essai philosophique sur les probabilités*, 5th edn (1825), prefaced by René Thom, Paris, Bourgois, 1986, p. 35. The first edition dates back to 1814 and the essentials of its content—in particular, the text cited here—had already been published in 1785 (Laplace, *Essai philosophique*, p. 240).

See also Lorraine Daston, *Classical Probability in the Enlightenment*, Princeton, Princeton University Press, 1988.

12. *Corresp.*, XXIV, no. 19028, p. 112, to Count Laplace, Vitebsk, 1 August 1812.

13. Auguste-Frédéric-Louis Viesse de Marmont, *De l'esprit des institutions militaires*, prefaced by Bruno Colson, Paris, ISC-FRS-Économica, 2001 (1st edn, 1845) p. 135. This superior perception on the part of Napoleon is captured well in Englund, *Napoleon*, p. 103.

14. *Corresp.*, XII, no. 10325, p. 442, to Joseph, King of Naples, Saint-Cloud, 6 June 1806.

15. [Pierre-Louis Roederer], *Autour de Bonaparte. Journal du comte P.-L. Roederer, ministre et conseiller d'État. Notes intimes et politiques d'un familier des Tuileries*, introduction and notes by Maurice Vitrac, Paris, Daragon, 1909, p. 250.

16. William Duggan, *Napoleon's Glance: The Secret of Strategy*, New York, Nation Books, 2002, pp. 3–4.

17. Clausewitz, *On War*, I, 3, p. 118.

18. Duggan, *Napoleon's Glance*, p. 6.

19. W. Duggan, *The Art of What Works: How Success Really Happens*, New York, McGraw-Hill, 2003, p. 6.

20. Nada Tomiche, *Napoléon écrivain*, Paris, Armand Colin, 1952, p. 17; Englund, *Napoleon*, pp. 41–2; Jacques Jourquin, 'Bibliothèques particulières de Napoléon', in Jean Tulard (ed.), *Dictionnaire Napoléon*, Paris, Fayard, 1987, pp. 214–15. Annie Jourdan speaks of a 'gorging on knowledge' in her *Napoléon. Héros, imperator, mécène*, Paris, Aubier, 1998, p. 22.

21. W. Duggan, *Coup d'œil: Strategic Intuition in Army Planning*, Carlisle, Pa., Strategic Studies Institute, 2005 (<http://www.StrategicStudiesInstitute. army.mil/>, accessed 2 December 2009), pp. v, 1–2.

22. Ulrike Kleemeier, 'Moral Forces in War', in Hew Strachan and Andreas Herberg-Rothe (eds), *Clausewitz in the Twenty-First Century*, Oxford, Oxford University Press, 2007, p. 114.

23. *Corresp.*, XVIII, no. 15144, p. 525, to Eugène Napoléon, Burghausen, 30 April 1809.

24. Chaptal (Jean-Antoine), *Mes Souvenirs sur Napoléon*, published by his great-grandson A. Chaptal, Paris, Plon, 1893, pp. 295–6. New edn presented and annotated by Patrice Gueniffey, Paris, Mercure de France, 2009.

25. *Corresp.*, XIV, no. 11658, p. 211, to the King of Naples, Warsaw, 18 January 1807.

26. François Antommarchi, *Derniers Momens de Napoléon, ou Complément du Mémorial de Sainte-Hélène*, 2 vols, Brussels, Tarlier, 1825, I, p. 321.

27. *Corresp.*, XVII, no. 14283, p. 480, notes on Spanish affairs, Saint-Cloud, 30 August 1808.

28. *Mémoires*, Montholon, II, p. 90.

29. *Mémoires*, Montholon, IV, p. 345.

30. 'Campagnes d'Égypte et de Syrie', *Corresp.*, XXX, p. 176.

31. Clausewitz, *On War*, I, 3, p. 121.
32. *Corresp.*, XII, no. 9810, p. 44, to Berthier, Paris, 14 February 1806.
33. [Auguste-Frédéric-Louis Viesse de Marmont], *Mémoires du duc de Raguse*, 8 vols, Paris, Perrotin, 1857, V, p. 256.
34. Jean-Baptiste Vachée, *Napoléon en campagne*, Paris, Berger-Levrault, 1913; Bernard Giovanangeli Éditeur, 2003, p. 54.
35. Bonnal, 'La Psychologie militaire de Napoléon', p. 434.
36. [Roederer], *Œuvres*, III, p. 537.
37. Riley, *Napoleon as a General*, pp. 11-12 and 177.
38. I am grateful to Martin Motte for his comments on this point.
39. *Mémoires*, Montholon, V, p. 213.
40. Clausewitz, *On War*, I, 3, p. 117.
41. AN, 314 AP 30, MS 50 (Gourgaud, *Journal*, II, p. 347).
42. Tomiche, *Napoléon écrivain*, pp. 9, 11-13.
43. F. G. Healey, 'La Bibliothèque de Napoléon à Sainte-Hélène. Documents inédits trouvés parmi les *Hudson Lowe Papers*', *Revue de l'Institut Napoléon*, nos. 73-4, 1959-60, p. 174.
44. Barry E. O'Meara, *Napoleon in Exile*, II, pp. 146-7; italics in original.
45. Eugène de Beauharnais, Viceroy of Italy.
46. Las Cases, *Mémorial*, vol. 3, pp. 244-7 (Dunan edn, I, pp. 278-9); italics in the original.
47. Charles-François-Tristan de Montholon, *Récits de la captivité de l'empereur Napoléon à Sainte-Hélène*, 2 vols, Paris, Paulin, 1847, II, pp. 240-1.
48. Clausewitz, *On War*, I, 3, p. 118.
49. *Corresp.*, X, no. 8832, p. 474, to Jérôme Bonaparte, Milan, 13 prairial an XIII—2 June 1805.
50. *Corresp.*, XIII, no. 10558, p. 9, to the King of Naples, Saint-Cloud, 28 July 1806.
51. Husband of the celebrated 'Madame Sans-Gêne', Marshal Lefebvre was uneducated, unlike General Mathieu Dumas, author of several books. Lefebvre knew how to command and conduct his men on a battlefield. Dumas essentially followed a career in the general staff, political assemblies, ministries, and diplomacy.
52. AN, 314 AP 30, MS 40 (Gourgaud, *Journal*, II, pp. 325-6).
53. AN, 390 AP 25, MS of 1819, p. 160 (Bertrand, *Cahiers*, II, p. 445, where the text is placed in 1820).
54. *Corresp.*, XV, no. 12511, p. 178, to Prince Jérôme, Finkenstein, 2 May 1807.
55. Jean Guitton, *La Pensée et la guerre*, Paris, Desclée de Brouwer, 1969, pp. 76-7.
56. With 41,000 men, Charles de Rohan, Prince of Soubise, was defeated by Frederick II's 22,000 Prussians at Rossbach on 5 November 1757. This did not prevent him from becoming Marshal of France the following year. He had the reputation of a sycophant who owed his positions to Louis XV's favourites.
57. AN, 314 AP 30, MS 39 (Gourgaud, *Journal*, II, p. 80).

58. AN, 390 AP 25, MS of 1818 (November), p. 75 (Bertrand, *Cahiers*, II, p. 197).

59. *Corresp.*, XV, no. 12641, p. 264, note for the major general, Finkenstein, 24 May 1807.

60. *Corresp.*, IX, no. 7766, pp. 368–9, to Rear-Admiral Ver Huell, commander of the Batavian flotilla, Saint-Cloud, 1 prairial an XII—21 May 1804.

61. [Roederer], *Œuvres*, III, p. 536.

62. *Corresp.*, IX, no. 7818, p. 399, to Vice-Admiral Ganteaume, Saint-Cloud, 4 messidor an XII—23 June 1804.

63. *Corresp.*, XIII, no. 10646, p. 72, to General Mouton, Saint-Cloud, 14 August 1806. General Mouton was made Count of Lobau after his heroic conduct during the 1809 campaign in Austria. He commanded an army corps in 1813 and 1815. Louis-Philippe appointed him Marshal of France. Zacharie Allemand distinguished himself in the Indies in Suffren's fleet, but was reprimanded for abuse of power in 1797. He was disgraced in 1813 on account of his harsh, irascible character.

64. *Corresp.*, XII, no. 10350, p. 458, to Prince Eugène, Saint-Cloud, 11 June 1806.

65. AN, 390 AP 25, MS of 1821 (9 March), p. 21 (Bertrand, *Cahiers*, III, p. 94).

66. [Roederer], *Autour de Bonaparte*, pp. 133–4.

67. Laurent Gouvion Saint-Cyr, *Mémoires pour servir à l'histoire militaire sous le Directoire, le Consulat et l'Empire*, 4 vols, Paris, Anselin, 1831, III, pp. 48–9.

68. *Corresp.*, XXV, no. 20090, p. 363, to Bertrand, Liegnitz, 6 June 1813.

69. In 1795, Barthélemy-Louis-Joseph Schérer commanded the army of the Pyrénées-Orientales and then that of Italy, to the head of which Napoleon succeeded. Minister of War from 1797 to 1799, he was again in command in Italy in 1799, but was defeated there.

70. 'Précis des événements militaires arrivés pendant les six premiers mois de 1799', *Corresp.*, XXX, p. 266. This clearly refers to Prince Eugène of Savoy, not Eugène de Beauharnais.

71. 'Campagnes d'Italie (1796–1797)', *Corresp.*, XXIX, p. 187.

72. Baron Jean-Pierre Beaulieu de Marconnay commanded the Austrian troops at the start of the first Italian campaign and was beaten at Lodi.

73. *Corresp.*, I, no. 366, p. 251, to Citizen Carnot, Plaisance, 20 floréal an IV—9 May 1796.

74. AN, 314 AP 30, MS 35 (Gourgaud, *Journal*, II, p. 51).

75. Montholon, *Récits de la captivité*, II, p. 361.

76. Alexander Suvorov initially distinguished himself against the Turks and forged a great reputation. He commanded the coalition army that invaded Italy in 1799 and won several victories, including Novi (15 August 1799). After the defeat of part of his forces at Zurich, he had to retreat in bad conditions and died shortly afterwards.

77. 'Précis des événements militaires arrivés pendant les six premiers mois de 1799', *Corresp.*, XXX, p. 269.

78. Montholon, *Récits de la captivité*, II, p. 106.

79. Jean Rapp, cavalry general and Napoleon's aide de camp. He led the charge of the Guard's cavalry at Austerlitz against his Russian counterpart and came to present Prince Repnin as a prisoner to Napoleon, with the captured flags. The scene was immortalized by the painter Gérard.

80. AN, 390 AP 30, MS 39 (Gourgaud, *Journal*, II, p. 60).

81. AN, 390, AP 25, MS of June 1817 (15 June), p. 7 (Bertrand, *Cahiers*, I, p. 234).

82. General Jean-Louis Reynier commanded the 2nd corps of the army of Naples.

83. *Corresp.*, XII, no. 10325, p. 440, to the King of Naples, Saint-Cloud, 6 June 1806.

84. Don Joachim, a Spanish general of Irish origin, during the surrender of Valencia, 9 January 1812.

85. Alexander I, Tsar of Russia. The sentence that follows refers to the meetings at Tilsit in 1807.

86. AN, 390 AP 25, MS of January–September 1819, pp. 52–3 (Bertrand, *Cahiers*, II, pp. 227–8).

87. Maurice de Saxe, *Mes Rêveries . . .* , presented by Jean-Pierre Bois, Paris, CFHM-ISC-Économica, 2002 (1st edn, 1756), p. 159.

88. Alexander the Great, King of Macedonia.

89. *Mémoires*, Gourgaud, II, pp. 189–92.

90. Generals François Haxo and Joseph Rogniat commanded, respectively, the engineers of the imperial guard and that of the Grande Armée in 1813. Pierre Fontaine was the great architect of the First Empire. Among other works, he rearranged the Tuileries and erected the Arc de Triomphe du Carrousel.

91. At the request of Napoleon, who wanted a manual on fortification for the École de l'artillerie et du génie de Metz, Lazare Carnot had published a treatise entitled *De la défense des places fortes* in 1810. The work was not well received and the Emperor did not have it distributed (Lazare Carnot, *Révolution et mathématique*, introduction by Jean-Paul Charnay, 2 vols, Paris, L'Herne, 1984–5, I, pp. 132–3).

92. An artillery officer, Louis Évain spent most of his career in the offices of the War Ministry.

93. AN, 314, AP 30, MS 17 (Gourgaud, *Journal*, I, pp. 212–14).

CHAPTER 4

1. Blank in the text.

2. AN, 390, AP 25, MS of January–April 1821 (20 February), p. 9 (Bertrand, *Cahiers*, III, pp. 66–7).

3. *Mémoires*, Montholon, V, p. 106.

4. AN, 390, AP 25, MS of 1818 (October), p. 62 (Bertrand, *Cahiers*, II, p. 177).

5. *Corresp.*, XI, no. 9738, p. 573, to Prince Joseph, Paris, 2 February 1806.

6. [Louis Marchand], *Mémoires de Marchand, premier valet de chambre et exécuteur testamentaire de l'Empereur*, ed. J. Bourguignon and H. Lachouque, 2 vols,

Paris, Plon, 1952–5, I, pp. 97–8. New edn in one volume, Paris, Tallandier, 2003.

7. *Mémoires*, Gourgaud, II, pp. 188–9.
8. Jean Tulard and Louis Garros, *Itinéraire de Napoléon au jour le jour 1769–1821*, Paris, Tallandier, 1992, pp. 52–3.
9. Tulard and Garros, *Itinéraire de Napoléon*, p. 314.
10. Tulard and Garros, *Itinéraire de Napoléon*, p. 319.
11. Henry Houssaye, *1814*, Paris, Perrin, 1888; Étrépilly, Presses du Village and Christian de Bartillat, 1986, pp. 308–9.
12. Louis Chardigny, *L'Homme Napoléon*, Paris, Perrin, 1987, p. 183 (new edn, 2010).
13. He had been previously—in particular, at the Battle of Abukir in Egypt.
14. AN, 390 AP 25, MS of January–September 1819, p. 108 (Bertrand, *Cahiers*, I, pp. 340–1).
15. AN, 390 AP 25, MS of July 1817 (5 July), p. 4 (Bertrand, *Cahiers*, I, p. 243).
16. David Chandler, 'Napoleon and Death', *Napoleonic Scholarship: The Journal of the International Napoleonic Society*, vol. 1, no. 1, April 1997, <http://www.napoleon-series.org/ins/scholarship97/c_death.html> (accessed 22 October 2008).
17. *Corresp.*, XIV, no. 11800, p. 297, note, Eylau, 12 February 1807.
18. *Corresp.*, XIV, no. 11813, p. 304, to the Empress, Eylau, 14 February 1807.
19. Clément-Wenceslas-Lothaire de Metternich, *Mémoires, documents et écrits divers laissés par le prince de Metternich, Chancelier de Cour et d'État*, published by his son Richard de Metternich, categorized and assembled by M. A. de Klinkowstroem, 8 vols, Paris, Plon, 1880–4, I, pp. 151–2.
20. AN, 314 AP 30, MS 35 (Gourgaud, *Journal*, II, p. 37).
21. King Charles XII of Sweden was one of the most renowned warlords of the early eighteenth century. He triumphed over Peter the Great's Russians at Narva (1700), but imprudently invaded Russia and was defeated at Poltava (1709). He was killed at the siege of Fredrikshald on 11 December 1718.
22. AN, 390 AP 25, MS of September 1817 (17 September), p. 7 (Bertrand, *Cahiers*, I, p. 273).
23. *Corresp.*, II, no. 1198, p. 120, to General Clarke, Verona, 29 brumaire an V—19 November 1796.
24. *Corresp.* IV, no. 3046, pp. 361–2, to Mme Brueys (wife of the admiral, killed at Abukir), Cairo, 2 fructidor an VI—19 August 1798.

CHAPTER 5

1. *Corresp.*, III, no. 222, p. 357, to the Minister of Foreign Relations, Passariano, 10 vendémiaire an VI—1 October 1797.
2. *Corresp.*, XIV, no. 11806, p. 300, to Lannes, Eylau, 12 February 1807.

3. With victory in this battle on 21 June 1813, the Duke of Wellington ejected Joseph Bonaparte from the Spanish throne for good.

4. From 17 to 23 April 1809, Napoleon restored a problematic situation in Bavaria by forcing Archduke Charles of Austria to pull back. His skilful manœuvres culminated in the victory of Eckmühl, but the Austrian army was able to escape and it required the tough battles of Essling and Wagram to impose peace.

5. AN, 314 AP 30, MS 39 (Gourgaud, *Journal*, II, p. 157).

6. [Caulaincourt, Armand-Louis-Augustin de], *Mémoires du général de Caulaincourt, duc de Vicence, grand écuyer de l'Empereur*, introduction and notes by Jean Hanoteau, 3 vols, Paris, Plon, 1933, II, p. 315.

7. *Corresp.*, VI, no. 4449, p. 38, to French soldiers, Paris, 4 nivôse an VIII— 25 December 1799.

8. Las Cases, *Mémorial*, vol. 10, pp. 21–2 (Dunan edn, I, p. 681).

9. Clausewitz, *On War*, I, 5, p. 134.

10. Mathieu-Louis Molé, state counsellor, director general of Ponts-et-Chaussées, Minister of Justice in 1813.

11. [Mathieu Molé and] Marquis de Noailles, *Le Comte Molé (1781–1855). Sa vie, ses mémoires*, 6 vols, Paris, Champion, 1922–30, I, pp. 131–2.

12. The published text reads 'fifty', but this must be a copy error from a confusion between 30 and 50 in reading the manuscript.

13. 'Dix-huit notes sur l'ouvrage intitulé *Considérations sur l'art de la guerre*', *Corresp.*, XXXI, p. 304.

14. *Corresp.*, V, no. 3424, p. 33, to Berthier (for putting in army orders), Cairo, 14 vendémiaire an VII—5 October 1798.

15. *Corresp.*, VIII, no. 6400, p. 79, to Murat, commander-in-chief of French troops in Italy, Saint-Cloud, 6 brumaire an XI—28 October 1802.

16. *Corresp.*, XI, no. 9105, p. 110, decision, camp of Boulogne, 4 fructidor an XIII—22 August 1805.

17. *Corresp.*, XIII, no. 10709, pp. 117–18, to Prince Eugène, Saint-Cloud, 30 August 1806.

18. *Corresp.*, XIV, no. 11412, p. 47, to Clarke, Posen, 8 December 1806.

19. *Corresp.*, XVIII, no. 14748, p. 246, to Prince Camille Borghese, governor general of the departments beyond the Alps, Paris, 27 January 1809.

20. André Corvisier, *La Guerre. Essais historiques*, Paris, Presses Universitaires de France, 1995, pp. 167–9. The ratio was only reversed in the 1870 war for Germany and the First World War for France.

21. *Corresp.*, IX, no. 7139, p. 7, to Davout, commander of the camp at Bruges, Paris, 5 vendémiaire an XII—28 September 1803.

22. *Corresp.*, XXII, no. 18041, p. 411, to Davout, commander of the army of Germany in Hamburg, Saint-Cloud, 16 August 1811.

23. *Corresp.*, XXIII, no. 18723, p. 431, to General Clarke, Dresden, 26 May 1812.

CHAPTER 6

1. Gérald Arboit, *Napoléon et le renseignement*, Paris, Centre français de recherche sur le renseignement, Historical Note no. 27, August 2009 (<www.cf2r.org>).
2. *Corresp.*, XIII, no. 10672, p. 87, to Joseph, King of Naples, Rambouillet, 20 August 1806.
3. *Guerre d'Espagne. Extrait des souvenirs inédits du général Jomini (1808–1814)*, published by Ferdinand Lecomte, Paris, L. Baudouin, 1892, p. 83.
4. Archduke Charles was commander-in-chief of the Austrian forces in the last phase of the first Italian campaign in 1797. Napoleon found himself facing him again in 1809, from Abensberg to Wagram.
5. Very difficult to decipher at this point, the manuscript probably refers to the general of engineers Simon Bernard, aide de camp to the Emperor and present at Ligny on 16 June 1815.
6. AN, 390, AP 25, MS of January–April 1821, p. 1 (Bertrand, *Cahiers*, III, pp. 29–30).
7. *Corresp.*, XVII, no. 14347, to Joseph Napoleon, King of Spain, Kaiserslautern, 24 September 1808.
8. Count Dagobert Sigismond von Wurmser succeeded Beaulieu at the head of the Austrian forces in Italy. He was beaten at Castiglione and Bassano, before being trapped in Mantua and surrendering.
9. *Corresp.*, II, no. 1632, p. 418, to the executive Directory, Goritz, 5 germinal an V—25 March 1797.
10. Bonaparte and Josephine had met again on 25 July 1796 and gone to Brescia. On 28 July, they left Brescia at 22.00 hrs for Castelnuovo. On the 29th, they were at Peschiera early in the morning (Tulard and Garros, *Itinéraire de Napoléon*, pp. 83–4). These operations resulted in the battles of Lonato and Castiglione.
11. Paul Guillaume, brigadier-general, commander at Peschiera. He was blockaded there by the Austrians on 30 July 1796 and released by Masséna on 6 August.
12. Colonel and aide de camp to the Tsar, Prince Czernitchev (or Tchernitchev) was sent to Paris to attend the marriage of Napoleon and Marie-Louise in 1810. His mission was also that of a spy. He was able to get his hands on an inventory of French forces and sent regular reports, thanks to his relations with an assistant at the War Ministry, whose identity was belatedly discovered by the police (Thierry Lentz, *L'Effondrement du système napoléonien 1810–1814*, Paris, Fayard, 2004, pp. 231, 245–7).
13. AN, 390 AP 25, MS of 1818 (October), pp. 63–4 (Bertrand, *Cahiers*, II, pp. 178–9).
14. *Corresp.*, IV, no. 2540, p. 811, to Vice-Admiral Brueys, Paris, 3 floréal an VI—22 April 1798.

15. *Corresp.*, X, no. 8787, p. 444, to Vice-Admiral Decrès, Milan, 5 prairial an XIII—25 May 1805.

16. *Corresp.*, XXIII, no. 18503, p. 230, to Marmont via Berthier, Paris, 18 February 1812.

17. *Corresp.*, XVII, no. 14276, pp. 470–1, observations on Spanish affairs, Saint-Cloud, 27 August 1808.

18. *Corresp.*, XVII, no. 14283, p. 479, observations on Spanish affairs, Saint-Cloud, 30 August 1808.

19. *Corresp.*, XIX, no. 15388, p. 141, to Eugène, Schönbrunn, 20 June 1809, 10.00 hrs.

20. José Fernández Vega, 'War as "Art": Aesthetics and Politics in Clausewitz's Social Thinking', in Strachan and Herberg-Rothe (eds), *Clausewitz in the Twenty-First Century*, p. 130.

21. *Corresp.*, V, no. 3605, p. 128, to Berthier, Cairo, 21 brumaire an VII—11 November 1798.

22. *Corresp.*, XXIII, no. 18727, p. 436, to Jérôme Napoléon, Dresden, 26 May 1812.

CHAPTER 7

1. *Corresp.*, III, no. 2292, p. 370, to Talleyrand, Minister of Foreign Relations, Passariano, 16 vendémiaire an VI—7 October 1797.

2. *Corresp.*, XVII, no. 14283, pp. 447–8, note on Spanish affairs, Saint-Cloud, 30 August 1808.

3. *Corresp.*, XXI, no. 17389, p. 420, to Decrès, Navy Minister, Paris, 26 February 1811.

4. *Mémoires*, Montholon, V, p. 21.

5. Alan D. Beyerchen, 'Clausewitz, Nonlinearity and the Unpredictability of War', *International Security*, vol. 17, no. 3, Winter 1992–3, pp. 59–90: here pp. 77–80.

6. *Corresp.*, IX, no. 8018, p. 524, to Vice-Admiral Decrès, La Hague, 25 fructidor an XII—12 September 1804.

7. *Corresp.*, X, no. 8897, p. 529 to Vice-Admiral Decrès, Verona, 27 prairial an XIII—16 June 1805.

8. *Corresp.*, XVI, no. 13652, p. 418, to the Grand Duke of Berg, Paris, 14 March 1808.

9. *Corresp.*, XII, no. 9997, p. 204, to Prince Joseph, Paris, 20 March 1806.

10. 'Quatre notes sur l'ouvrage intitulé *Mémoires pour servir à l'histoire de la révolution de Saint-Domingue*', *Corresp.*, XXX, p. 536.

11. *Corresp.*, XXVI, no. 20612, p. 229, to Marshal Macdonald, Dresden, 22 September 1813, 10.00 hrs.

12. Thierry Widemann, *L'Antiquité dans la guerre au siècle des Lumières. Représentation de la guerre et référence antique dans la France du XVIIIe siècle*, doctoral thesis in

history supervised by François Hartog, Paris, École des hautes études en sciences sociales, 2009, pp. 251–2.

13. Beyerchen, 'Clausewitz...', p. 77.
14. Jacques de Lauriston had served in the artillery, general staffs, and diplomacy before commanding an army corps in 1813. A brilliant cavalry officer, Jean-Toussaint Arrighi de Casanova was a cousin of Napoleon's by marriage. He commanded the 3rd cavalry corps and was also governor of Leipzig in 1813.
15. *Corresp.*, XVII, no. 14328, p. 515, notes for Joseph Napoléon, King of Spain, Saint-Cloud, 15 September 1808.
16. *Corresp.*, XXV, no. 19776, p. 133, to Lauriston, commanding the 5th corps of the Grande Armée at Koenigsborn, Paris, 27 March 1813.
17. *Corresp.*, XXVI, no. 20676, pp. 278–9, to Berthier, Dresden, 2 October 1813.
18. Clausewitz, *On War*, I, 6, p. 136.
19. AN, 390 AP 25, MS of January–September 1819, p. 127 (Bertrand, *Cahiers*, II, p. 322).
20. Las Cases, *Mémorial*, vol. 18, p. 124 (Dunan edn, II, p. 578).
21. *Corresp.*, III, no. 2259, p. 342, to Talleyrand, Passariano, 5 vendémiaire an VI—26 September 1797.
22. [Rémusat], *Mémoires*, I, p. 270.
23. Beyerchen, 'Clausewitz...', p. 74.
24. Guitton, *La Pensée et la guerre*, p. 90.
25. *Mémoires*, Montholon, V, pp. 116–17.
26. Las Cases, *Mémorial*, vol. 5, p. 4 (Dunan edn, I, p. 367).
27. *Corresp.*, XIX, no. 15322, p. 89, to Jérôme, King of Westphalia, commanding the 10th corps of the army of Germany, Schönbrunn, 9 June 1809.
28. *Corresp.*, XVIII, no. 15144, pp. 524–5, to Eugène Napoléon, Burghausen, 30 April 1809.
29. *Corresp.*, II, no. 1582, p. 388, to General Joubert, Sacile, 25 ventôse an V—15 March 1797.
30. *Corresp.*, XIX, no. 15381, pp. 136–7, to Marmont, commanding the army of Dalmatia, Schönbrunn, 19 June 1809, midday.
31. General Louis Baraguey d'Hilliers served in the army of Catalonia from 22 August 1810.
32. *Corresp.*, XXI, no. 16965, pp. 157–8, to Clarke, Fontainebleau, 29 September 1810.
33. *Corresp.*, XXIII, no. 18784, p. 484, to Berthier, Prince of Neuchâtel and Wagram, Danzig, 11 June 1812.
34. *Corresp.*, XXIV, no. 18936, p. 40, the Prince of Neuchâtel to the Duke of Tarente, Vilna, 9 July 1812.

CONCLUSION TO BOOK I

1. Of the works on these campaigns written by Clausewitz, one has existed in English translation since the mid-nineteenth century and is available in several

current editions: *The Campaign in Russia*, London, J. Murray, 1843. Two translations of Clausewitz's 1815 campaign have recently been published: *On Waterloo: Clausewitz, Wellington, and the Campaign of 1815*, trans. and ed. Christopher Bassford, Daniel Moran, and Gregory W. Pedlow, Charleston, SC, Clausewitz.com, 2010; *On Wellington: A Critique of Waterloo*, trans. and ed. Peter Hofschröer, Norman, University of Oklahoma Press, 2010. Extracts from Clausewitz's critique of the 1806 and 1814 campaigns appear in Carl von Clausewitz, *Historical and Political Writings*, ed. and trans. Peter Paret and Daniel Moran, Princeton, Princeton University Press, 1992.

2. Riley, *Napoleon as a General*, pp. 4–5 and 15–17.
3. Clausewitz, *On War*, I, 6, p. 136.
4. Clausewitz, *On War*, I, 8, p. 141.

BOOK II

CHAPTER I

1. The mathematician Louis de Lagrange had in particular chaired the commission that established the system of weights and measures in 1790.
2. AN, 314 AP 30, MS 39 (Gourgaud, *Journal*, II, pp. 77–8).
3. Clausewitz, *On War*, II, 1, p. 146.
4. Paret, *The Cognitive Challenge of War*, p. 114.
5. Hervé Coutau-Bégarie, *Traité de stratégie*, 5th edn, Paris, ISC-Économica, 2006, p. 106.
6. 'Dix-huit notes sur l'ouvrage intitulé *Considérations sur l'art de la guerre*', *Corresp.*, XXXI, pp. 321–2.
7. *Corresp.*, XXXI, p. 410.
8. Wallace P. Franz, 'Grand Tactics', *Military Review*, 61, 1981–12, pp. 32–9; Claus Telp, *The Evolution of Operational Art, 1740–1813: From Frederick the Great to Napoleon*, London, Frank Cass, 2005, pp. 3–4, 96–7.
9. 'Dix-huit notes sur l'ouvrage intitulé *Considérations sur l'art de la guerre*', *Corresp.*, XXXI, p. 380.
10. 'Notes sur le *Précis des événements militaires ou Essais historiques sur les campagnes de 1799 à 1814*', *Corresp.*, XXX, p. 496.
11. *Corresp.*, XXX, p. 500.
12. The Battle of Moskowa or Borodino on 7 September 1812.
13. Napoleon's target is General Rogniat, whose book he is criticizing.
14. 'Dix-huit notes sur l'ouvrage intitulé *Considérations sur l'art de la guerre*', *Corresp.*, XXXI, pp. 337–8.
15. *Mémoires*, Montholon, I, p. 296.
16. *Mémoires*, Montholon, II, pp. 201–3.
17. Clausewitz, *On War*, I, 3, p. 130. Another allusion can be found in II, 2, p. 170: 'Experience, with its wealth of lessons, will never produce a *Newton* or an

Euler, but it may well bring forth the higher calculations of a *Condé* or a *Frederick*.' Clausewitz refers to the passage again in connection with the difficulty of establishing the objective of war (VIII, 3, p. 708): 'Bonaparte was quite right when he said that Newton himself would quail before the algebraic problems it could pose.'

18. Ami-Jacques Rapin, *Jomini et la stratégie. Une approche historique de l'œuvre*, Lausanne, Payot, 2002, p. 27.

19. [Caulaincourt], *Mémoires*, II, p. 66.

20. Charles of Austria, *Principes de la stratégie, développés par la relation de la campagne de 1796 en Allemagne*, trans. from the German and annotated by General Jomini and J.-B.-F. Koch, 3 vols, Paris, Magimel, Anselin et Pochard, 1818.

21. An allusion to Jomini, who was struck by Leuthen's account as regards tactics and Castiglione's as regards strategy (Jean-Jacques Langendorf, *Faire la guerre: Antoine-Henri Jomini*, 2 vols, Geneva, Georg, 2001–4, II, pp. 202–3).

22. AN, 390 AP 25, MS of 1818 (December), pp. 85–6 (Bertrand, *Cahiers*, II, pp. 212–13).

23. Clausewitz, *On War*, II, 1, p. 146.

24. Clausewitz, *On War*, III, 1, p. 207.

25. Coutau-Bégarie, *Traité de stratégie*, p. 59. Paul-Gédéon Joly de Maizeroy served in the War of the Austrian Succession and the Seven Years War. The author of several books, he introduced the substantive 'stratégique', and then 'stratégie', which he situated between politics and grand tactics. Heinrich Dietrich von Bülow belonged to an illustrious family of Brandenburg. Cultivated but eccentric, he had an eventful life. His *Spirit of the System of Modern Warfare* (1801) had a major impact.

26. AN, 390 AP 25, MS of January–September 1819 (20 October), p. 136 (Bertrand, *Cahiers*, II, p. 405).

CHAPTER 2

1. *Mémoires*, Montholon, II, pp. 51–2.

2. Spenser Wilkinson, *The Rise of General Bonaparte*, Oxford, Clarendon Press, 1930, p. 144.

3. Henry Lloyd, *The History of the Late War in Germany between the King of Prussia and the Empress of Germany and her Allies*; re-edition of the original text in *War, Society and Enlightenment: The Works of General Lloyd*, ed. Patrick J. Speelman, Boston and Leiden, Brill, 2005, p. 14.

4. A military engineer, Pierre de Bourcet took part in the War of the Austrian Succession and the Seven Years War. In 1775, he wrote *Principes de la guerre de montagnes*, published a century later (Paris, Imprimerie nationale, 1888). Of Hanoverian origin, Gerhard von Scharnhorst was the great reformer of the Prussian army after the defeat at Jena (1806). He was also Clausewitz's spiritual

father (Azar Gat, *The Origins of Military Thought: From the Enlightenment to Clausewitz*, Oxford, Clarendon Press, 1989, p. 165).

5. C. von Clausewitz, 'Considérations sur l'art' (*c*.1820–25), in C. von Clause-witz, *De la Révolution à la Restauration. Écrits et lettres*, ed. M.-L. Steinhauser, Paris, Gallimard, 1976, p. 131.

6. 'Dix-huit notes sur l'ouvrage intitulé *Considérations sur l'art de la guerre*', *Corresp.*, XXXI, p. 365.

7. Duggan, *Coup d'œil*, p. 4.

8. *Corresp.*, XV, no. 12416, pp. 108 and 110, observations on a plan to establish a special school of literature and history at the Collège de France, Finkenstein, 19 April 1807.

9. Gat, *The Origins of Military Thought*, p. 166.

10. [Roederer], *Autour de Bonaparte*, p. 6.

11. *Corresp.*, XV, no. 12465, p. 145, to Prince Jérôme, Finkenstein, 24 April 1807.

12. Jean Colin, *L'Éducation militaire de Napoléon*, Paris, Chapelot, 1900; Teissèdre, 2001, pp. 118–26, 137. Antoine du Pas, Marquis de Feuquière was a lieuten-ant-general under Louis XIV and made his *Mémoires* a veritable treatise on the art of war, based on a critical history of recent wars. Chevalier Jean du Teil published *De l'usage de l'artillerie nouvelle dans la guerre de campagne* in 1778.

13. For a brief presentation of these authors and their writings, readers are referred to my *L'Art de guerre de Machiavel à Clausewitz*, Namur, Presses Universitaires de Namur, 1999.

14. Englund, *Napoleon*, p. 147.

15. Machiavelli, *The Art of War*, trans. Ellis Farneworth with an introduction by Neal Wood, Indianapolis, Bobbs-Merrill, 1965.

16. Bertrand, *Cahiers*, I, p. 158.

17. Jean Chagniot, *Le Chevalier de Folard. La stratégie de l'incertitude*, Paris and Monaco, Éditions du Rocher, 1997.

18. AN, 390 AP 25, MS of January 1817 (12 January), p. 8 (Bertrand, *Cahiers*, I, p. 181).

19. Daniel Reichel, *Davout et l'art de la guerre. Recherches sur la formation, l'action pendant la Révolution et les commandements du maréchal Davout, duc d'Auerstaedt, prince d'Eckmühl (1770–1823)*, preface by André Corvisier, Neuchâtel and Paris, Delachaux and Niestlé, 1975, pp. 243–52.

20. Antommarchi, *Derniers Momens de Napoléon*, I, p. 115.

21. Antommarchi, *Derniers Momens de Napoléon*, II, p. 30.

22. M. de Saxe, *Mes Rêveries . . .*, presentation by J.-P. Bois, Paris, CFHM-ISC-Économica, 2002.

23. AN, 390 AP 25, MS of January 1817 (12 January), p. 8 (Bertrand, *Cahiers*, I, p. 181).

24. *Corresp.*, XXIII, no. 18418, pp. 159–60, to Marshal Berthier, Paris, 6 January 1812.

25. AN, 314 AP 30, MS 40 (Gourgaud, *Journal*, II, p. 267).

26. AN, 314 AP 30, MS 40 (Gourgaud, *Journal*, II, pp. 268–9).
27. For the origins and set of Frederick II's military writings, readers can consult Christopher Duffy, *Frederick the Great: A Military Life*, London and New York, Routledge, 1985, pp. 76–80 and 403 (index).
28. Jeroom Vercruysse, *Bibliographie descriptive des écrits du prince de Ligne*, Paris, Honoré Champion, 2008, p. 69.
29. A copy bound with the general's name, belonging to a private collector, was seen at the exhibition devoted to the general at the town hall in Binche (Belgium), on 3 November 2000. A veteran of Italy and Egypt, André-Joseph Boussart commanded a brigade of dragoons in 1805. He served in Spain between 1808 and 1813.
30. Colin, *L'Éducation militaire*, p. 123.
31. He mentions him in a manuscript dating from February 1789 (Napoleon Ier, *Manuscrits inédits, 1786–1791*, published in accordance with the original autograph manuscripts by Frédéric Masson and Guido Biagi, Paris, P. Ollendorff, 1907, p. 265).
32. Matti Lauerma, *Jacques-Antoine-Hippolyte de Guibert (1743–1790)*, Helsinki, Suomalainen Tiedeakatemia, 1989, p. 269.
33. We know that the library on Saint Helena contained the complete works of Guibert, including, obviously, the *Essai général de tactique*: *Œuvres militaires de Guibert*, published by his widow based on the manuscripts and in accordance with the corrections of the author, 5 vols, Paris, Magimel, an XII—1803 (Healey, 'La Bibliothèque . . .', p. 174).
34. AN, 314 AP 30, MS 40 (Gourgaud, *Journal*, II, p. 286).
35. AN, 390 AP 25, MS of 1818, p. 49 (Bertrand, *Cahiers*, II, p. 152).
36. Jacques-Antoine-Hippolyte de Guibert, *Stratégiques*, introduction by Jean-Paul Charnay, Paris, L'Herne, 1977, p. 27; *Écrits militaires 1772–1790*, preface and notes by General Ménard, Paris, Copernic, 1977, p. 7; *Essai général de tactique 1772*, presentation by Jean-Pierre Bois, Paris, Économica, 2004, p. xxxiv.
37. *Œuvres militaires de Guibert . . .*, I, pp. vii–viii.
38. Colin, *L'Éducation militaire*, p. 123; Lauerma, *Jacques-Antoine-Hippolyte de Guibert*, p. 272; Guglielmo Ferrero, *Bonaparte en Italie (1796–1797)*, trans. from the Italian, Paris, Éditions de Fallois, 1994 (1st edn, 1936), p. 83.
39. Jean Tulard, Jean-François Fayard, and Alfred Fierro, *Histoire et dictionnaire de la Révolution française 1789–1799*, Paris, Robert Laffont, 1987, p. 561.
40. Brother of the Duke of Bassano, indicates Fleuriot de Langle.
41. In 1787, Guibert formed part of an Administrative Council in the War Department.
42. Allusion to the regulation of 22 May 1781, known as the edict of Ségur, which required all candidates for a sub-lieutenancy in the infantry, cavalry, and dragoons to supply proof of four degrees of nobility. The Council Guibert belonged to tightened the provisions of this edict in 1788 (Jean Chagniot, 'Les

Rapports entre l'armée et la société à la fin de l'Ancien Régime', in André
Corvisier [ed.], *Histoire militaire de la France*, 4 vols, Paris, Presses Universitaires
de France, 1992–4, II, pp. 118–19).

43. This order of attack, whereby one concentrates one's forces on one wing,
while declining the other, so as to overwhelm a wing of the enemy army by an
oblique march, had in fact been practised by Epaminondas at the battles of
Leuctra and Mantinea (371 and 362 BCE). See Thierry Widemann, 'Référence
antique et "raison stratégique" au XVIIIe siècle', in Bruno Colson and Hervé
Coutau-Bégarie (eds), *Pensée stratégique et humanisme. De la tactique des Anciens à
l'éthique de la stratégie*, Paris, FUNDP-ISC-Économica, 2000, pp. 147–56.

44. AN, 390 AP 25, MS of 1818 (November), pp. 75–6 (Bertrand, *Cahiers*, II,
pp. 195–8).

45. At the Battle of Kolin in Bohemia on 18 June 1757, Frederick launched his
army into the attack by having it make a flanking march under the eyes of
Daun's Austrians, who occupied a very good defensive position. All the
Prussian attacks failed.

46. Bertrand, *Cahiers*, II, p. 446.

47. Telp, *The Evolution of Operational Art*, p. 33.

48. Henry Lloyd, *Histoire des guerres d'Allemagne*, Paris, ISC-FRS-Économica, 2001
(1st edn, 1784); Patrick J. Speelman, *Henry Lloyd and the Military Enlightenment
of Eighteenth-Century Europe*, Westport, Conn., Greenwood Press, 2002.

49. Two engagements in the French campaign of 1814. At Nangis or Mormant,
on 17 February, the French cavalry turned the two wings of the position
occupied by the Russians. At Champaubert the previous week, the French
cuirassiers had pushed the Russian infantry into pools, broken down its squares,
and put it to flight.

50. AN, 390 AP 25, MS of April 1817 (13 April), p. 8. (Bertrand, *Cahiers*, I,
pp. 216–17).

51. Gourgaud, *Journal*, II, p. 70.

52. Adam Heinrich Dietrich von Bülow, *Histoire des campagnes de Hohenlinden et de
Marengo* [trans. from the German by C. L. Sevelinges], containing the notes
that Napoleon made on this book in 1819 on Saint Helena, arranged and
published by Brevet Major Emmett, London, Whittaker, Treacher & Arnot,
1831, 2nd title page.

53. *Corresp.*, XXVI, no. 20382, p. 62, to Prince Cambacérès, Bautzen, 16 August
1813.

54. Allusion to Napoleon's victory over the allies on 26 and 27 August 1813.

55. AN, 390 AP 25, MS of January–September 1819, p. 59 (Bertrand, *Cahiers*, II,
pp. 240–1).

56. The reference is to the first work for which Jomini was known and which
went through numerous editions: the *Traité de grande tactique* (3 vols, 1805–6),
which subsequently became the *Traité des grandes opérations militaires* (5 vols,
1807–9; 3rd edn, 8 vols, 1811–16); continued by the *Relation critique des*

campagnes des Français contre les Coalisés depuis 1792 (1806), which subsequently became *Histoire critique et militaire des campagnes de la Révolution* (2 vols, 1817), etc. See Langendorf, *Faire la guerre: Antoine-Henri-Jomini*, II, pp. 435–6, and Rapin, *Jomini et la stratégie*, pp. 303–4.

57. AN, 314 AP 30, MS 50 (Gourgaud, *Journal*, II, p. 347).
58. AN, 390 AP 25, MS of 1818 (October), p. 62 (Bertrand, *Cahiers*, II, p. 168).
59. AN, 314 AP 30, MS 39 (Gourgaud, *Journal*, II, p. 104).
60. AN, 390 AP 25, MS of 1818 (October), p. 58 (Bertrand, *Cahiers*, II, p. 168).
61. While remaining Ney's titular aide de camp, Jomini had been attached to the Emperor's general staff for the duration of the campaign against Prussia (*Corresp. gén*, no. 1300, p. 879, to Marshal Berthier, Saint-Cloud, 20 September 1806).
62. AN, 390 AP 25, MS of 1818 (October), p. 62 (Bertrand, *Cahiers*, II, p. 176).
63. AN, 390 AP 25, MS of 1818 (December), p. 84 (Bertrand, *Cahiers*, II, pp. 209–10).
64. Bertrand, *Cahiers*, II, p. 362.
65. AN, 390 AP 25, MS of 1818 (December), p. 87 (Bertrand, *Cahiers*, II, p. 214).
66. Bertrand, *Cahiers*, II, p. 288.
67. Bertrand, *Cahiers*, III, p. 169.
68. Bruno Colson, *Le Général Rogniat, ingénieur et critique de Napoléon*, Paris, ISC-Économica, 2006.
69. 'Dix-huit notes sur l'ouvrage intitulé *Considérations sur l'art de la guerre*', *Corresp.*, XXXI, p. 302.
70. In all likelihood this is an allusion to the 'Essai sur la fortification de campagne', *Corresp.*, XXXII, pp. 462–84.
71. Rogniat was not present at Austerlitz and it is not certain that he attended the Battle of Jena.
72. Like many officers in the engineers, Rogniat had never been the head of a corps, whether a company or a battalion. Engineering regiments were only created in 1814, by Rogniat.
73. AN, 390 AP 25, MS of 1818 (December), pp. 80–2 (Bertrand, *Cahiers*, II, pp. 203–5).
74. Jean-Baptiste-Antoine-Marcellin de Marbot, *Remarques critiques sur l'ouvrage de M. le Lieutenant Général Rogniat, intitulé:* Considérations sur l'art de la guerre, Paris, Anselin et Pochard, September 1820.
75. The 'Dix-huit notes sur l'ouvrage intitulé *Considérations sur l'art de la guerre*', a refutation of Rogniat, were only published in 1823, in the *Mémoires*.
76. AN, 390 AP 25, MS of January–April 1821 (15 March), p. 23 (Bertrand, *Cahiers*, III, pp. 99–100).
77. *Corresp.*, V, no. 4300, pp. 528–9, to General Marmont, El-Rahmânyeh, 3 thermidor an VII—21 July 1799.
78. AN, 314 AP, MS 28 (Gourgaud, *Journal*, I, pp. 192–3). The book by Colonel Simon-François Gay de Vernon, *Traité élémentaire d'art militaire et de fortification*

à l'usage des élèves de l'École polytechnique et des élèves des écoles militaires, 2 vols and an atlas, Paris, Allais, an XIII—1805, is recorded in the library at Longwood (Healey, 'La Bibliothèque...', p. 174).

79. AN, 390 AP 25, MS of January–April 1821 (19 April), p. 33 (Bertrand, *Cahiers*, III, p. 130).
80. AN, 390 AP 25, MS of January 1817 (12 January), p. 8 (Bertrand, *Cahiers*, I, p. 181).

CHAPTER 3

1. *Corresp.*, XXVI, no. 20678, p. 281, to Marshal Macdonald, Dresden, 2 October 1813.
2. AN, 390 AP 25, MS of 1818, p. 54 (Bertrand, *Cahiers*, II, pp. 159–60).
3. AN, 390 AP 25, MS of November 1816 (13 November), pp. 4–5 (Bertrand, *Cahiers*, I, p. 146).
4. *Corresp.*, X, no. 8716, p. 399, to Vice-Admiral Decrès, Milan, 21 floréal an XIII—11 May 1805.
5. *Corresp.*, II, no. 1976, p. 163, note following the response to M. Dunan, presumed to be from Mombello, 13 messidor an V—1 July 1797. This note derives from the *Mémoires* of Bourrienne, who is not to be trusted.
6. *Corresp.*, XXIII, no. 18503, p. 228, by Berthier for Marmont, commanding the army of Portugal, Paris, 18 February 1812.
7. AN, 314 AP 30, MS 40 (Gourgaud, *Journal*, II, p. 324).
8. Placed at the head of three army corps and a cavalry corps, Ney was to march on Berlin and be reinforced by other units led by Napoleon in person. But the latter had to repel two unanticipated offensives by Blücher and Schwarzenberg. Left without news, Ney succumbed to his impetuousness and was beaten by Bülow's and Tauentzien's Prussians on 6 September 1813.
9. Gouvion Saint-Cyr, *Mémoires*, IV, pp. 149–50.
10. AN, 390 AP 25, MS of 1818, p. 42 (Bertrand, *Cahiers*, II, p. 139).

CHAPTER 4

1. 'Campagnes d'Égypte et de Syrie', *Corresp.*, XXX, p. 171.
2. Las Cases, *Mémorial*, vol. 18, pp. 117–18 (Dunan edn, II, pp. 576–7).
3. Feuquière and Villars have already been mentioned. Marshal Jean-Baptiste de Maillebois had fought the Austrians in Germany and the Austro-Sardinians in Provence and Lombardy during the War of the Austrian Succession. His good advice was insufficiently followed by his Spanish allies and he skilfully directed the evacuation of Italy by French troops in 1746. Marshal Nicolas de Catinat was one of Louis XIV's best generals, the victor of Staffarde (1690) and Marsaille (1693). Duke Henri de Rohan was one of the leaders of the Protestant

party. Drawing on the example of Caesar, he wrote *Le Parfait Capitaine* (1636), a work that went through numerous editions and translations.

4. Arthur de Ganniers, 'Napoléon chef d'armée. Sa formation intellectuelle, son apogée, son déclin', *Revue des questions historiques*, 73, 1903, pp. 518–21.

5. 'Dix-huit notes sur l'ouvrage intitulé *Considérations sur l'art de la guerre*', *Corresp.*, XXXI, pp. 353–4.

6. *Corresp.*, X, no. 8209, p. 69, to General Lauriston, Paris, 21 frimaire an XIII— 12 December 1804.

7. *Mémoires*, Montholon, V, p. 20; italics in the original.

8. *Mémoires*, Montholon, IV, pp. 350–1; italics in the original, 354.

9. AN, 390 AP 25, MS of January–September 1819, p. 107 (Bertrand, *Cahiers*, II, pp. 339–40).

10. Variant for the end of this sentence in *Mémoires*, Montholon, II, p. 193: 'and with an objective; it must be composed of forces proportionate to the obstacles one anticipates.'

11. Peter the Great, at Poltava, on 28 June 1709. Before invading Russia, Napoleon had asked his librarian what existed in French that was 'most detailed on Charles XII's campaign in Poland and Russia' (*Corresp.*, XXIII, no. 18348, p. 93, to M. Barbier, Paris, 19 December 1811).

12. Allusion to the Battle of Ankara (20 July 1402), where Tamerlane and his Mongols defeated Sultan Bayezid I or Bajazet and his Ottomans.

13. 'Dix-huit notes sur l'ouvrage intitulé *Considérations sur l'art de la guerre*', *Corresp.*, XXXI, p. 418.

14. *Corresp.*, XX, no. 16372, p. 284, to Marshal Berthier, Compiègne, 9 April 1810.

15. *Corresp.*, XII, no. 10325, p. 442, to the King of Naples, Saint-Cloud, 6 June 1806.

16. AN, 314 AP 30, MS 39 (Gourgaud, *Journal*, II, pp. 72–3).

17. *Mémoires*, Gourgaud, II, pp. 192–3.

18. Antulio J. Echevarria II, *Clausewitz and Contemporary War*, Oxford, Oxford University Press, 2007, pp. 154–6.

19. Hew Strachan, *Carl von Clausewitz's* On War: *A Biography*, London, Atlantic Books, 2007, p. 96.

20. Félix Lemoine, 'En relisant Clausewitz', *Revue militaire française*, December 1929, p. 266, cited by Benoît Durieux, *Clausewitz en France. Deux siècles de réflexion sur la guerre 1807–2007*, Paris, ISC-CID-Fondation Saint-Cyr-Économica, 2008, p. 459.

21. AN, 314 AP 30, MS 40 (Gourgaud, *Journal*, II, p. 269).

22. 'Dix-huit notes sur l'ouvrage intitulé *Considérations sur l'art de la guerre*', *Corresp.*, XXXI, p. 412.

23. AN, 390 AP 25, MS of 1818 (December), p. 86 (Bertrand, *Cahiers*, II, p. 214).

24. Beaten at Castiglione, south of Lake Garda, on 5 August 1796, Wurmser went back up the course of the Adige, abandoning Verona to the French. Bonaparte

pursued him in the direction of Trento and Tyrol, waiting for Wurmser to redescend on Bassano via the valley of Brenta, and then on to Verona and Mantua. Wurmser made this move, believing that Bonaparte was continuing towards Tyrol. On 8 September, Wurmser was trapped at Bassano. Part of his army fled towards Friuli, while he shut himself up in Mantua with the remainder.

25. Bertrand, *Cahiers*, II, pp. 368–9, in italics in the original.
26. AN, 314 AP 30, MS 40 (Gourgaud, *Journal*, II, pp. 324–5).
27. Clausewitz, *On War*, II, 2, p. 163.

CHAPTER 5

1. Tomiche, *Napoléon écrivain*, p. 15.
2. *Corresp. gén.*, I, no. 207, to Captain Andréossy, Nice, 17 messidor an II—5 July 1794.
3. Probably the *Histoire du prince François Eugène de Savoie* by Mauvillon (Amsterdam, 1740) and the *Histoire militaire du prince Eugène de Savoye* by Dumont and Rousset de Missy (The Hague, 1729).
4. *Corresp. gén.*, I, p. 1322, letter without text (lost) to General Calon, director of the Dépôt de la guerre, 7 March 1796. Spenser Wilkinson discovered in a Viennese bookshop the copy of Pezay's book forgotten by Bonaparte in Verona in late 1796 (Wilkinson, *The Rise of General Bonparte*, pp. 156–7). Louis-Joseph de Vendôme, Duke of Penthièvre, had fought and beaten Prince Eugène of Savoy in Piedmont in 1705, at Luzzara and Cassano. Maillebois has been referred to above.
5. *Corresp. gén.*, V, no. 10657, p. 618, to Marshal Berthier, Pont-de-Briques, 7 fructidor an XIII—25 August 1805.
6. Antoine Casanova, *Napoléon et la pensée de son temps: une histoire intellectuelle singulière*, Paris, La Boutique de l'Histoire, 2000, pp. 86–7; [Agathon-Jean-François Fain], *Mémoires du baron Fain, premier secrétaire du cabinet de l'Empereur*, introduction and notes by P. Fain, Paris, Plon, 1908, pp. 70–1.
7. AN, 390, AP 25, MS of 1819, p. 160 (Bertrand, *Cahiers*, II, p. 445, where the text is placed in 1820).
8. Colin, *L'Éducation militaire*, pp. 151–2.
9. AN, 314 AP 30, MS 39 (Gourgaud, *Journal*, II, p. 72).
10. AN, 314 AP 30, MS 40 (Gourgaud, *Journal*, II, p. 268). See Bk. II, Ch. 2 n. 45 above. Clearly, Napoleon did not know the Prince of Ligne's account, which was available but not widely diffused in France (*Mélanges militaires, littéraires et sentimentaires*, vols XIV, XV, and XVI, Dresden, Walther, 1796. See our critical re-edition: Charles-Joseph de Ligne, *Mon Journal de la guerre de Sept Ans*, unpublished texts introduced, established, and annotated by Jeroom Vercruysse and Bruno Colson, Paris, Honoré Champion, 2008).
11. AN, 314 AP 30, MS 50 (Gourgaud, *Journal*, II, p. 347).

12. *Corresp. gén.*, V, no. 11097, p. 848, to Marshal Berthier, Schönbrunn, 24 brumaire an XIV—15 November 1805.
13. Clausewitz, *On War*, II, 5, pp. 182, 185.
14. This is 'alternate history': Jonathan North (ed.), *The Napoleon Options: Alternate Decisions of the Napoleonic Wars*, London, Greenhill Books, 2000.
15. AN, 314 AP 30, MS 40 (Gourgaud, *Journal*, II, p. 275). Maurice-Étienne Gérard and Nicolas-Joseph Maison were two of the generals in whom Napoleon had most hopes in his last campaigns. They both distinguished themselves in Russia in 1812 and in Germany in 1813 and were made marshals of France after the Empire, Maison in 1829 and Gérard in 1830.
16. AN, 390 AP 25, MS of 1819, p. 151 (Bertrand, *Cahiers*, II, p. 433, where the text is placed in 1820).

CHAPTER 6

1. Colin, *L'Éducation militaire*, pp. 146–7.
2. Wilkinson, *The Rise of General Bonaparte*, p. 149.
3. Las Cases, *Mémorial*, vol. 18, pp. 221–5 (Dunan edn, II, pp. 614–15); in italics in the original.
4. Montholon, *Récits de la captivité*, II, p. 528.
5. [Rémusat], *Mémoires*, I, p. 268.
6. Annie Jourdan devotes a chapter to 'Napoléon et l'histoire' in *Napoléon. Héros*, pp. 19–56.
7. [Caulaincourt], *Mémoires*, II, p. 302.
8. 'Dix-huit notes sur l'ouvrage intitulé *Considérations sur l'art de la guerre*', *Corresp.*, XXXI, pp. 347, 414 and 418.
9. Gat, *The Origins of Military Thought*, p. 160.
10. Pierre Bergé and associates, *Manuscrits, autographes et très beaux livres anciens et modernes*, Drouot, 7 December 2004 (sales catalogue containing at no. 22 *Les Mémoires*, manuscript dictated by Napoleon to Generals Bertrand, Gourgaud, and Montholon, in-folio of 84 pp., approximately 40 pages of which are in Napoleon's hand in ink or pencil), p. 30.
11. 'Notes sur l'introduction à l'*Histoire de la guerre en Allemagne en 1756, entre le roi de Prusse et l'impératrice-reine et ses alliés, etc.*, par le général Lloyd', *Corresp.*, XXXI, p. 422.
12. [Feuquière], *Mémoires de M. le marquis de Feuquière . . . contenans ses maximes sur la guerre, et l'application des exemples aux maximes*, new edn, 4 books in 2 vols, Amsterdam, L. Honoré et fils, 1731; Clausewitz, *On War*, II, 6, p. 202.
13. Johann Marchese di Provera and Friedrich, Prince of Hohenzollern-Hechingen, Austrian generals.
14. A line of defence in the rear of the besieging army's camps to protect it against a relief army.

15. Arras was besieged in 1654 by the Spanish. On 24 August, Turenne and a relief army routed the Spanish. In May 1706, a French army commanded by the Duke of Feuillade began a siege of Turin. In September, Prince Eugène of Savoy led a relief army, which raised the siege and expelled the French from Italy.

16. *Mémoires*, Montholon, V, pp. 91–3; in italics in the original.

17. Second son of King George III of England, Frederick Duke of York commanded the Anglo-Hanoverian troops that had disembarked in the Austrian Low Countries, reconquered in 1793. He began a siege of Dunkerque. The French general Houchard marched to its aid and confronted York at Hondschoote 6–8 September. The battle was very confused, but the Duke of York eventually decided to lift the siege and withdraw.

18. AN, 314 AP 30, MS 40 (Gourgaud, *Journal*, II, pp. 320–1).

19. AN, 314 AP 30, MS 40 (Gourgaud, *Journal*, II, p. 283).

20. Probably Charles-Jean-Dominique de Lacretelle (1766–1855), author of a *Histoire de France pendant le XVIIIe siècle*, Paris, Treuttel and Wurtz, 1808–26.

21. AN, 390 AP 25, MS of 1818, p. 16 (Bertrand, *Cahiers*, II, p. 30).

22. L'École d'application de l'artillerie et du génie.

23. Jean II le Bon at Poitiers in 1356 and François I at Pavia in 1525.

24. Or Dürrenstein, 6 kilometres south-west of Krems. On 11 November 1805, attacked in the front and the rear by the Russians in a narrow space on the left bank of the Danube, Mortier and the Gazan division succeeded in emerging by riding over the body of their enemies to join the French forces.

25. *Corresp.*, XIX, no. 15889, pp. 540–3, to Clarke, Schönbrunn, 1 October 1809; italics in the original.

26. AN, 390 AP 25, MS of January–April 1821, p. 2 (Bertrand, *Cahiers*, III, p. 30).

CONCLUSION TO BOOK II

1. Bernard Druène, 'Der Feldherr Napoleon—Theorie und Praxis seiner Kriegskunst', and R. Van Roosbroeck, 'Der Einfluss Napoleons...', in von Groote and Müller (eds), *Napoleon I.*, pp. 49 and 200.

2. Echevarria II, *Clausewitz and Contemporary War*, p. 17.

3. Andreas Herberg-Rothe, *Clausewitz's Puzzle: The Political Theory of War*, trans. from the German, Oxford, Oxford University Press, 2007, p. 67.

4. 'The Most Important Principles for the Conduct of War to Complete my Course of Instruction of His Royal Highness The Crown Prince': see Carl von Clausewitz, *Principles of War*, trans. and ed. with an introduction by Hans W. Gatzke, Mineola, NY, Dover Publications, 2003.

5. Colin, *L'Éducation militaire*, pp. 376–8.

6. Rudolf van Caemmerer, *L'Évolution de la stratégie au XIXe siècle*, trans. from the German, Paris, Fischbacher, 1907, p. 301.

7. Echevarria II, *Clausewitz and Contemporary War*, pp. 168, 191–2.

BOOK III

CHAPTER I

1. AN, 390 AP 25, MS of 1818, p. 23 (Bertrand, *Cahiers*, II, p. 48).
2. 'Précis des événements militaires arrivés pendant les six premiers mois de 1799', *Corresp.*, XXX, p. 263.
3. Jules Lewal, *Introduction à la partie positive de la stratégie*, notes by A. Bernède, Paris, CFHM-ISC-Économica, 2002 (1st edn, 1892), p. 125.
4. 'Précis des événements militaires arrivés pendant les six derniers mois de 1799', *Corresp.*, XXX, p. 289.
5. Clausewitz, *On War*, III, 1, p. 208.
6. Clausewitz, *On War*, III, 1, p. 209.
7. Alan Beyerchen, 'Clausewitz and the Non-Linear Nature of Warfare: Systems of Organized Complexity', in Strachan and Herberg-Rothe (eds), *Clausewitz in the Twenty-First Century*, p. 56.
8. *Mémoires*, Montholon, II, p. 89. The word 'stratégie' is in French in Montholon's later text, *Récits de la captivité*, II, p. 462. Wellington was victorious at Talavera against Joseph Bonaparte on 28 July 1809, at Salamanca (or Arapiles) against Marmont on 22 July 1812, and at Vitoria against Joseph on 21 June 1813.
9. *Mémoires*, Montholon, II, p. 101.
10. Clausewitz, *On War*, III, 1, p. 210.
11. *Corresp.*, XIX, no. 15454, p. 185, to General Marmont, Schönbrunn, 28 June 1809, 10.00 hrs.
12. AN, 314 AP 30, MS 50 (Gourgaud, *Journal*, II, pp. 346–7).
13. *Corresp.*, II, no. 1402, p. 261, to the executive Directory, Verona, 1 pluviôse an V—20 January 1797.

CHAPTER 2

1. *Corresp.*, XVII, no. 1426, p. 42, observations on Spanish affairs, Saint-Cloud, 27 August 1808.
2. Chaptal, *Mes Souvenirs*, pp. 301–2.
3. Lacuna in the original; the minute contains the material in brackets.
4. *Corresp.*, XVII, n. 14343, p. 526, note for the King of Spain, Châlons-sur-Marne, 22 September 1808.
5. *Corresp.*, XVIII, p. 15144, p. 525, to Eugène, Burghausen, 30 April 1809.
6. *Inédits napoléoniens*, I, no. 1416, p. 383, to Marshal Davout, Paris, 26 March 1815.
7. [Thibaudeau], *Mémoires*, pp. 114–15; italics in the original. The French tried to execute an adroit enveloping manoeuvre in Prussian fashion, but were taken completely by surprise in the middle of it.

8. *Corresp.*, XIX, no. 15933, pp. 570–1, to Clarke, Schönbrunn, 10 October 1809.
9. AN, 314 AP 30, MS 39 (Gourgaud, *Journal*, II, p. 143).
10. Clausewitz, *On War*, I, 1, p. 95.
11. *Corresp.*, V, no. 3949, pp. 304–5, to Marmont, Cairo, 21 pluviôse an VII—9 February 1799.
12. *Œuvres de Napoléon Bonaparte*, 5 vols, Paris, Panckoucke, 1821, I, p. 451, to the head of the general staff, Milan, 25 messidor an V—13 July 1797.
13. *Corresp.*, X, no. 8628, p. 347, to Decrès, Stupinigi, 4 floréal an XIII—24 April 1805.
14. *Lettres inédites de Napoléon Ier*, Lecestre, I, no. 430, p. 301, to General Clarke, Minister of War, Paris, 27 March 1809.
15. *Dernières lettres inédites de Napoléon Ier*, published by Léonce de Brotonne, 2 vols, Paris, Honoré Champion, 1903, I, no. 1016, p. 468, to Fouché, General Police Minister, Paris, 13 February 1810.
16. *Corresp.*, XXVII, no. 21316, p. 206, to Savary, Police Minister, Château de Surville, 19 February 1814.
17. *Corresp.*, XXVII, no. 21360, p. 239, to King Joseph, village of Noës, Troyes, 24 February 1814, 07.00 hrs.
18. *Corresp.*, XXIV, no. 19056, p. 130, to the Prince of Neuchâtel, Vitebsk, 7 August 1812.
19. *Corresp.*, XXIV, no. 19100, p. 158, to Berthier for Oudinot, Smolensk, 19 August 1812.
20. *Corresp.*, XXV, no. 19688, pp. 46–7, to Eugène, Trianon, 9 March 1813.
21. *Corresp.*, XIII, no. 10599, p. 38, to Eugène, Saint-Cloud, 5 August 1806.
22. *Corresp.*, X, no. 8507, p. 279, to Marshal Moncey, Saint-Cloud, 10 germinal an XIII—31 March 1805.

CHAPTER 3

1. 'Dix-huit notes sur l'ouvrage intitulé *Considérations sur l'art de la guerre*', *Corresp.*, XXXI, p. 417.
2. Friedrich-Wilhelm von Seydlitz-Kurzbach was the best general in Frederick's II's heavy cavalry. He contributed to his principal victories—in particular, Rossbach (1757).
3. AN, 390 AP 25, MS of 1818 (October), p. 61 (Bertrand, *Cahiers*, II, p. 173).
4. AN, 390 AP 25, MS of September 1816 (10 September), pp. 4–5 (Bertrand, *Cahiers*, I, pp. 120–1).
5. Clausewitz, *On War*, III, 4, p. 218.

CHAPTER 4

1. *Dernières lettres inédites*, I, no. 711, p. 325, to Clarke, Minister of War, Bayonne, 20 May 1808.

2. *Corresp.*, IV, no. 2710, p. 183, proclamation to the land army, on board the *Orient*, 4 messidor an VI—22 June 1798, shortly before disembarking in Egypt.

3. *Corresp.*, XV, no. 12282, pp. 17–18, to the King of Naples (Joseph Bonaparte), Finkenstein, 3 April 1807.

4. *Corresp.*, XVIII, no. 14552, p. 112, army order, Chamartin, 12 December 1808.

5. Laurence Montroussier, *Éthique et commandement*, Paris, Économica, 2005, p. 182.

6. Las Cases, *Mémorial*, vol. 11, pp. 137–8 (Dunan edn, II, pp. 5–6).

7. *Corresp.*, XXI, no. 16973, p. 162, to Eugène Napoléon, Viceroy of Italy, Fontainebleau, 1 October 1810.

8. *Corresp.*, XXII, no. 17672, p. 126, to Minister of War Clarke, Saint-Cloud, 30 April 1811.

9. AN, 390 AP 25, MS of July 1817 (17 July), p. 7 (Bertrand, *Cahiers*, I, p. 247).

10. *Corresp.*, XIV, no. 12212, p. 566, to General Savary, Osterode, 29 March 1807, 05.00 hrs.

11. O'Meara, *Napoleon in Exile*, I, p. 202.

12. For example, in a report to Marshal Berthier, General Suchet wrote that on the occasion of the capture of a fort, the Italian sappers had rivalled the French in worth and 'equalled their elders'—i.e. the Romans (Colson, *Le Général Rogniat*, pp. 272, 278, and 279).

13. *Corresp.*, VI, no. 4660, p. 18, to Brune, commander-in-chief of the army of the West, Paris, 21 ventôse an VIII—12 March 1800.

14. *Corresp.*, XIV, no. 11906, p. 355, to Marshal Soult, Osterode, 28 February 1807, 18.00 hrs.

15. *Corresp.*, XIV, no. 12150, p. 516, to Marshal Lefebvre, Osterode, 24 March 1807, midday.

16. AN, 314 AP 30, MS 39 (Gourgaud, *Journal*, II, p. 83).

17. [Thibaudeau], *Mémoires*, pp. 83–4; in italics in the original.

18. John Lynn, 'Toward an Army of Honor: The Moral Evolution of the French Army, 1789–1815', *French Historical Studies*, 16, 1989–1, pp. 152–73; Bertaud, *Quand les enfants parlaient de gloire*, pp. 173–4.

19. Didier Le Gall, *Napoléon et le* Mémorial de Sainte-Hélène. *Analyse d'un discours*, Paris, Kimé, 2003, pp. 235–6 and 253.

20. *Corresp.*, IX, no. 7527, p. 239, decision, Paris, 16 pluviôse an XII—6 February 1804.

21. *Corresp.*, XI, no. 9235, p. 220, decision, Saint-Cloud, 1st complementary day an XIII—18 September 1805.

22. *Corresp.*, XIII, no. 10817, p. 216, to Berthier, note on the defence of Inn and occupation of Braunau, Saint-Cloud, 19 September 1806.

23. *Corresp.*, XVI, no. 13557, p. 326, to General Menou, Governor General of the Departments beyond the Alps, Paris, 13 February 1808.

24. *Corresp.*, XI, no. 9522, p. 434, order of the day, Brünn, 3 frimaire an XIV—24 November 1805.

25. *Corresp.*, XVII, no. 13836, p. 81, to Joachim Murat, Grand Duke of Berg, Lieutenant-General of the Kingdom of Spain, Bayonne, 9 May 1808, 17.00 hrs.

26. *Corresp.*, X, no. 8375, p. 181, note for the Police Minister, Paris, 10 ventôse an XIII—1 March 1805.

27. 'Notes sur l'introduction à l'*Histoire de la guerre en Allemagne en 1756, entre le roi de Prusse et l'impératrice-reine et ses alliés, etc.*, par le général Lloyd', *Corresp.*, XXXI, pp. 424–6.

28. AN, 390 AP 25, MS of October 1817 (24 October), p. 12 (Bertrand, *Cahiers*, I, p. 289).

29. Bertrand, *Cahiers*, II, p. 289.

30. *Corresp.*, XII, no. 10086, p. 276, to the King of Naples, La Malmaison, 11 April 1806.

31. *Corresp.*, XIII, no. 10630, p. 65, to the King of Naples, Saint-Cloud, 10 August 1806.

32. *Corresp.*, XV, no. 12511, p. 178, to Prince Jérôme, Finkenstein, 2 May 1807.

33. *Corresp.*, XIV, no. 12094, p. 479, to General Dejean, Osterode, 20 March 1807.

34. *Corresp.*, XIV, no. 12106, pp. 488–9, to the King of Naples, Osterode, 20 March 1807.

35. *Corresp.*, XIV, no. 12107, p. 489, to the King of Naples, Osterode, 20 March 1807.

36. *Corresp.*, IX, no. 7616, p. 287, to General Marmont, commanding the camp of Utrecht, La Malmaison, 26 ventôse an XII—12 March 1804.

37. *Corresp.*, X, no. 8446, p. 232, to Marshal Bernadotte, La Malmaison, 26 ventôse an XIII—17 March 1805.

38. *Corresp.*, X, no. 8785, p. 442, to Berthier, Milan, 5 prairial an XIII—25 May 1805.

39. *Corresp.*, XIII, no. 11172, p. 477, to Prince Eugène, Berlin, 4 November 1806.

40. *Corresp.*, XIV, no. 12174, p. 534, to Eugène, Osterode, 25 March 1807.

41. *Corresp.*, XII, no. 10032, p. 244, official report of the Battle of Austerlitz, presented to Emperor Alexander by General Kutuzov, and the observations of a French officer, Braunau, 28 March 1806.

42. *Corresp.*, IX, no. 8001, p. 511, to Fouché, Aix-la-Chapelle, 22 fructidor an XII—9 September 1804.

43. Our correction for 'and' in the published text.

44. 'Dix-huit notes sur l'ouvrage intitulé *Considérations sur l'art de la guerre*', *Corresp.*, XXXI, pp. 416–17.

45. During the many engagements pitting the army of Italy against the Austrian columns that descended from both sides of Lake Garda at the start of August 1796, the 32nd battle semi-brigade, 2,626 men strong, played a key role in the Masséna division at Lonato.
46. AN, 314 AP 30, MS 39 (Gourgaud, *Journal*, II, p. 127).
47. *Mémoires*, Montholon, X, p. 237.
48. *Corresp.*, XIII, no. 10709, p. 118, to Eugène de Beauharnais, Saint-Cloud, 30 August 1806.
49. *Correspondance inédite de Napoléon Ier conservée aux Archives de la Guerre*, published by Ernest Picard and Louis Tuetey, 5 vols, Paris, Charles Lavauzelle, 1912–25, I, no. 247, p. 145, order of the day, Schönbrunn, 2 nivôse an XIV— 23 December 1805.
50. O'Meara, *Napoleon in Exile*, I, p. 479.

CHAPTER 5

1. On 2 August 47 BCE, Julius Caesar was attacked by Pharnaces, King of Pontus, while his legions were constructing their camp. The Romans formed up for battle very rapidly and completely routed their opponents. It was when announcing this victory to Rome that Caesar uttered his famous 'Veni, vidi, vici'.
2. 'Précis des guerres de Jules César', *Corresp.*, XXXII, p. 69.
3. *Corresp.*, I, no. 537, p. 345, to the executive Directory, Peschiera, 13 prairial an IV—1 June 1796.
4. *Corresp.*, XI, no. 9405, p. 343, proclamation, Elchingen, 29 vendémiaire an XIV—21 October 1805.
5. *Lettres inédites*, I, no. 536, pp. 370–1, to General de Wrède, commander-in-chief of the Bavarian troops, Schönbrunn, 8 October 1809; italics in the original.
6. 'Campagne de 1815', *Corresp.*, XXXI, pp. 206–7.
7. *Corresp.*, XIII, no. 11325, p. 588, to Marshal Mortier, Posen, 29 November 1806, 22.00 hrs.
8. *Mémoires*, Montholon, V, p. 272.
9. *Corresp.*, X, no. 8469, p. 257, to General Lauriston, La Malmaison, 1 germinal an XIII—22 March 1805.
10. *Corresp.*, XI, no. 9160, p. 161, to Decrès, camp of Boulogne, 11 fructidor an XIII—29 August 1805.

CHAPTER 6

1. Clausewitz, *On War*, III, 7, p. 227.
2. *Corresp.*, VI, no. 4450, p. 39, proclamation to the army of Italy, Paris, 4 nivôse an VIII—25 December 1799.

3. *Corresp.*, VII, no. 5403, p. 40, proclamation to the army of the East, Paris, 1 ventôse an IX—20 February 1801.

4. *Corresp.*, X, no. 8237, pp. 91–2, report on the situation of the French Empire, Paris, 6 nivôse an XIII—27 December 1804.

5. *Correspondance inédite*, I, p. 112, order of the day, Elchingen, 23 vendémiaire an XIV—15 October 1805.

6. *Corresp.*, XIV, no. 12100, p. 484, message to the Senate, Osterode, 20 March 1807.

7. *Corresp.*, VII, no. 6080, p. 460, order of the day, Saint-Cloud, 22 floréal an X—12 May 1802.

8. Englund, *Napoleon*, p. 105.

CHAPTER 7

1. Clausewitz, *On War*, III, 8, p. 228.

2. AN, 390 AP 25, MS of 1818 (January), p. 18 (Bertrand, *Cahiers*, II, pp. 34–5).

3. Paddy Griffith, *The Art of War of Revolutionary France 1789–1802*, London, Greenhill Books, 1998, p. 200.

4. AN, 390 AP 25, MS of 1818 (January), pp. 27–8 (Bertrand, *Cahiers*, II, p. 40).

5. Jean-Paul Bertaud and Daniel Reichel, *L'Armée et la guerre*, 3rd instalment of Serge Bonin and Claude Langlois (eds), *Atlas de la Révolution française*, Paris, École des hautes études en sciences sociales, 1989, p. 53.

6. *Corresp.*, XIII, no. 10656, p. 77, to the King of Naples, Saint-Cloud, 16 August 1806.

7. *Corresp.*, XIV, no. 11579, p. 161, to Eugène, Warsaw, 7 January 1807.

8. *Corresp.*, XV, no. 12530, p. 189, to the King of Naples, 4 May 1807.

9. *Corresp.*, XVIII, no. 14846, p. 308, to Joseph Napoléon, King of Spain, Paris, 4 March 1809.

10. *Lettres inédites*, I, no. 155, p. 97, to the Landammann of Switzerland, Finkenstein, 18 May 1807.

11. *Corresp.*, XX, no. 16090, p. 89, to Joachim Napoléon, King of the Two Sicilies, Paris, 27 December 1809.

12. *Corresp.*, XXI, no. 16894, p. 102, to Jérôme Napoléon, King of Westphalia, Saint-Cloud, 11 September 1810.

13. AN, 390 AP 25, MS of January–April 1821 (21 April), p. 34 (Bertrand, *Cahiers*, III, p. 133). On 2 August 216 BCE, Hannibal, with an army of 40,000–50,000 men, defeated 80,000 Romans, killing 45,000 of them and taking 20,000 prisoners (Yann Le Bohec, *Histoire militaire des guerres puniques*, Paris and Monaco, Éditions du Rocher, 1996, pp. 190–2).

14. 'Précis des guerres de Jules César', *Corresp.*, XXXII, p. 70.

15. 'Précis des événements militaires arrivés pendant les six derniers mois de 1799', *Corresp.*, XXX, p. 291.

16. Clausewitz, *On War*, III, 8, p. 229.

17. Bertrand, *Cahiers*, II, p. 287.
18. [Gohier], *Mémoires de Louis-Jérôme Gohier, président du Directoire au 18 Brumaire*, 2 vols, Paris, Bossange, 1824, I, p. 204.
19. Clausewitz, *On War*, III, 8, pp. 230–1.
20. It is difficult to assess these figures, given that they depend on the moment and position at which they are counted. Overall, Napoleon had almost 40,000 men and 60 guns at his disposal for his operations, facing around 25,000 of Colli's Sardinians and 31,000 of Beaulieu's Austrians. In the initial engagements around Montenotte, Bonaparte succeeded in achieving local numerical superiority of more than 2 to 1 with his manoeuvres.
21. 'Campagnes d'Italie (1796–1797)', *Corresp.*, XXIX, pp. 83–4.
22. Clausewitz, *On War*, III, 8, p. 232.
23. 'Campagnes d'Italie (1796–1797)', *Corresp.*, XXIX, p. 131.
24. AN, 390 AP 25, MS of April 1817 (26 April), p. 12 (Bertrand, *Cahiers*, I, p. 219).

CHAPTER 8

1. 'Campagnes d'Italie (1796–1797)', *Corresp.*, XXIX, p. 102.
2. *Corresp.*, IV, no. 2724, p. 193, to General Desaix, Alexandria, 15 messidor an VI—3 July 1798.
3. AN, 390 AP 25, MS of 1818 (October), p. 63 (Bertrand, *Cahiers*, II, pp. 178–9).
4. AN, 390 AP 25, MS of 1818 (January), p. 4 (Bertrand, *Cahiers*, II, pp. 46–7). Allusion to the disembarkation in Egypt. Bonaparte and his army disembarked on the beach at Marabout, near Alexandria, in the night of 1–2 July 1798.
5. 'Campagnes d'Égypte et de Syrie', *Corresp.*, XXX, pp. 168–9.
6. *Corresp.*, VI, no. 4711, p. 216, to General Masséna, Paris, 19 germinal an VIII—9 April 1800.

CHAPTER 9

1. Ganniers, 'Napoléon chef d'armée . . .', p. 532.
2. Clausewitz, *On War*, III, 10, p. 238.
3. *Corresp. gén.*, I, no. 457, p. 321, to Masséna, 15 germinal an IV—4 April 1796.
4. *Corresp.*, IV, no. 2632, p. 135, to General Vaubois, on board the *Orient*, 23 prairial an VI—11 June 1798.
5. *Corresp.*, V, no. 3972, p. 323, to Reynier, before El-Arich, 29 pluviôse an VII—1 February 1799.
6. 'Dix-huit notes sur l'ouvrage intitulé *Considérations sur l'art de la guerre*', *Corresp.*, XXXI, p. 366.
7. *Lettres inédites*, I, no. 537, p. 371, to Clarke, Minister of War, Schönbrunn, 10 October 1809.

8. 'Dix-huit notes sur l'ouvrage intitulé *Considérations sur l'art de la guerre*', *Corresp.*, XXXI, p. 366.
9. That of Blücher, who escaped Napoleon's attacks that day.
10. *Corresp.*, XXVI, no. 20781, p. 343, to Joachim Napoléon at Wachau, Düben, 12 October 1813, 20.00 hrs.
11. *Corresp.*, XXVII, no. 21393, p. 259, to Berthier, Troyes, 27 February 1814, 09.00 hrs.
12. 'Campagnes d'Italie (1796–1797)', *Corresp.*, XXIX, p. 186.
13. *Corresp.*, XI, no. 9496, p. 413, 24th bulletin of the Grande Armée, Schönbrunn, 24 brumaire an XIV—15 November 1805.

CHAPTER 10

1. Stéphane Béraud, *La Révolution militaire napoléonienne*, I. *Les Manœuvres*, Paris, Bernard Giovanangeli, 2007, p. 11.
2. Clausewitz, *On War*, III, 11, p. 240.
3. 'Dix-huit notes sur l'ouvrage intitulé *Considérations sur l'art de la guerre*', *Corresp.*, XXXI, p. 418.
4. Jean Colin, *Les Transformations de la guerre*, Paris, Flammarion, 1911, p. 208.
5. *Corresp.*, XIII, no. 10941, p. 310, to Soult, Würzburg, 5 October 1806, 11.00 hrs.
6. Béraud, *La Révolution militaire napoléonienne*, I, pp. 143–4.
7. *Corresp.*, XXV, no. 19916, p. 236, to Eugène, Erfurt, 28 April 1813, 03.00 hrs.
8. *Corresp.*, XIX, no. 15736, pp. 412–13, to Clarke, Schönbrunn, 2 September 1809.
9. Field-Marshal Curt Christoph von Schwerin had stressed the risks of marching in two columns that were so far apart on difficult terrain. Even so, luck and the quality of their army vouchsafed the Prussians victory before Prague on 6 May 1757. Schwerin was killed during the battle, while leading his infantry into the attack flag in hand (Dennis E. Showalter, *The Wars of Frederick the Great*, London and New York, Longman, 1996, pp. 149–56).
10. Charles of Lorraine, brother-in-law of the Empress Maria-Theresa, an Austrian field-marshal.
11. AN, 390 AP 25, MS of April 1817 (13 April), p. 7 (Bertrand, *Cahiers*, I, p. 216).
12. *Mémoires*, Montholon, IV, pp. 323–4. The second paragraph is in italics in the original.
13. More exactly, Alvinczy von Borberek, Joseph, baron. Aged 61, he had already been pushed back to Arcole by Bonaparte in November 1796.
14. Peter Vitius Quosdanovich, Austrian general of Croatian origin.
15. *Mémoires*, Montholon, IV, pp. 335–40; several passages in italics in the original.
16. *Corresp.*, VI, no. 4642, p. 166, to Masséna, Paris, 14 ventôse an VIII—5 March 1800.
17. Clausewitz, *On War*, III, 11, p. 240.

18. Montholon, *Récits de la captivité*, II, p. 363.
19. Claude-Jacques Lecourbe led a division of the army of Switzerland in several engagements in 1799 with brio. A friend of Moreau, he was disgraced in 1802, but rallied to Napoleon during the Hundred Days.
20. AN, 314 AP 30, MS 50 (Gourgaud, *Journal*, II, p. 345).
21. *Corresp.*, III, no. 1975, p. 160, reply to M. Dunan, note dictated at Mombello and presumed to be of 13 messidor an V—1 July 1797.
22. AN, 314 AP 30, MS 40 (Gourgaud, *Journal*, II, p. 320). Archduke Jean's army did not reach the battlefield of Wagram.
23. AN, 390 AP 25, MS of January–April 1821 (12 February), p. 7 (Bertrand, *Cahiers*, III, p. 62).
24. *Mémoires*, Montholon, V, p. 65; italics in the original.
25. AN, 390 AP 25, MS of 1818, p. 49 (Bertrand, *Cahiers*, II, p. 153). Having dispersed his forces and then concentrated his cavalry too close to the imperial forces, Turenne was surprised and defeated by them at Marienthal or Mergentheim (Swabia) on 5 May 1645 (Jean Bérenger, *Turenne*, Paris, Fayard, 1987, pp. 213–15).

CHAPTER 11

1. AN, 390 AP 25, MS of 1818 (October), p. 62 (Bertrand, *Cahiers*, II, pp. 175–6).
2. *Mémoires*, Montholon, V, pp. 272–3; italics in the original after the colon.
3. *Corresp.*, XXIII, no. 18312, p. 59, to Berthier, Major General of the army of Spain, Paris, 6 December 1811.
4. *Mémoires*, Montholon, V, p. 311.
5. Clausewitz, *On War*, III, 12, pp. 241–6.
6. The French remained masters of the battlefield. Utterly exhausted, however, they were incapable of pursuing the Russians, who retreated during the night. The casualties were enormous on both sides and are difficult to establish precisely: between 1,500 and 3,000 killed, including 7 generals, and from 4,300 to 7,000 wounded on the French side; unquestionably 7,000 dead and 23,000 wounded on the Russian side.
7. AN, 390 AP 25, MS of 1818 (January), p. 20 (Bertrand, *Cahiers*, II, pp. 44–5).
8. *Corresp.*, XI, no. 9275, p. 253, to Berthier, Strasbourg, 5 vendémiaire an XIV—27 September 1805.
9. *Corresp.*, XIII, no. 11251, p. 530, to Davout, Berlin, 13 November 1806, 16.00 hrs.
10. *Corresp.*, XV, no. 12465, pp. 144–5, to Prince Jérôme, Finkenstein, 24 April 1807.
11. *Corresp.*, XV, no. 12605, p. 247, to Prince Jérôme, Finkenstein, 18 May 1807.
12. *Corresp.*, XI, no. 9665, p. 535, to Prince Joseph, Munich, 12 January 1806.
13. *Corresp.*, XI, no. 9738, p. 573, to Prince Joseph, Paris, 2 February 1806.
14. *Corresp.*, XII, no. 9789, p. 29, to Prince Joseph, Paris, 9 February 1806.

15. *Corresp.*, XII, no. 9808, pp. 41–2, to Prince Joseph, Paris, 14 February 1806.
16. *Corresp.*, XIII, no. 10554, p. 5, to the King of Naples, Saint-Cloud, 26 July 1806.
17. *Corresp.*, XIII, no. 10573, p. 21, to the King of Naples, Saint-Cloud, 30 July 1806.
18. *Corresp.*, XII, no. 10368, p. 468, to Eugène, Saint-Cloud, 14 June 1806; XIX, no. 15305, p. 76, to Eugène, Schönbrunn, 6 June 1809, 09.00 hrs; XXVI, no. 20516, p. 171, to Macdonald, Dresden, 3 September 1813.
19. *Corresp.*, XIX, no. 15340, p. 100, to Clarke, Schönbrunn, 12 June 1809.

CHAPTER 12

1. Together with Soubise, the Count of Clermont embodied the incompetence of the French high command during the Seven Years War. At Crefeld, on 23 June 1758, he objected to cutting short his meal to put his army in a position to repel the attack by Duke Ferdinand of Brunswick. Fifty thousand French were utterly defeated by 32,000 Hanoverians, Hessians, and Brunswickians.
2. 'Dix-huit notes sur l'ouvrage intitulé *Considérations sur l'art de la guerre*', *Corresp.*, XXXI, pp. 342 and 345.
3. Clausewitz, *On War*, III, 13, p. 249.
4. Field-Marshal Duke Charles of Brunswick commanded the main Prussian army, in the presence of King Frederick-William II, at the Battle of Auerstaedt and was mortally wounded there. Field-Marshal Wichard von Möllendorff was another veteran of Frederick II's wars. He was 82 in 1806. Wounded, he took refuge at Erfurt after Jena, and to his great shame could not prevent its surrender. General Ernst von Rüchel took a long time to arrive on the battlefield of Jena on 14 October. Seriously wounded, he bravely remained at his post.
5. 'Dix-huit notes sur l'ouvrage intitulé *Considérations sur l'art de la guerre*', *Corresp.*, XXXI, p. 419.

CHAPTER 13

1. AN, 390 AP 25, MS of March 1818, p. 101 (Bertrand, *Cahiers*, II, p. 86).
2. AN, 390 AP 25, MS of January–September 1819 (May), p. 122 (Bertrand, *Cahiers*, II, p. 369).
3. 'Campagnes d'Italie (1796–1797)', *Corresp.*, XXIX, p. 137.
4. *Lettres inédites*, I, no. 513, p. 352, to Fouché, General Police Minister, Schönbrunn, 22 August 1809.
5. Marshal Victor, who commanded the siege of Cadiz from 5 February 1810.
6. General André Perreimond or Perreymond commanded a brigade of light cavalry.

7. General Édouard-Jean-Baptiste Milhaud, ex-deputy in the Convention, at the time commanded a division of dragoons.
8. General Deo-Gratias-Nicolas Godinot commanded an infantry brigade in the army of Andalusia.
9. *Corresp.*, XXI, no. 17531, p. 526, to the Prince of Neuchâtel and Wagram, Paris, night of 29–30 March 1811.
10. Clausewitz, *On War*, III, 14, p. 258.

CHAPTER 14

1. Clausewitz, *On War*, III, 17, p. 258.
2. *Corresp.*, III, no. 1800, p. 47, to the national guards of the Cisalpine Republic, draft proclamation presumed to be from Milan, 25 floréal an V—14 May 1797.
3. *Corresp.*, VII, no. 6068, p. 452, words of the First Consul at the State Council during the session of 14 floréal an X—4 May 1802.
4. *Corresp.*, VII, no. 6213, message to the legislative body of the Italian Republic, Paris, 9 thermidor an X—28 July 1802.
5. *Corresp.*, VIII, no. 6483, p. 129, short speech to the five deputies from Switzerland, Saint-Cloud, 20 frimaire an XI—11 December 1802.
6. *Corresp.*, X, no. 8204, p. 62, minutes of the reception for the presidents of the electoral colleges, prefects, presidents of the courts of appeal, etc., Paris, 15 frimaire an XIII—6 December 1804.

CONCLUSION TO BOOK III

1. Riley, *Napoleon as a General*, p. 60.

BOOK IV

1. Carl von Clausewitz, *Théorie du combat*, trans. from the German with a preface by Thomas Lindemann, Paris, ISC-Économica, 1998.

CHAPTER I

1. 'Dix-huit notes sur l'ouvrage intitulé *Considérations sur l'art de la guerre*', *Corresp.*, XXXI, p. 331.
2. 'Précis des guerres de Jules César', *Corresp.*, XXXII, pp. 82–3.
3. Clausewitz, *On War*, IV, 2, pp. 266–7.
4. Robert M. Epstein, *Napoleon's Last Victory and the Emergence of Modern War*, Lawrence, University Press of Kansas, 1994.
5. Clausewitz, *On War*, IV, 2, p. 266.
6. 'Dix-huit notes sur l'ouvrage intitulé *Considérations sur l'art de la guerre*', *Corresp.*, XXXI, p. 311.

7. Doubtless a writing error for Hohenlinden or Ettlingen.
8. *Corresp.*, XXXI, pp. 328–9.
9. AN, 314 AP 30, MS 40 (Gourgaud, *Journal*, II, p. 267).
10. 'Notes tirées du mémoire de M. le marquis de Vallière inséré dans les *Mémoires de l'Académie*, année 1772' (Napoleon I, *Manuscrits inédits, 1768–1791*, published in accordance with the original autographs by Frédéric Masson and Guido Biagi, Paris, P. Ollendorff, 1907, p. 53). At Raucoux (Rocourt), near Liège, in 1746, the infantry of Marshal de Saxe was preceded by artillery pieces and followed by cavalry. The villages held by the Austro-Dutch-British allies of Prince Charles of Lorraine were taken after suffering heavy shelling. Three years earlier, the British, commanded by their king in battle for the last time, passed over the body of the Duke of Noailles's French at Dettingen. The artillery enabled George II to throw the French into the Main. In 1757, Marshal d'Estrées, with 60,000 men, crushed the Duke of Cumberland's 36,000 Hanoverians and British at Hastembeck. French numbers had as much of an impact as their artillery.
11. *Corresp.*, XXVI, no. 20929, p. 458, to Eugène, Saint-Cloud, 20 November 1813.
12. Las Cases, *Mémorial*, vol. 18, pp. 125–6 (Dunan edn, II, p. 579).
13. Clausewitz, *Théorie du combat*, nos. 232 and 235, pp. 57 and 58.

CHAPTER 2

1. Clausewitz, *On War*, IV, 3, pp. 270–1.
2. *Notes inédites de l'empereur Napoléon Ier sur les Mémoires militaires du général Lloyd*, published by Ariste Ducaunnès-Duval, Bordeaux, Imprimerie G. Gounouilhou, 1991 (extracts from Vol. XXXV of the *Archives historiques de la Gironde*), p. 13.
3. Clausewitz, *Théorie du combat*, nos. 62 and 66, p. 33.
4. The Parisian toise was equivalent to 1.949 metres.
5. 'Essai sur la fortification de campagne', *Corresp.*, XXXI, pp. 464–5.
6. See Bk. III, Ch. 12 n. 1.
7. *Mémoires*, Montholon, V, p. 242; in italics in the original.
8. AN, 390 AP 25, MS of June 1817 (2 June), p. 4 (Bertrand, *Cahiers*, I, p. 231).
9. Clausewitz, *Théorie du combat*, nos. 350 and 356–61, pp. 72–3.
10. Clausewitz, *Théorie du combat*, nos. 401–2 and 432, pp. 78 and 82.
11. *Corresp.*, XII, no. 10032, pp. 231–3, official report of the Battle of Austerlitz, presented to Emperor Alexander by General Kutuzov, and observations by a French officer [Napoléon], Braunau, 28 March 1806.
12. Clausewitz, *Théorie du combat*, no. 505, pp. 91–2.
13. 'Précis des guerres de Jules César', *Corresp.*, XXXII, p. 83.
14. *Corresp.*, XXXII, pp. 58–9.
15. Antommarchi, *Derniers Momens de Napoléon*, I, p. 185.

16. Clausewitz, *On War*, IV, 4, pp. 274–5.
17. Clausewitz, *On War*, IV, 4, p. 276.
18. AN, 390 AP 25, MS of 1819, p. 143 (Bertrand, *Cahiers*, II, p. 440, where the text is placed in 1820).
19. Clausewitz, *Théorie du combat*, nos. 170 and 171, p. 48.
20. Las Cases, *Mémorial*, vol. 11, p. 50 (Dunan edn, I, p. 779).
21. 'Dix-huit notes sur l'ouvrage intitulé *Considérations sur l'art de la guerre*', *Corresp.*, XXXI, pp. 330–1.
22. AN, 390 AP 25, MS of 1819, pp. 161–2 (Bertrand, *Cahiers*, II, p. 448, where the text is placed in 1820).
23. 'Dix-huit notes sur l'ouvrage intitulé *Considérations sur l'art de la guerre*', *Corresp.*, XXXI, p. 413.
24. *Corresp.*, XXXI, p. 413.
25. *Notes inédites de l'empereur Napoléon Ier sur les Mémoires militaires du général Lloyd*, p. 14.
26. 'Dix-huit notes sur l'ouvrage intitulé *Considérations sur l'art de la guerre*', *Corresp.*, XXXI, p. 414.
27. *Corresp.*, XXXI, p. 415.
28. Gunther E. Rothenberg, *The Art of Warfare in the Age of Napoleon*, London, Batsford, 1977, pp. 152–3.
29. Clausewitz, *Théorie du combat*, nos. 104 and 105, pp. 39 and 40. See also no. 537, p. 95.
30. *Corresp.*, XIII, no. 10900, p. 277, to Marshal Soult, Mainz, 29 September 1806.
31. *Corresp.*, XXV, no. 19643, pp. 12–13, to General Bertrand, Paris, 2 March 1813. Another letter of the same kind can be found in *Corresp.*, XXV, no. 19868, p. 201, to Marshal Marmont, Mainz, 17 April 1813.
32. 'Campagnes d'Égypte et de Syrie', *Corresp.*, XXX, p. 53.
33. 'Précis des guerres de Jules César', *Corresp.*, XXXII, p. 69.

CHAPTER 3

1. Clausewitz, *On War*, IV, 7, p. 284.
2. Clausewitz, *Théorie du combat*, no. 156, p. 47.
3. 'Précis des guerres de Jules César', *Corresp.*, XXXII, p. 82.
4. Las Cases, *Mémorial*, vol. 3, p. 242 (Dunan edn, II, p. 277); in italics in the original.
5. Antommarchi, *Derniers Momens de Napoléon*, I, pp. 188–9. This episode from the Battle of Arcole occurred on 17 November 1796 and in fact contributed to the winning of the battle, with the presentation of 800 infantrymen on another point. Clausewitz estimated that 'the retreat of the Austrians [should be] attributed to their general situation, and to the news that a column was arriving from Legnago' (Carl von Clausewitz, *La Campagne de 1796 en Italie*, trans. from German, Paris, 1901; Pocket, 1990, p. 190).

6. Antommarchi, *Derniers Momens de Napoléon*, p. 187; the end of the sentence is in italics in the original.
7. *Corresp.*, XXV, no. 19951, p. 260, report of the Grande Armée, Lützen, 2 May 1813.
8. Gouvion Saint-Cyr, *Mémoires*, IV, p. 41.
9. Chaptal, *Mes Souvenirs*, pp. 294–5.
10. Clausewitz, *Théorie du combat*, nos. 127 and 184, pp. 43–4 and 50.
11. 'Campagne de 1815', *Corresp.*, XXXI, p. 187.
12. Guitton, *La Pensée et la guerre*, pp. 89–90.
13. Clausewitz, *Théorie du combat*, nos. 115a–117, p. 41.
14. Clausewitz, *On War*, IV, 7, p. 285.
15. 'Dix-huit notes sur l'ouvrage intitulé *Considérations sur l'art de la guerre*', *Corresp.*, XXXI, p. 398.

CHAPTER 4

1. Clausewitz, *On War*, IV, 8, p. 292.
2. *Corresp.*, XV, no. 12747, p. 329, 78th bulletin of the Grande Armée, Heilsberg, 12 June 1807.
3. Jourdan had decided to take the offensive in Germany on 25 March 1799 at Masséna's entreaties in order to release the latter's army of Switzerland. He committed his divisions on an excessive front without ensuring their liaison and against the much more numerous Austrian forces of Archduke Charles. The latter won one of his finest victories.
4. 'Précis des événements militaires arrivés pendant les six premiers mois de 1799', *Corresp.*, XXX, p. 263.
5. AN, 314 AP 30, MS 50 (Gourgaud, *Journal*, II, p. 347).
6. *Corresp.*, XI, no. 9532, p. 440, to Talleyrand, in the bivouac two leagues from Brünn, 9 frimaire an XIV—30 November 1805, 16.00 hrs.
7. *Corresp.*, XIX, no. 15694, p. 379, to Clarke, Schönbrunn, 21 August 1809.

CHAPTER 5

1. 'Campagnes d'Italie (1796–1797)', *Corresp.*, XXIX, p. 192.
2. 'Campagnes d'Égypte et de Syrie', *Corresp.*, XXX, p. 170.
3. *Corresp.*, XIII, no. 11001, p. 348, to M. de La Marche, n.p. n.d. but delivered 13 October 1806.
4. *Corresp.*, XVIII, no. 14445, p. 40, to Marshal Victor, Vitoria, 6 November 1808, midnight.
5. *Mémoires*, Montholon, V, pp. 268–72.
6. *Corresp.*, V, pp. 22–3.
7. David Chandler, *The Military Maxims of Napoleon*, trans. G. C. D'Aguilar, London, Greenhill Books, 2002, p. 127.

8. Clausewitz, *Théorie du combat*, no. 604, pp. 103–4.

9. *Corresp.*, XXIII, no. 18503, p. 231, to Marmont, commanding the army of Portugal, Paris, 18 February 1812; italics in the original.

10. Auguste-Frédéric-Louis Viesse de Marmont, *De l'esprit des institutions militaires*, preface by Bruno Colson, Paris, ISC-FRS-Économica, 2001 (1st edn, 1845), p. 14.

11. Chaptal, *Mes Souvenirs*, p. 301.

12. Owen Connelly, *Blundering to Glory: Napoleon's Military Campaigns*, 3rd edn, Lanham, Md., Rowman & Littlefield Publishers, 2006, pp. 44–5.

13. AN, 314 AP 30, MS 40 (Gourgaud, *Journal*, II, p. 325).

14. Clausewitz, *On War*, IV, 9, p. 294.

CHAPTER 6

1. Clausewitz, *On War*, IV, 11, p. 306.

2. *Corresp.*, XI, no. 9405, pp. 342–3, proclamation, Elchingen, 29 vendémiaire an XIV—21 October 1805.

3. *Corresp.*, XIII, no. 10983, p. 337, to Murat, Auma, 12 October 1806, 04.00 hrs.

4. Clausewitz, *On War*, IV, 11, p. 307.

5. Clausewitz, *On War*, IV, 11, p. 309.

6. Clausewitz, *On War*, IV, 11, p. 309.

7. *Corresp.*, XIII, no. 10977, p. 333, to Soult, Ebersdorf, 10 October 1806, 08.00 hrs.

8. *Corresp.*, XXVI, no. 20360, p. 35, instructions for the Prince of Moskowa and the Duke of Raguse, Dresden, 12 August 1813.

9. *Corresp.*, XXVI, no. 20437, p. 112, to Maret, Löwenberg, 22 August 1813.

10. Having redeemed a situation that was turning out badly, thanks to the arrival of Desaix, Napoleon pushed the Austrians back to the River Bormida. But the bulk of their army remained intact. They were not destroyed on a tactical level. In terms of operations, however, their position was jeopardized.

11. In the complex match that occurred in late April 1809 in Bavaria, Napoleon did not have enough good troops to do battle with Archduke Charles immediately. His system consisted in depriving the Archduke of his line of communication with Vienna, in expelling him from Bavaria into Bohemia, and himself driving on to the Austrian capital. He sought to achieve a success in the campaign before obtaining another one in a major battle.

12. General Levin von Bennigsen was commander-in-chief of the Russian army.

13. A letter from Berthier to Bernadotte fell into Cossack hands. It revealed to Bennigsen the position of the Grande Armée's corps and Napoleon's plan, which sought to cut off the Russians' line of retreat (F. Loraine Petre, *Napoleon's Campaign in Poland 1806–1807*, London, Greenhill Books, 2001 (1st edn, 1901), pp. 147–9).

14. Napoleon sought to cut Bennigsen off from his communication with the Niemen and Russia in order to push him westwards, to the town of Elbing, very close to the Baltic Sea, in the Gulf of Danzig.

15. AN, 390 AP 25, MS of 1819, p. 161 (Bertrand, *Cahiers*, II, p. 447, where the text is placed in 1820).

16. Epstein, *Napoleon's Last Victory*, p. 17.

17. On 28 November 1805, Napoleon deliberately allowed his advance guard, installed at Wischau, to be attacked by Bagration's Russians and had it withdraw so that the latter became emboldened and pursued the French army to the battlefield he had chosen (Jacques Garnier, *Austerlitz, 2 décembre 1805*, Paris, Fayard, 2005, pp. 191–203).

18. *Corresp.*, XII, no. 10032, pp. 230–1 and 233, official report of the Battle of Austerlitz, presented to Emperor Alexander by General Kutuzov, with observations by a French officer, Braunau, 28 March 1806.

19. Telp, *The Evolution of Operational Art*, pp. 1–2.

CHAPTER 7

1. *Corresp.*, XVIII, no. 14460, p. 51, to Joseph, Cubo, 10 November 1808, 20.00 hrs.

2. *Mémoires*, Montholon, V, p. 197.

3. AN, 390 AP 25, MS of 1819, p. 161 (Bertrand, *Cahiers*, II, p. 446: placed in 1820).

4. *Corresp.*, XI, no. 9386, p. 329, to Murat, Elchingen Abbey, 25 vendémiaire an XIV—17 October 1805, 14.00 hrs.

5. Frédéric-Henri Walther commanded the 2nd division of dragoons in the cavalry reserve. On 16 November 1805, he had participated in the hard fighting at Hollabrunn (or Schöngraben) against the Russian rearguard, commanded by Bagration.

6. *Corresp.*, XI, no. 9509, p. 425, to Lannes, Znaym, 27 brumaire an XIV—18 November 1805, 21:00 hrs.

7. Clausewitz, *On War*, IV, 12, p. 313; italics in original.

8. *Corresp.*, XIII, no. 11030, p. 372, 12th bulletin of the Grande Armée, Halle, 19 October 1806.

9. *Corresp.*, XIII, no. 11053, p. 385, 14th bulletin of the Grande Armée, Dessau, 22 October 1806.

10. AN, 390 AP 25, MS of April 1817 (13 April), p. 8 (Bertrand, *Cahiers*, I, p. 217).

11. 'Diplomatie.—Guerre', *Corresp.*, XXX, p. 442.

12. *Corresp.*, I, no. 1000, p. 616, to the executive Directory, Due-Castelli, 30 fructidor an IV—16 September 1796.

13. Maurice de Saxe, *Mes Rêveries*, presented by Jean-Pierre Bois, Paris, CFHM-ISC-Économica, 2002 (1st edn, 1756), p. 224; Vegetius, *De re militari*, III, 21, Paris, Corréard, 1859, p. 151. I owe this point to Thierry Widemann.

14. 'Dix-huit notes sur l'ouvrage intitulé *Considérations sur l'art de la guerre*', *Corresp.*, XXXI, p. 343.
15. Clausewitz, *On War*, IV, 12, p. 312.

CHAPTER 8

1. 'Précis des événements militaires arrivés pendant les six derniers mois de 1799', *Corresp.*, XXX, p. 302.
2. Clausewitz, *On War*, IV, 13, p. 322.
3. 'Précis des guerres du maréchal de Turenne', *Corresp.*, XXXII, p. 117.
4. O'Meara, *Napoleon in Exile*, I, p. 479.
5. 'Dix-huit notes sur l'ouvrage intitulé *Considérations sur l'art de la guerre*', *Corresp.*, XXXI, p. 345.
6. Las Cases, *Mémorial*, vol. 17, p. 262 (Dunan edn, II, p. 538).
7. Jean-Baptiste Lemonnier-Delafosse, *Campagnes de 1810 à 1815 ou Souvenirs militaires*, Le Havre, A. Lemale, 1850, pp. 163–4.
8. 'Précis des événements militaires arrivés pendant les six derniers mois de 1799', *Corresp.*, XXX, p. 298.

CONCLUSION TO BOOK IV

1. Georges Lefebvre, *Napoléon*, Paris, Félix Alcan, 1935, p. 203.

BOOK V

CHAPTER 1

1. 'Dix-huit notes sur l'ouvrage intitulé *Considérations sur l'art de la guerre*', *Corresp.*, XXXI, pp. 303–4.
2. Chaptal, *Mes Souvenirs*, p. 299.
3. Bertaud, *Quand les enfants parlaient de gloire*, pp. 127 and 130.
4. Las Cases, *Mémorial*, vol. 4, p. 193 (Dunan edn, I, p. 355).
5. Frederick II.
6. *Lettres inédites*, I, no. 419, pp. 290–1, to Caulaincourt, Paris, 6 March 1809.

CHAPTER 2

1. Clausewitz, *On War*, V, 4, p. 338. On the distinction between fire and clash, see the valuable brochures by Daniel Reichel, 'Le Feu' (I, II, and III), *Études et documents*, Berne, Département militaire fédéral, 1982-I, 2 and 1983; 'Le Choc', *Études et documents*, Berne, Département militaire fédéral, 1984.
2. *Mémoires*, Montholon, V, p. 120.
3. *Mémoires*, Montholon, I, pp. 277–8.

4. 'Dix-huit notes sur l'ouvrage intitulé *Considérations sur l'art de la guerre*', *Corresp.*, XXXI, p. 323.
5. 'Projet d'une nouvelle organisation de l'armée', *Corresp.*, XXXI, p. 453.
6. 'Dix-huit notes sur l'ouvrage intitulé *Considérations sur l'art de la guerre*', *Corresp.*, XXXI, pp. 410–11.
7. *Corresp.*, XIX, no. 15678, p. 361, to Clarke, Schönbrunn, 18 August 1809.
8. Clausewitz, *On War*, V, 4, p. 346.
9. Clausewitz, *On War*, V, 4, p. 339.
10. This over-simplifies matters somewhat. While Wellington's infantry was not supported by cavalry at the start of the engagement, it nevertheless had fifteen cannons. When the French cuirassiers charged, the allies had a little over 2,000 cavalry—much fewer, it is true, than their opponents (Alain Arcq, *La Bataille des Quatre-Bras, 16 juin 1815*, Annecy-le-Vieux, Historic'one Éditions, 2005, pp. 53–83).
11. 'Campagne de 1815', *Corresp.*, XXXI, p. 211.
12. Clausewitz, *On War*, V, 4, pp. 340–1.
13. Colin, *L'Éducation militaire*, pp. 73–4.
14. *Mémoires*, Montholon, I, pp. 271–2.
15. Las Cases, *Mémorial*, vol. 11, pp. 48–9 (Dunan edn, I, p. 779).
16. Bonnal, 'La Psychologie militaire de Napoléon', p. 422.
17. In directing the artillery of the imperial guard with skill and vigour, Antoine Drouot contributed to the victories of Wagram (6 July 1809) and Hanau (30 October 1813), in particular.
18. Jean-Barthélemy Sorbier had in fact started out as a captain in a horse artillery company in 1792. He was commander-in-chief of the artillery of the Grande Armée in 1813. Elzéar-Auguste Cousin de Dommartin had a shorter career. He died of tetanus in Egypt in 1799, where he was commander-in-chief of the artillery of the army of the Orient. He had commanded the light artillery of the army of Italy in 1796 and distinguished himself at Castiglione and Roveredo.
19. AN, 390 AP 25, MS of January–April 1821 (21 February), p. 11 (Bertrand, *Cahiers*, III, p. 72).
20. Jean-Jacques-Basilien de Gassendi had had Lieutenant Bonaparte under his command in the La Fère artillery regiment in 1785. Brigadier-general in 1800, retired in 1803, and inspector-general of artillery in 1805, he did not participate in the Empire's campaigns. He is above all known for having written an *Aide-mémoire à l'usage des officiers d'artillerie*, which never possessed official status, but went through many editions down to the 1840s. It amounted to a veritable summa of technical military knowledge.
21. AN, 314 AP 30, MS 39 (Gourgaud, *Journal*, II, p. 83).
22. 'Campagnes d'Italie (1796–1797)', *Corresp.*, XXIX, p. 191.
23. Paddy Griffith, *The Art of War of Revolutionary France 1789–1802*, London, Greenhill Books, 1998, pp. 236–42; John Lynn, *The Bayonets of the Republic:*

Motivation and Tactics in the Army of Revolutionary France, 1791–94, 2nd edn, Boulder, Colo., Westview Press, 1996, pp. 212–13.

24. *Corresp.*, XIV, no. 11417, p. 52, to Prince Eugène, Posen, 8 December 1806.
25. *Corresp.*, XIV, no. 11896, p. 346, to Bernadotte, Osterode, 27 February 1807, 17.30 hrs.
26. *Corresp.*, XXII, no. 17603, p. 58, to Jérôme Napoléon, Paris, 12 April 1811.
27. *Corresp.*, XXVI, no. 20929, p. 458, to Eugène Napoléon, Saint-Cloud, 20 November 1813.
28. *Corresp.*, XIX, no. 15530, p. 248, to Clarke, Schönbrunn, 15 July 1809.
29. AN, 390 AP 25, MS of 1818 (January), p. 4 (Bertrand, *Cahiers*, II, p. 47).
30. AN, 314 AP 30, MS 50 (Gourgaud, *Journal*, II, p. 347).
31. *Corresp.*, XXII, no. 18113, p. 463, to Clarke, Compiègne, 4 September 1811.
32. Clausewitz, *On War*, V, 4, pp. 343–5.
33. 'Campagnes d'Italie (1796–1797)', *Corresp.*, XXIX, p. 191. The same idea is to be found in Bertrand, *Cahiers*, III, p. 29.
34. At Vauchamps, on 14 February 1814, on Grouchy's orders the French cavalry broke several Prussian regiments formed into squares, cornered others in a wood, took numerous prisoners, and created utter disarray among Blücher's troops. Three days later, at Nangis, the French cavalry turned the position of the Russians and broke their ranks.
35. AN, 314 AP 30, MS 17 (Gourgaud, *Journal*, I, p. 214).
36. 'Dix-huit notes sur l'ouvrage intitulé *Considérations sur l'art de la guerre*', *Corresp.*, XXXI, p. 320.
37. 'Notes sur l'introduction à l'*Histoire de la guerre en Allemagne en 1756, entre le roi de Prusse et l'impératrice-reine et ses alliés, etc.*, par le général Lloyd', *Corresp.*, XXXI, pp. 426–8.
38. *Corresp.*, XXXI, pp. 428–9.
39. General Horace Sébastiani at the time commanded the Grande Armée's 2nd cavalry corps and Latour-Maubourg the 1st.
40. *Lettres inédites*, II, no. 967, p. 217, to Prince Eugène, Trianon, 15 March 1813.
41. *Corresp.*, XII, no. 9966, p. 183, to Prince Eugène, 13 March 1806.
42. *Corresp.*, XII, no. 10104, p. 288, to Prince Eugène, Saint-Cloud, 15 April 1806.
43. *Corresp.*, XVII, no. 13751, p. 14, to Maréchal Bessières, Bayonne, 16 April 1808.
44. *Corresp.*, XXIII, no. 18248, p. 4, to Minister of War Clarke, Saint-Cloud, 12 November 1811.
45. *Corresp.*, XXIII, no. 18366, p. 106, to Clarke, Paris, 25 December 1811.
46. 'Projet d'une nouvelle organisation de l'armée', *Corresp.*, XXXI, p. 456.
47. O'Meara, *Napoleon in Exile*, I, p. 131.
48. Commanding the 2nd corps of the army of Naples, charged with conquering Calabria, General Reynier had been beaten by the British at the engagement of Maida or Sant'Eufemia on 4 July 1806.
49. *Corresp.*, XIII, no. 10629, pp. 63–4, to the King of Naples, Saint-Cloud, 9 August 1806.

50. *Corresp.*, XIX, no. 15274, p. 58, to General Dejean, Ebersdorf, 29 May 1809 and no. 15530, p. 249, to Clarke, Schönbrunn, 15 July 1809.

51. 'Dix-huit notes sur l'ouvrage intitulé *Considérations sur l'art de la guerre*', *Corresp.*, XXXI, p. 309.

52. *Corresp.*, VII, no. 6061, p. 449, to Berthier, Minister of War, Paris, 9 floréal an X—29 April 1802.

53. *Corresp.*, IV, no. 3220, p. 454, order, Cairo, 17 fructidor an VI—3 September 1798.

54. 'Dix-huit notes sur l'ouvrage intitulé *Considérations sur l'art de la guerre*', *Corresp.*, XXXI, p. 314.

55. *Corresp.*, XXVI, no. 20791, p. 350, to Maréchal Marmont, Düben, 13 October 1813, 10.00 hrs.

56. *Mémoires*, Montholon, V, p. 120.

57. *Corresp.*, II, no. 1311, p. 195, army order, Verona, 1er nivôse an V—21 December 1796.

58. *Corresp.*, XI, no. 9522, p. 435, order of the day, Brünn, 3 frimaire an XIV—24 November 1805.

59. [Marchand], *Mémoires*, I, p. 97.

60. General Jean-Ambroise Baston, Count of Lariboisière, was commander-in-chief of the Grande Armée's artillery in 1812 and died from illness after the retreat from Russia.

61. Chaptal, *Mes Souvenirs*, pp. 296–9.

62. [Barante, Amable-Guillaume-Prosper de], *Souvenirs du baron de Barante de l'Académie française (1782–1866)*, published by Cl. de Barante, 3 vols, Paris, Calmann-Lévy, 1890, I, p. 72.

63. *Corresp.*, V, no. 4323, p. 542, to the executive Directory, Alexandria, 10 thermidor an VII—28 July 1799.

64. *Corresp.*, XVI, no. 13166, p. 38, decision, Rambouillet, 16 September 1807.

65. AN, 314 AP 30, MS 39 (Gourgaud, *Journal*, II, p. 167).

66. AN, 390, AP 25, MS of January 1817 (12 January), p. 8 (Bertrand, *Cahiers*, I, p. 181).

67. *Corresp.*, XII, no. 9820, p. 50, to General Dejean, Paris, 15 February 1806.

68. AN, 390 AP 25, MS of April 1817 (5 April), p. 3 (Bertrand, *Cahiers*, I, p. 210).

69. Jean-François Lemaire, *Les Blessés dans les armées napoléoniennes*, Paris, Lettrage Distribution, 1999, pp. 271–5.

70. AN, 390 AP 25, MS of August 1817 (10 August), p. 3 (Bertrand, *Cahiers*, I, p. 256).

CHAPTER 3

1. Clausewitz, *On War*, V, 5, p. 348.

2. AN, 314 AP 30, MS 39 (Gourgaud, *Journal*, II, pp. 143–4). Virtually the same text is to be found in Montholon, *Récits de la captivité*, II, p. 133.

3. Clausewitz, *On War*, V, 5, p. 347.
4. AN, 314 AP 30, MS 40 (Gourgaud, *Journal*, II, p. 268).
5. AN, 314, AP 30, MS 35 (Gourgaud, *Journal*, II, p. 50). The victory of Maurice de Saxe's French army over the Duke of Cumberland's allies at Fontenoy, on 11 May 1745, was incontestable. It would have been more decisive if the armies of the time had been integrated, which would have given them more flexibility in their operations. The vanquished were not pursued (Jean-Pierre Bois, *Fontenoy 1745. Louis XV, arbitre de l'Europe*, Paris, Économica, 1996).
6. Clausewitz, *On War*, V, 5, p. 348.
7. *Corresp.*, VI, no. 4552, p. 107, to General Berthier, Minister of War, Paris 5 pluviôse an VIII—25 January 1800.
8. 'Dix-huit notes sur l'ouvrage intitulé *Considérations sur l'art de la guerre*', *Corresp.*, XXXI, p. 412.
9. Clausewitz, *On War*, V, 5, pp. 349–50.
10. Béraud, *La Révolution militaire napoléonienne*, 1, pp. 11–12.
11. François-Étienne-Christophe Kellermann, the victor at Valmy, Marshal of the Empire in 1804.
12. *Corresp.*, I, no. 420, pp. 277–8, to the executive Directory, Lodi, 25 floréal an IV—14 May 1796.
13. *Corresp.*, I, no. 421, p. 279, to Carnot, Lodi, 25 floréal an IV—14 May 1796.
14. *Corresp.*, II, no. 1637, p. 420, to Carnot, Goritz, 5 germinal an V—25 March 1797.
15. *Corresp.*, V, no. 4188, p. 461, to the executive Directory, Cairo, 1 messidor an VII—19 June 1799.
16. *Corresp.*, VI, no. 4744, p. 245, to Bernadotte, commander-in-chief of the army of the West, Paris, 11 floréal an VIII—1 May 1800.
17. 'Dix-huit notes sur l'ouvrage intitulé *Considérations sur l'art de la guerre*', *Corresp.*, XXXI, p. 418.
18. *Mémoires*, Montholon, II, p. 51.

CHAPTER 4

1. Clausewitz, *On War*, V, 6, pp. 353–6.
2. AN, 390 AP 25, MS of March 1818, p. 102 (Bertrand, *Cahiers*, II, pp. 87–8).
3. 'Campagnes d'Italie (1796–1797)', *Corresp.*, XXIX, p. 189.
4. Clausewitz, *On War*, V, 6, p. 357.
5. Clausewitz, *On War*, V, 6, pp. 357–8.
6. 'Dix-huit notes sur l'ouvrage intitulé *Considérations sur l'art de la guerre*', *Corresp.*, XXXI, p. 412.
7. AN, 314 AP 30, MS 39 (Gourgaud, *Journal*, II, p. 144, with an inversion of the words 'arithmetical' and 'geometrical'). Montholon gives virtually the same text, in conformity with Gourgaud's manuscript (Montholon, *Récits de la captivité*, II, p. 134).

CHAPTER 5

1. Clausewitz, *On War*, V, 7, p. 359.
2. AN, 390 AP 25, MS of 1818 (October), p. 64 (Bertrand, *Cahiers*, II, p. 181).
3. *Correspondance inédite*, I, no. 217, p. 124, order of the day, Vienna, 23 brumaire an XIV—14 November 1805.
4. 'Dix-huit notes sur l'ouvrage intitulé *Considérations sur l'art de la guerre*', *Corresp.*, XXXI, pp. 320–2.
5. Guillaume-Marie-Anne Brune, Marshal of France in 1804, assassinated during the white terror in Avignon in 1815.
6. *Corresp.*, VI, no. 4989, p. 407, to Carnot, Paris, 26 messidor an VIII—15 July 1800.
7. *Lettres inédites*, I, no. 444, p. 309, to Prince Eugène, Saint-Polten, 10 May 1809, 05.00 hrs.
8. *Corresp.*, XXIII, no. 18411, p. 154, to Berthier, Paris, 2 January 1812.
9. *Corresp.*, XXVI, no. 20595, pp. 219–20, to Berthier, Pirna, 19 September 1813.
10. 'Dix-huit notes sur l'ouvrage intitulé *Considérations sur l'art de la guerre*', *Corresp.*, XXXI, p. 323.

CHAPTER 6

1. Clausewitz, *On War*, V, 8, pp. 369–71.
2. *Corresp.*, XIII, no. 10572, p. 20, to the King of Naples, Saint-Cloud, 30 July 1806.
3. On 15 April 1809, the 35th of the line lost its eagle and more than 2,000 prisoners in the engagement of Pordenone.
4. Aide de camp to Napoleon, General Lauriston had taken charge of a division from Baden in the mountains and seized the Semmering Pass before joining the army of Italy.
5. The young Austrian archdukes were often joined by a more experienced officer to advise them. Count Laval Nugent von den Grafen von Westmeath was already a general in 1809 and became a field-marshal in 1849.
6. *Corresp.*, XIX, no. 15310, pp. 81–2, to Eugène Napoléon, Schönbrunn, 7 June 1809, 02.30 hrs.
7. Béraud, *La Révolution militaire napoléonienne*, 1, p. 168.
8. *Corresp.*, XXVI, no. 20492, p. 155, note on the general situation of my affairs, Dresden, 30 August 1813.

CHAPTER 7

1. Clausewitz, *On War*, V, 9, p. 372.
2. 'Dix-huit notes sur l'ouvrage intitulé *Considérations sur l'art de la guerre*', *Corresp.*, XXXI, p. 415.

3. Brother of Frederick II, he was likewise an excellent general, who understood the development of the art of war, despite this remark by Napoleon. He was the victor at the Battle of Freiberg (1762).

4. *Mémoires*, Montholon, V, p. 330. The second part of the sentence is in italics in the original.

5. Clausewitz, *On War*, V, 9, pp. 373–4.

6. Charles Esdaile, *The Peninsular War: A New History*, London, Penguin, 2003, p. 441.

7. 'Dix-huit notes sur l'ouvrage intitulé *Considérations sur l'art de la guerre*', *Corresp.*, XXXI, p. 315.

8. *Corresp.*, XVII, no. 13762, p. 23, to Eugène Napoléon, Bayonne, 18 April 1808.

9. Telp, *The Evolution of Operational Art*, p. 89.

10. *Corresp.*, XVII, to Murat, Bayonne, 19 May 1808, 14.00 hrs.

11. *Corresp.*, XIV, no. 11579, p. 161, to Eugène, Warsaw, 7 January 1807.

12. *Corresp.*, XII, no. 10003, p. 208, to Eugène, Paris, 21 March 1806.

13. Other letters to Prince Eugène about camps in Dalmatia and Italy: *Corresp.*, XII, no. 10324 and 10343, Saint-Cloud, 6 and 10 June 1806.

14. A friend from Napoleon's youth, Andoche Junot commanded the army charged with invading Portugal to enforce the blockade of British exports decreed in Berlin in 1806.

15. *Corresp.*, XVI, no. 13416, p. 215, to Junot, commander of the army of Portugal, Milan, 23 December 1807.

CHAPTER 8

1. Clausewitz, *On War*, V, 10, p. 375.

2. 'Dix-huit notes sur l'ouvrage intitulé *Considérations sur l'art de la guerre*', *Corresp.*, XXXI, p. 411.

3. Clausewitz, *On War*, V, 12, pp. 386–7.

4. Clausewitz, *On War*, V, 10, pp. 375–8.

5. 'Note sur le *Traité des grandes opérations militaires* par le général baron Jomini', *Corresp.*, XXIX, p. 349.

6. *Mémoires*, Montholon, IV, p. 316.

7. Béraud, *La Révolution militaire napoléonienne*, 1, p. 12.

8. Carolyn Shapiro, 'Napoleon and the Nineteenth-Century Concept of Force', *Journal of Strategic Studies*, 11, 1988–4, pp. 509–19.

9. Joachim Fischer, *Napoleon und die Naturwissenschaften*, Wiesbaden and Stuttgart, F. Steiner, 1988.

10. Strachan, *Carl von Clausewitz's* On War, pp. 88 and 143.

11. *Corresp.*, XI, no. 9225, p. 210, to Prince Eugène, Saint-Cloud, 29 fructidor an XIII—16 September 1805.

12. War artist, officer in the engineers, aide de camp to Marshal Berthier in 1805.

13. [Louis-François Lejeune], *Mémoires du général Lejeune*, publiés par Germain Bapst, 2 vols, Paris, Firmin-Didot, 1895–6; Éditions du Grenadier, 2001, pp. 22–3.

14. *Corresp.*, XI, no. 9374, p. 318, to Marshal Soult, Augsburg, 20 vendémiaire an XIV—12 October 1805.

15. The engagements of Wertingen (8 October 1805), Günzburg (9 October), and Elchingen (14 October) made it possible to engage with Mack's Austrian army and forced it to shut itself up in Ulm. At Memmingen on 14 October, Marshal Soult cut him off from any possibility of retreating towards the Tyrol. During the engagements of Albeck (15 and 16 October), Langenau (16 October), and Neresheim (17 October), Ney's and Murat's troops intercepted other Austrian elements that were trying to escape the French grip and took them prisoner.

16. *Corresp.*, XI, no. 9392, pp. 335–6, 6th bulletin of the Grande Armée, Elchingen, 26 vendémiaire an XIV—18 October 1805.

17. *Corresp.*, XI, no. 9393, p. 336, to the Empress Josephine, Elchingen, 27 vendémiaire an XIV—19 October 1805.

18. Connelly, *Blundering to Glory*, pp. 81–2; Garnier, *Austerlitz*, pp. 89–93.

19. Anne-Jean-Marie-René Savary was a gendarmerie and, secondarily, a cavalry general and an aide de camp to the Emperor. He accomplished several kinds of mission, in particular diplomatic ones. Provisionally placed at the head of the army of Spain after Murat's departure on 15 June 1808, he left this command the following month.

20. The victor of Fleurus (26 June 1794).

21. *Lettres inédites*, I, no. 324, p. 221, to Joseph Napoléon, Bayonne, 18 July 1808.

22. *Corresp.*, XXVII, no. 21417, p. 275, to Berthier, La Ferté-sous-Jouarre, 2 March 1814, 18.00 hrs.

23. Clausewitz, *On War*, V, 12, pp. 385–6.

24. Murat exhausted his cavalry at the start of the campaign. The horses died in their thousands. In letting him take the head of the Grande Armée's advance guard, Napoleon imposed an unsustainable pace on its columns. Davout was furious about this and quarrelled with Murat. The Emperor affected to agree with Davout, but secretly ordered Murat to continue (Connelly, *Blundering to Glory*, pp. 168 and 171).

25. *Corresp.*, XXIII, no. 18869, p. 541, to Davout, Kovno, 26 June 1812, 03.30 hrs.

26. *Corresp.*, XIII, no. 11100, p. 428, to Lannes, Berlin, 28 October 1806, midnight.

27. *Corresp.*, XVII, no. 14223, p. 409, notes on the current position of the army in Spain, Bayonne, 21 July 1808.

CHAPTER 9

1. *Corresp.*, XIV, no. 11897, p. 347, to Talleyrand, Osterode, 27 February 1807.
2. *Corresp.*, XIV, no. 12004, p. 425, to General Lemarois, Osterode, 12 March 1807.
3. Clausewitz, *On War*, V, 14, p. 396.
4. Clausewitz, *On War*, V, 14, p. 398.
5. *Corresp.*, XII, no. 9944, p. 166, to Prince Joseph, Paris, 8 March 1806.
6. Clausewitz, *On War*, V, 14, p. 402.
7. *Corresp.*, XVIII, no. 14909, p. 359, to Prince Eugène, Paris, 16 March 1809; *Corresp.*, XXV, no. 19706, p. 70, to Jérôme Napoléon, King of Westphalia, Trianon, 12 March 1813.
8. *Corresp.*, XX, no. 16521, p. 388, to Berthier, Major General of the army of Spain, Le Havre, 29 May 1810.
9. Las Cases, *Mémorial*, Vol. 18, p. 128 (Dunan edn, II, p. 579).
10. AN, 390 AP 25, MS of November 1816 (24 November), p. 10 (Bertrand, *Cahiers*, I, p. 151).
11. *Mémoires*, Montholon, II, p. 51.
12. Clausewitz, *On War*, V, 14, p. 404.
13. *Corresp.*, XVI, no. 13327, p. 439, to Clarke, Minister of War, Fontainebleau, 5 November 1807.
14. Clausewitz, *On War*, V, 14, p. 405.
15. Georges Lefebvre, *Napoléon*, Paris, Félix Alcan, 1935, pp. 198–200.

CHAPTER 10

1. *Corresp. gén.*, VI, no. 11279, p. 37 to Joseph, commander-in-chief of the army of Naples, Munich, 12 January 1806; Hubert Camon, *La Guerre napoléonienne. Les systèmes d'opérations. Théorie et technique*, Paris, Chapelot, 1907; ISC-Économica, 1997, p. 57.
2. *Corresp.*, XIX, no. 15373, p. 126, to Eugène Napoléon, Schönbrunn, 18 June 1809, 09.00 hrs.
3. *Lettres inédites*, I, no. 461, p. 318, to Prince Eugène, Schönbrunn, 20 June 1809.
4. An allusion to several eighteenth-century thinkers, such as Maurice de Saxe, the Prince de Ligne, and Guibert (Colson, *L'Art de la guerre de Machiavel à Clausewitz*, pp. 165, 195, and 214).
5. Joseph II, Emperor of Austria had dismantled the post in the Low Countries (today's Belgium), of which he was the sovereign.
6. *Corresp.*, XIII, no. 10726, p. 131, to General Dejean, Saint-Cloud, 3 September 1806.
7. A tributary of the Danube.

8. 'Dix-huit notes sur l'ouvrage intitulé *Considérations sur l'art de la guerre*', *Corresp.*, XXXI, pp. 357–8.
9. Camon, *La Guerre napoléonienne*, p. 58.
10. *Corresp.*, XIII, no. 10920, pp. 292–3, to the King of Holland, Mainz, 30 September 1806.
11. AN, 314 AP 30, MS 50 (Gourgaud, *Journal*, II, p. 347).
12. Jérémie Benoît and Bernard Chevallier, *Marengo. Une victoire politique*, Paris, Réunion des Musées nationaux, 2000, pp. 122–4; Thierry Lentz, *Le Grand Consulat 1799–1804*, Paris, Fayard, 1999, p. 235.
13. Connelly, *Blundering to Glory*, pp. 67–8.
14. Bourcet, *Principes de la guerre de montagnes*, p. 110; Wilkinson, *The Rise of General Bonaparte*, p. 147.
15. AN, 314 AP 30, MS 39 (Gourgaud, *Journal*, II, p. 81).
16. *Mémoires*, Montholon, V, p. 338; italics in the original.
17. *Mémoires*, Montholon, V, pp. 216–17; in italics in the original.
18. *Mémoires*, Montholon, V, pp. 173 and 190–1.
19. A palace and arsenal in Madrid.
20. *Corresp.*, XVII, no. 14343, pp. 524–5, note for the King of Spain, Châlons-sur-Marne, 22 September 1808.
21. Partisans whom the troops of Louis XIV had to confront. The barbets were Waldensian Protestants from the Piedmontese side of the Alps; the miquelets were Pyreneans, defenders of the cause of the Hapsburgs during the War of the Spanish Succession.
22. *Corresp.*, XVII, no. 14317, p. 528, to Joseph Napoléon, King of Spain, Kaiserslautern, 24 September 1808.

CHAPTER 11

1. Clausewitz, *On War*, V, 17, p. 416.
2. *Corresp.*, XI, no. 9174, p. 172, to Prince Eugène, camp of Boulogne, 14 fructidor an XIII—1 September 1805.
3. Provinces of southern Austria.
4. *Corresp.*, XIX, no. 15305, p. 77, to Prince Eugène, Schönbrunn, 6 June 1809, 09.00 hrs.
5. *Œuvres de Napoléon Bonaparte*, 5 vols, Paris, Panckoucke, 1821, I, p. 46.
6. 'Campagnes d'Italie (1796–1797)', *Corresp.*, XXIX, pp. 75–7.

CONCLUSION TO BOOK V

1. Bonnal, 'La Psychologie militaire de Napoléon', p. 435.
2. Riley, *Napoleon as a General*, pp. 116–29.

BOOK VI

CHAPTER 1

1. Clausewitz, *On War*, VI, 1, p. 427.
2. 'Dix-huit notes sur l'ouvrage intitulé *Considérations sur l'art de la guerre*', *Corresp.*, XXXI, p. 347.
3. *Corresp.*, XIII, no. 10572, pp. 20–1, to the King of Naples, Saint-Cloud, 30 July 1806.
4. Clausewitz, *On War*, VI, 1, p. 430.
5. *Corresp.*, XIX, no. 15667, p. 351, to Clarke, Schönbrunn, 16 August 1809. At the start of August 1809, the British laid siege to Vlissingen, a fortified port on the island of Walcheren. The Franco-Dutch garrison surrendered on 15 August.
6. *Corresp.*, XIX, no. 15678, pp. 359–61, to Clarke, Schönbrunn, 18 August 1809.
7. *Corresp.*, XIX, 15698, pp. 382–4, to Clarke, Schönbrunn, 22 August 1809, 16.00 hrs.
8. 'Précis des événements militaires arrivés pendant l'année 1798', *Corresp.*, XXX, p. 245.

CHAPTER 2

1. Clausewitz, VI, 2, p. 431.
2. AN, 390 AP 25, MS of November 1816 (13 November), pp. 4–5 (Bertrand, *Cahiers*, I, p. 146).
3. *Notes inédites de l'empereur Napoléon Ier sur les Mémoires militaires du général Lloyd*, p. 8.
4. AN, 390 AP 25, MS of 1818 (October), p. 64 (Bertrand, *Cahiers*, II, p. 180).

CHAPTER 3

1. 'Dix-huit notes sur l'ouvrage intitulé *Considérations sur l'art de la guerre*', *Corresp.*, XXXI, p. 342 n. 2.
2. 'Campagnes d'Italie (1796–1797)', *Corresp.*, XXIX, pp. 84–5.
3. 'Précis des événements militaires arrivés pendant les six derniers mois de 1799', *Corresp.*, XXX, p. 291.

CHAPTER 4

1. Clausewitz, *On War*, VI, 4, pp. 439–41.
2. SHD/DAT, 17 C 2, 'Note sur la position politique et militaire de nos armées de Piémont et d'Espagne', n.p. n.d., but no doubt in the headquarters at Loano, 25 messidor an II—13 July 1794, pp. 4–5.

3. Bourcet, *Principes de la guerre de montagnes*, p. 125.
4. *Corresp.*, I, no. 49, pp. 64–5, mémoire on the army of Italy, Paris, early thermidor an III—July 1795.
5. 'Précis des opérations de l'armée d'Italie pendant les années 1792, 1793, 1794 and 1795', *Corresp.*, XXIX, p. 34.
6. *Corresp.*, XVII, no. 14192, p. 382, notes for General Savary, aide de camp to the Emperor on mission in Madrid, Bayonne, 13 July 1808. Commanding the elements of the imperial guard in Spain and the observation division of the Pyrénées-Orientales at this time, Marshal Bessières clashed with General La Cuesta's Spanish at Medina del Rio Seco on 14 July and then entered Madrid with King Joseph.

CHAPTER 5

1. Barthélemy-Catherine Joubert served throughout the first Italian campaign. He was a major general from 7 December 1796 and was an architect of the victory at Rivoli on 14 January 1797.
2. *Corresp.*, II, no. 1501, p. 337, to Joubert, Tolentino, 29 pluviôse an V—17 February 1797.
3. Clausewitz, *On War*, VI, 5, p. 443.
4. Readers will have understood that it is a question here of countering British landings.
5. *Corresp.*, XIII, no. 10558, pp. 9–10, to the King of Naples, Saint-Cloud, 28 July 1806.

CHAPTER 6

1. Clausewitz, *On War*, VI, 9, p. 468.
2. *Corresp.*, XIII, no. 10941, p. 310, to Marshal Soult, Würzburg, 5 October 1806, 11.00 hrs.

CHAPTER 7

1. Clausewitz, *On War*, VI, 10, pp. 471–9.
2. *Mémoires*, Montholon, V, p. 76.
3. AN, 314 AP 30, MS 17 (Gourgaud, *Journal*, I, p. 205).
4. 'Dix-huit notes sur l'ouvrage intitulé *Considérations sur l'art de la guerre*', *Corresp.*, XXXI, p. 420.
5. *Correspondance inédite*, IV, no. 5888, p. 524, extract from the Emperor's notes on Erfurt, applied to several posts on the northern border, presumed to be August 1811.
6. *Corresp.*, XII, no. 10419, p. 492, to General Dejean, Saint-Cloud, 27 June 1806.

7. Marshal Masséna employed to good effect Verona's Castel Vecchio, positioning in it batteries that covered the crossing of the Adige by his troops on 19 October 1805 (James Marshall Cornwall, *Marshal Massena*, London, Oxford University Press, 1965, pp. 136–7).

8. *Corresp.*, XII, no. 10419, pp. 494–5, to General Dejean, Saint-Cloud, 27 June 1806.

9. On the Tagliamento, north-west of Udine.

10. *Corresp.*, XIV, no. 11667, p. 218, to Prince Eugène, Warsaw, 20 January 1807.

11. *Corresp.*, XII, no. 10419, p. 492, to General Dejean, Saint-Cloud, 27 June 1806.

12. AN, 314 AP 30, MS 17 (Gourgaud, *Journal*, I, p. 205).

13. *Mémoires*, Montholon, II, p. 50.

14. 'Campagne de 1815', *Corresp.*, XXXI, pp. 150–1.

15. A reference to the campaigns of 1711 and 1712, at the end of the War of the Spanish Succession, when Prince Eugène of Savoy lingered over sieges and did not manage to invade France.

16. The Austrians of the Prince of Coburg laid siege to several locations rather than forging ahead to Paris.

17. Archduke Leopold William of Hapsburg, governor of the Spanish Low Countries. Probably an allusion to the campaign of 1654, when Turenne manoeuvred skilfully and won a great victory at Arras. The Grand Condé then transferred to the service of the Spanish.

18. AN, 390 AP 25, MS of January–September 1819 (May), p. 115 (Bertrand, *Cahiers*, II, pp. 357–8).

19. 'Campagnes d'Égypte et de Syrie', *Corresp.*, XXX, p. 180.

20. *Corresp.*, XIX, no. 15889, pp. 541–2, to Clarke, Schönbrunn, 1 October 1809.

21. *Corresp.*, II, no. 1059, p. 31, to the executive Directory, Milan, 11 vendémiaire an V—2 October 1796.

22. *Corresp.*, XXII, no. 17732, p. 166, to Clarke, note on the defence of Corfu, Rambouillet, 19 May 1811.

23. *Corresp.*, XX, no. 16387, pp. 292–3, orders concerning the places of Italy, Compiègne, 19 April 1810.

24. *Corresp.*, XXII, no. 17557, pp. 36 and 39, to Clarke, Paris, 6 April 1811.

25. *Œuvres de Napoléon Bonaparte*, 5 vols, Paris, Panckoucke, 1821, I, p. 40.

26. AN, 314 AP 30, MS 39 (Gourgaud, *Journal*, II, p. 80).

27. *Corresp.*, XXIII, no. 18308, p. 53, to Marshal Davout, commanding the observation corps of Elbe, at Hamburg, Paris, 5 December 1811.

28. *Supplément à la correspondance de Napoléon Ier: lettres curieuses omises par le comité de publication, rectifications*, publié par Albert Du Casse, Paris, E. Dentu, 1887, pp. 206–7, to Joseph, Reims, 14 March 1814.

CHAPTER 8

1. Clausewitz, *On War*, VI, 12, pp. 486 and 488.
2. 'Précis des guerres de Jules César', *Corresp.*, XXXII, p. 30.
3. Montholon, *Récits de la captivité*, II, p. 362.
4. *Corresp.*, V, no. 4083, p. 390, to Murat, before Acre, 21 germinal an VII— 10 April 1799.
5. *Corresp.*, XXV, no. 20065, p. 338 to Berthier, Neumarkt, 31 May 1813, 23.30 hrs.
6. AN, 390 AP 25, MS of 1818 (January), p. 27 (Bertrand, *Cahiers*, II, p. 40).
7. *Corresp.*, II, p. 152.
8. 'Dix-huit notes sur l'ouvrage intitulé *Considérations sur l'art de la guerre*', *Corresp.*, XXXI, p. 416.
9. AN, 390 AP 25, MS of 1818 (November), p. 74 (Bertrand, *Cahiers*, II, p. 196).
10. *Mémoires*, Montholon, V, pp. 93–6.
11. *Corresp.*, XIII, no. 10797, p. 197, to Berthier, Saint-Cloud, 16 September 1806.
12. *Corresp.*, XIII, 10941, p. 310, to Soult, Würzburg, 5 October 1806, 11.00 hrs.
13. *Corresp.*, XIV, no. 11906, p. 354, to Soult, Osterode, 28 February 1807, 18.00 hrs.
14. *Corresp.*, XIV, no. 11939, pp. 380–1, to Soult, Osterode, 5 March 1807, 16.00 hrs.
15. *Corresp.*, XVIII, no. 14942, p. 382, to Bertrand, La Malmaison, 22 March 1809.
16. Colson, *Le Général Rogniat*, pp. 339–48, 354–9, 382–90.
17. Bertrand, *Cahiers*, II, p. 334.
18. Field fortification was used much more during the Crimean War (1854–6) and especially during the American Civil War (1861–5).
19. 'Essai sur la fortification de campagne', *Corresp.*, XXXI, p. 467.
20. Clausewitz, *Théorie du combat*, no. 528, p. 94.

CHAPTER 9

1. Clausewitz, *On War*, VI, 13, pp. 491 and 497.
2. Lazare Carnot, *De la défense des places fortes*, Paris, Courcier, 1810.
3. 'Notes sur la fortification dictées par Napoléon à Sainte-Hélène' [à Gourgaud], *Revue du génie militaire*, Vol. 14, July 1897, pp. 13–14.
4. 'Essai sur la fortification de campagne', *Corresp.*, XXXI, p. 466.
5. The siege began on 12 August 1708 and Marshal de Boufflers, having repelled several attacks, surrendered the post on 25 October.
6. In late May 1712, when the large army of Prince Eugène of Savoy covered the investment of Landrecies, Marshal de Villars conceived a daring manoeuvre and through a concealed march took the entrenched camp of Denain, where Eugène was supporting his base, by surprise. On 27 July, the camp of Denain

was taken and the allies' lines of communication were cut. Eugène de Savoie lifted the siege of Landrecies.

7. 'Dix-huit notes sur l'ouvrage intitulé *Considérations sur l'art de la guerre*', *Corresp.*, XXXI, pp. 335–6.
8. *Corresp.*, XXXI, p. 337.
9. *Corresp.*, XXXI, p. 338.
10. The role of 'national recess' was exactly that conferred on Antwerp by the government of Belgium from the 1850s onwards and which it was to play in August–September 1914.
11. Las Cases, *Mémorial*, Vol. 17, pp. 123–4 (Dunan edn, II, p. 492).

CHAPTER 10

1. Clausewitz, *On War*, VI, 15, pp. 502–3.
2. AN, 390 AP 25, MS of February 1817 (14 February), p. 10 (Bertrand, *Cahiers*, I, p. 198).
3. AN, 314 AP 30, MS 40 (Gourgaud, *Journal*, II, pp. 267–8). In May 1800, after crossing the Great Saint Bernard Passage, Napoleon and his Reserve army were blocked in front of the fort of Bard, which had the only road viable for artillery under its fire. A few French cannons with their wheels covered in straw succeeded in crossing one night unnoticed. The following day, they made a breach in the wall of the fort, which surrendered. Its garrison consisted of a mere 400 Austrians with 18 cannons.
4. Clausewitz, *On War*, VI, 15, pp. 503–4.
5. This confirms Clausewitz's intellectual identification with Napoleon, at least in part, to the point of sometimes proving more 'Napoleonic' than him (René Girard, *Achever Clausewitz. Entretiens avec Benoît Chantre*, Paris, Carnets Nord, 2007, pp. 249–51).

CHAPTER 11

1. 'Campagnes d'Italie (1796–1797)', *Corresp.*, XXIX, p. 72.
2. *Corresp.*, II, no. 11125, instructions for General Joubert, Verona, 8 brumaire an V—29 October 1796.
3. *Corresp.*, XVIII, no. 14707, p. 218, notes on the defence of Italy, Valladolid, 14 January 1809.
4. Clausewitz, *Théorie du combat*, no. 288, p. 64.
5. *Mémoires*, Montholon, V, pp. 24–5.
6. *Corresp.*, XVII, no. 1426, p. 42, observations on Spanish affairs, Saint-Cloud, 27 August 1808.
7. *Corresp.*, XIV, no. 11961, p. 397, to Bernadotte, Osterode, 6 March 1807, midnight.

8. *Corresp.*, XVII, no. 14283, p. 480, notes on Spanish affairs, Saint-Cloud, 30 August 1808.
9. *Corresp.*, XXV, no. 19721, pp. 88–9, to Eugène Napoléon, Trianon, 15 March 1813.
10. Clausewitz, *On War*, VI, 18, p. 522.

CHAPTER 12

1. Armand-Charles Guilleminot, brigadier-general in 1808, major general in 1813, topographer, one of the most highly cultivated of generals of the First Empire.
2. *Corresp.*, XV, no. 12321, p. 41, order, Finkenstein, 6 April 1807.
3. Clausewitz, *On War*, VI, 20, pp. 541–2.
4. 'Précis des événements militaires arrivés pendant les six derniers mois de 1799', *Corresp.*, XXX, p. 283.

CHAPTER 13

1. Clausewitz, *On War*, VI, 22, p. 547.
2. Clausewitz, *On War*, VI, 22, pp. 547–9.
3. *Mémoires*, Montholon, IV, pp. 310–12.
4. *Corresp.*, VI, no. 4662, p. 181, to Masséna, commander-in-chief of the army of Italy, Paris, 21 ventôse an VIII—12 March 1800.
5. *Corresp.*, XII, no. 10285, pp. 407–8, to Joseph, Saint-Cloud, 2 May 1806.
6. *Corresp.*, XII, no. 10329, p. 444, to Joseph, Saint-Cloud, 7 June 1806.
7. *Corresp.*, XIII, no. 10726, p. 131, to General Dejean, Saint-Cloud, 3 September 1806.
8. *Corresp.*, XVII, no. 14253, p. 438, note for the Prince of Neuchâtel, Saint-Cloud, 16 August 1808.
9. *Corresp.*, XVII, no. 14328, p. 515, notes for Joseph Napoleon, King of Spain, Saint-Cloud, 15 September 1808.
10. *Corresp.*, XXVII, no. 21120, p. 58, to General Maison, Paris, 20 January 1814.

CHAPTER 14

1. AN, 390 AP 25, MS of March 1818, p. 101 (Bertrand, *Cahiers*, II, p. 86).
2. 'Campagnes d'Italie (1796–1797)', *Corresp.*, XXIX, p. 214.
3. *Corresp.*, XX, no. 16495, p. 31, to Clarke, Minister of War, Ostend, 21 May 1810.
4. Clausewitz, *On War*, VI, 23, pp. 551–2.

CHAPTER 15

1. *Mémoires*, Montholon, IV, p. 310.
2. *Notes inédites de l'empereur Napoléon Ier sur les Mémoires militaires du général Lloyd*, p. 10.
3. Clausewitz, *On War*, VI, 25, pp. 566, 569.
4. Chaptal, *Mes Souvenirs*, pp. 304–5.

CHAPTER 16

1. *Mémoires*, Montholon, IV, pp. 347–8.
2. 'Précis des guerres de Jules César', *Corresp.*, XXXII, p. 14.
3. [Thibaudeau], *Mémoires*, p. 108.
4. Lentz, *L'Effondrement du système napoléonien*, pp. 520–1.
5. *Corresp.*, XXVIII, no. 21861, pp. 150–1, to Davout, Paris, 1 May 1815.
6. Englund, *Napoleon*, p. 434.
7. AN, 390 AP 25, MS of January 1818, pp. 93–4 (Bertrand, *Cahiers*, II, p. 53).
8. Esdaile, *The Peninsular War*, p. 280.
9. 'Dix-huit notes sur l'ouvrage intitulé *Considérations sur l'art de la guerre*', *Corresp.*, XXXI, p. 420.
10. Clausewitz, *On War*, VI, 26, p. 578.
11. Clausewitz, *On War*, VI, 26, p. 580.

CHAPTER 17

1. *Corresp.*, VI, no. 422, p. 224, proclamation to French youth, Paris, 1 floréal an VIII—21 April 1800.
2. Clausewitz, *On War*, VI, 27, p. 588.
3. *Corresp.*, XII, no. 10526, pp. 563–4, to General Dejean, Saint-Cloud, 20 July 1806.
4. 'Campagnes d'Égypte et de Syrie', *Corresp.*, XXX, p. 10.
5. Today Zadar (Croatia).
6. *Corresp.*, XIII, no. 10726, pp. 132–4, to General Dejean, Saint-Cloud, 3 September 1806.
7. *Corresp.*, XIII, no. 11075, p. 411, to Louis, Potsdam, 25 October 1806.
8. *Corresp.*, XIII, no. 1052, p. 20, to the King of Naples, Saint-Cloud, 30 July 1806.
9. *Corresp.*, XIII, no. 10629, p. 62, to the King of Naples, Saint-Cloud, 9 August 1806.
10. *Corresp.*, XIII, no. 10672, pp. 86–8, to the King of Naples, Rambouillet, 20 August 1806.

11. 'Projet pour la défense du golfe de Saint-Florent', Corsica, early 1793 (*Napoléon inconnu: papiers inédits (1786–1793)*, published by Frédéric Masson and Guido Biagi, 2 vols, Paris, Ollendorff, 1895, II, pp. 451–2).

12. SHD/DAT, 17 C 2, 'Note sur la position politique et militaire de nos armées de Piémont et d'Espagne', n.p. n.d., but doubtless the headquarters at Loano, 25 messidor an II—13 July 1794, p. 4.

13. *Corresp.*, XII, no. 10176, p. 339, to Prince Eugène, Saint-Cloud, 30 April 1806.

14. *Corresp.*, XVIII, no. 1407, pp. 217–18, notes on the defence of Italy, Valladolid, 14 January 1809.

BOOK VII

CHAPTER I

1. AN, 314 AP 30, MS 1 (Gourgaud, *Journal*, I, p. 211). Machiavelli had minimized the significance of artillery. Napoleon's notes on artillery were discovered in Gourgaud's papers and published. They are highly technical: 'Notes sur l'artillerie dictées par Napoléon à Sainte-Hélène au baron Gourgaud' [notes inédites communiquées par le vicomte de Grouchy], *Revue d'artillerie*, Vol. 50, June 1897, pp. 213–29.

2. *Corresp.*, XIII, no. 11009, p. 354, 5th bulletin of the Grande Armée, Jena, 15 October 1806.

3. AN, 314 AP 30, MS 40 (Gourgaud, *Journal*, II, p. 267).

CHAPTER 2

1. Béraud, *La Révolution militaire napoléonienne*, 1, pp. 18 and 25.
2. Clausewitz, *On War*, VII, 2, p. 634.
3. 'Campagnes d'Italie (1796–1797)', *Corresp.*, XXIX, p. 184.
4. Clausewitz, *On War*, VII, 4 and 5, pp. 638–9.
5. Bertrand, *Cahiers*, II, pp. 134–5.

CHAPTER 3

1. Clausewitz, *On War*, VII, 7, p. 641.
2. *Mémoires*, Montholon, V, p. 215.
3. A crushing defeat of Frederick at the hands of Daun on 18 June 1757. Here, from the heights they occupied, the Austrians saw the Prussians' flanking march against their right wing perfectly.
4. Left blank in the manuscript. Probable allusion to Lieutenant-General Gerd Heinrich von Manteuffel at the Battle of Zorndorf (25 August 1758). Manteuffel led the advance guard of the Prussians' oblique attack and his troops suffered heavy casualties.

5. AN, 390 AP 25, MS of 1818 (October), pp. 58–9 and 61 (Bertrand, *Cahiers*, II, pp. 168–9, 173, and 175).

6. *Mémoires*, Montholon, V, p. 297.

7. Robert M. Citino, *The German Way of War: From the Thirty Years' War to the Third Reich*, Lawrence, University Press of Kansas, 2005, p. 89.

8. *Corresp.*, II, no. 1022, p. 5, to General Kilmaine, Milan, 3 vendémiaire an V—24 September 1796.

9. *Corresp.*, IV, no. 3233, p. 462, to Desaix, Cairo, 18 fructidor an VI—4 September 1798.

10. AN, 390 AP 25, MS of October 1817 (24 October), p. 12 (Bertrand, *Cahiers*, I, p. 290).

11. *Corresp.*, II, p. 138.

12. 'Essai sur la fortification de campagne', *Corresp.*, XXXI, p. 467.

13. *Corresp.*, XIX, no. 15310, p. 82, to Eugène Napoléon, Schönbrunn, 7 June 1809, 02.30 hrs.

14. *Corresp.*, XIX, no. 15458, p. 116, to Eugène Napoléon, Schönbrunn, 16 June 1809, 17.00 hrs.

15. *Corresp.*, VII, no. 5514, p. 111, to Citizen Forfait, Minister of the Navy and the Colonies, Paris, 17 germinal an IX—7 April 1801.

16. Pierre de Villeneuve commanded the right wing of Admiral Bruyès's fleet at Abukir (1798). He escaped the disaster with two vessels and two frigates. He was defeated by Nelson at Trafalgar in 1805.

17. *Mémoires*, Gourgaud, II, pp. 193–4.

CHAPTER 4

1. Clausewitz, *On War*, VII, 8, pp. 643–4.

2. 'Campagnes d'Italie (1796–1797)', *Corresp.*, XXIX, pp. 97–8.

3. *Corresp.*, XXIX, p. 188.

4. *Corresp.*, II, no. 1735, p. 491, to the executive Directory, Leoben, 27 germinal an V—16 April 1797.

5. AN, 314 AP 30, MS 40 (Gourgaud, *Journal*, II, p. 268).

6. Clausewitz, *On War*, VII, 8, p. 644.

7. 'Diplomatie.—Guerre', *Corresp.*, XXX, p. 438.

8. *Corresp.*, XXX, pp. 447–8. On 16 August 1705, the Duke of Vendôme succeeded with 10,000 men in preventing the 24,000 imperial troops of Prince Eugène of Savoy from crossing the Adda. However, the French lost more than half their number. Chevalier Folard is said to have encouraged the Duke to have the old castle of Cassano, from which the French could fire on their enemies massed on the other side of the river, occupied (Chagniot, *Le Chevalier de Folard*, p. 43; Archer Jones, *The Art of War in the Western World*, Oxford, Oxford University Press, 1987, p. 286).

9. 16,000 for Alain Pigeard, *Dictionnaire des batailles de Napoléon*, Paris, Tallandier, 2004, p. 488.

10. The Lesmont bridge, the only one on the Aube, over which flowed the French army defeated at La Rothière on 1 February 1814.

11. AN, 390 AP 25, MS of January–September 1819 (11 September), pp. 130–1 (Bertrand, *Cahiers*, II, pp. 393–4). In connection with the Lesmont bridge, Bertrand specifies in a note: 'I remember this circumstance very well and all the details which I saw unfold before the very eyes of the Emperor.'

CHAPTER 5

1. Clausewitz, *On War*, VII, 9, p. 646.

2. *Mémoires*, Montholon, V, p. 100; italics in the original.

3. By mistake, Montholon's text has 'fulminating'.

4. *Mémoires*, Montholon, V, p. 195; italics in the original.

5. Christopher Duffy, *Frederick the Great: A Military Life*, London and New York, Routledge, 1985, pp. 124–31; Simon Millar, *Kolin 1757: Frederick the Great's First Defeat*, Oxford, Osprey, 2001, pp. 87–9.

6. Bonnal, 'La Psychologie militaire de Napoléon', p. 437.

7. *Corresp.*, XIII, no. 10941, p. 310, to Soult, Würzburg, 5 October 1806, 11.00 hrs.

8. *Corresp.*, XIII, no. 10982, p. 337, to Lannes, Auma, 12 October 1806, 04.00 hrs.

9. *Corresp.*, XV, no. 12755, p. 334, to the Grand Duke of Berg, Preussisch-Eylau, 14 June 1807, 03.30 hrs.

10. *Corresp.*, XIX, no. 15310, p. 82, to Eugène, Schönbrunn, 7 June 1809, 02.30 hrs.

11. *Corresp.*, XIX, no. 15694, pp. 379–80, to Clarke, Schönbrunn, 21 August 1809.

12. Montholon, *Récits de la captivité*, II, p. 169.

13. Montholon, *Récits de la captivité*, II, pp. 217–18.

14. Alexander Mikaberidze, *The Battle of Borodino: Napoleon against Kutuzov*, Barnsley, Pen and Sword, 2007, pp. 69–70.

15. AN, 314 AP 30, MS 39 (Gourgaud, *Journal*, II, p. 81).

16. *Corresp.*, XXVI, no. 20365, p. 38, to Marshal Oudinot, Dresden, 12 August 1813.

17. *Dernières Lettres inédites*, II, no. 2128, pp. 443–4, to Bertrand, commanding the 4th corps, Dresden, 13 August 1813.

CHAPTER 6

1. AN, 390 AP 25, MS of 1818 (January), p. 28 (Bertrand, *Cahiers*, II, p. 41).

2. 'Précis des opérations de l'armée d'Italie pendant les années 1792, 1793, 1794 et 1795', *Corresp.*, XXIX, p. 30.

3. *Mémoires*, Montholon, V, p. 9; italics in the original.
4. 'Précis des événements militaires arrivés pendant les six premiers mois de 1799', *Corresp.*, XXX, pp. 262–3.
5. Clausewitz, *On War*, VI, 16, pp. 514–15.
6. Clausewitz, *On War*, VII, 11, p. 651.

CHAPTER 7

1. *Corresp.*, I, no. 1196, p. 117, to the executive Directory, Verona, 29 brumaire an V—19 November 1796.
2. *Corresp.*, VI, no. 521, p. 561, to General Moreau, Paris, 19 nivôse an IX—9 January 1801.
3. *Corresp.*, I, no. 27, p. 38, plan for the second operation preparatory to the opening of the Piedmont campaign, Colmar, 2 prairial an II—21 May 1794.
4. *Corresp.*, XI, no. 9277, p. 253, to M. Otto, Strasbourg, 6 vendémiaire an XIV—28 September 1805.
5. *Corresp. gén.*, V, no. 10885, p. 741, to Fouché, Strasbourg, 6 vendémiaire an XIV—28 September 1805.
6. *Corresp.*, XI, no. 9325, p. 283, to Brune, Ludwigsburg, 11 vendémiaire an XIV—3 October 1805.
7. Bruno Colson, 'La place et la nature de la manœuvre dans l'art de la guerre napoléonien', in Christian Malis (ed.), *Guerre et manœuvre. Héritages et renouveau*, Paris, Fondation Saint-Cyr-Économica, 2009, pp. 118–40.
8. *Corresp.*, XI, no. 9372, p. 315, to Prince Murat, Augsburg, 20 vendémiaire an XIV—12 October 1805, 09.00 hrs.
9. 'Précis des guerres de Jules César', *Corresp.*, XXXII, p. 51.
10. Jones, *The Art of War*, pp. 75–9.
11. *Corresp.*, I, no. 337, p. 236, to the executive Directory, Tortone, 17 floréal an IV—6 May 1796.
12. *Corresp.*, XIII, no. 10970, p. 329, to Marshal Soult, Kronach, 8 October 1806, 15.30 hrs.
13. What is today called a manoeuvre of deception.
14. Telp, *The Evolution of Operational Art*, pp. 69 and 127; Béraud, *La Révolution militaire napoléonienne*, I, pp. 237, 291–3.
15. AN, 390 AP 25, MS of 1818 (October), pp. 59–60 (Bertrand, *Cahiers*, II, pp. 170–1).
16. Béraud, *La Révolution militaire napoléonienne*, I, p. 293.
17. 'Précis des événements militaires arrivés pendant les six premiers mois de 1799', *Corresp.*, XXX, p. 289.
18. AN, 314 AP 30, MS 50 (Gourgaud, *Journal*, II, pp. 34–8).
19. AN, 390 AP 25, MS of 1818 (January), p. 20 (Bertrand, *Cahiers*, II, p. 44).

20. *Corresp.*, IV, no. 3234, p. 463, to Berthier, Cairo, 18 fructidor an VI—4 September 1798.
21. Clausewitz, *On War*, VII, 13, pp. 654–5.
22. Hubert Camon, 'Essai sur Clausewitz', *Journal des sciences militaires*, 1900, p. 117.

CHAPTER 8

1. Clausewitz, *On War*, VII, 17, p. 666.
2. In sieges, the line of contravallation is a line of works, continuous or placed at intervals, designed to shield the camps of the besieging army from sorties by the besieged garrison. The line of circumvallation is of the same type, but is established in the rear of the camps of the besieging army to protect it from the attacks of the relief army.
3. *Mémoires*, Montholon, V, pp. 245–7.
4. Clausewitz, *On War*, VII, 17, p. 668.
5. The attack on the city of Extremadura, which was held by 4,000 French, on the night of 6–7 April 1812, cost Wellington 3,713 men (Esdaile, *The Peninsular War*, p. 387).
6. Montholon, *Récits de la captivité*, II, p. 171.
7. AN, 314 AP 30, MS 40 (Gourgaud, *Journal*, II, p. 222).
8. *Corresp.*, V, no. 4009, p. 346, to General Caffarelli, before Jaffa, 16 ventôse an VII—6 March 1799.
9. *Corresp.*, XVI, no. 13733, p. 487, to the Grand Duke of Berg, Bordeaux, 10 April 1808, midday.
10. *Corresp.*, XVII, no. 14223, p. 410, notes on the current position of the army in Spain, Bayonne, 21 July 1808. On 5 July 1807, 9,000 (rather than 12,000) British soldiers attacked Buenos Aires, entered the streets, and suffered heavy casualties as they came under fire from defenders firing from houses and terraces.

CHAPTER 9

1. Clausewitz, *On War*, VII, 21, p. 683; italics in the original.
2. 'Dix-huit notes sur l'ouvrage intitulé *Considérations sur l'art de la guerre*', *Corresp.*, XXXI, p. 417.
3. Clausewitz, *On War*, VII, 21, p. 683.
4. Gourgaud, *Journal*, I, p. 231.
5. 'Dix-huit notes sur l'ouvrage intitulé *Considérations sur l'art de la guerre*', *Corresp.*, XXI, p. 360.
6. 'Observations sur le plan de campagne des puissances belligérantes' [in 1799], copy of a sample of Napoleon's writing, Bibliothèque Thiers, Fonds Masson, quoted in Tomiche, *Napoléon écrivain*, pp. 309–10.

7. Clausewitz, *On War*, VI, 5, p. 444.

8. *Corresp.*, II, no. 1431, pp. 287–8, to General Joubert, Verona, 10 pluviôse an V—29 January 1797.

9. *Corresp.*, IV, no. 2710, p. 183, proclamation to the land army, on board *L'Orient*, 4 messidor an VI—22 June 1798.

10. *Corresp.*, IV, no. 2778, pp. 224–5, to General Kléber, Alexandria, 19 messidor an VI—7 July 1798.

11. Jacques-François de Boussay, Baron de Menou, was commander-in-chief of the army of the Orient following the assassination of General Kléber in June 1800. Beaten at Canope by the British in 1801, he surrendered in Alexandria and was able to return to France. He took the concern to conciliate the Egyptians very far, to the point of converting to Islam. He was a better administrator than general.

12. *Corresp. gén.*, II, no. 2602, p. 182, to General Menou, Alexandria, 19 messidor an VI—7 July 1798.

13. *Lettres inédites*, I, no. 441, p. 307, to Jérôme Napoléon, Burghausen, 29 April 1809.

14. AN, 390 AP 25, MS of 1818 (December), p. 88 (Bertrand, *Cahiers*, II, p. 218).

15. Ali (Louis-Étienne Saint-Denis, known as), *Souvenirs sur l'empereur Napoléon*, presentation and notes by C. Bourrachot, Paris, Arléa, 2000 (1st edn, 1926), p. 238.

16. *Corresp.*, IV, no. 2967, p. 316, to General Dupuy, Cairo, 17 thermidor an VI—4 August 1798.

17. O'Meara, *Napoleon in Exile*, I, p. 284.

18. *Corresp.*, XII, no. 10086, p. 276, to the King of Naples, Malmaison, 11 April 1806.

19. Guillaume-Joseph-Nicolas de Lafon-Blaniac was at the time equerry of King Joseph and a colonel in the 6th regiment of the horse chasseurs. He became a brigadier-general on 12 September 1806 and then a major general in 1813.

20. *Corresp.*, XII, no. 10156, p. 321, to the King of Naples, Saint-Cloud, 27 April 1806.

21. *Corresp.*, XVIII, no. 14730, p. 236, to Joseph Napoléon, Valladolid, 16 January 1809, evening.

22. *Corresp.*, XVII, no. 13875, p. 110, to Marshal Berthier, Bayonne, 12 May 1808.

23. *Corresp.*, XVII, 13749, pp. 9–10, notes for the Prince of Neuchâtel, Major General of the Grande Armée, Bayonne, 16 April 1808.

24. Major General Jean-Baptiste Drouet d'Erlon commanded the 9th corps of the army of Spain from 30 August 1810. This unit was transferred to the army of Portugal under Masséna on 10 September.

25. *Corresp.*, XXI, no. 16921, p. 126, to Berthier, Major General of the army of Spain, Fontainebleau, 18 September 1810.

26. *Corresp.*, XVII, no. 14192, p. 382, notes for General Savary, Bayonne, 13 July 1808.

27. *Corresp.*, XII, no. 10118, p. 298, to the King of Naples, Saint-Cloud, 21 April 1806.

28. *Lettres inédites*, I, no. 273, p. 187, to Prince Murat, Bayonne, 29 April 1808, 10.00 hrs.

29. In November 1811, Jean-Marie Lepaige Doursenne, known as Dorsenne, was commander-in-chief of the army of the North in Spain; Marie-Françoise-Auguste Caffarelli du Falga commanded the French forces in Biscay and Navarre; Pierre Thouvenot commanded the province and post of San Sebastián.

30. *Corresp.*, XXIII, no. 18276, p. 27, to Berthier, Major General of the army of Spain, Saint-Cloud, 20 November 1811.

31. *Corresp.*, IV, no. 2901, p. 284, to General Zajonchek, Cairo, 12 thermidor an VI—30 July 1798.

32. *Corresp.*, IV, no. 3200, p. 444, to General Dugua, Cairo, 14 fructidor an VI—31 August 1798.

33. Honoré Vial commanded the provinces of Damiette and Mansourah in late July 1798. He repulsed the Arabs from them in August and September.

34. *Corresp.*, V, no. 3436, p. 38, to General Vial, Cairo, 15 vendémiaire an VII—6 October 1798.

35. *Corresp.*, VI, no. 4523, p. 87, to General Brune, Paris, 24 nivôse an VIII—14 January 1800.

36. This village in Lombardy, whose peasants had rebelled, had been razed on Bonaparte's orders in May 1796, during the first Italian campaign.

37. *Corresp.*, XI, no. 9678, p. 543, to General Junot, Stuttgart, 19 January 1806.

38. *Corresp.*, XII, no. 9744, p. 5, to Junot, Paris, 4 February 1806.

39. *Corresp.*, XII, no. 9844, p. 62, to Junot, Paris, 18 February 1806.

40. *Corresp.*, XII, no. 9852, p. 66, to Junot, Paris, 19 February 1806.

41. *Corresp.*, XII, no. 9951, p. 121, to Prince Joseph, Paris, 2 March 1806.

42. *Corresp.*, XIII, no. 10657, p. 78, to Joseph, King of Naples, Rambouillet, 17 August 1806.

43. *Corresp.*, XII, no. 9944, pp. 165–6, to Prince Joseph, Paris, 8 March 1806.

44. *Corresp.*, XII, no. 10042, p. 249, to Prince Joseph, Paris, 31 March 1806.

45. 'Campagnes d'Italie (1796–1797)', *Corresp.*, XXIX, p. 113.

46. *Corresp.*, XVII, no. 14265, pp. 456–7, to General Clarke, Saint-Cloud, 22 August 1808.

47. *Corresp.*, XVII, no. 14273, p. 46, to General Clarke, Saint-Cloud, 27 August 1808.

48. Colson, *Le Général Rogniat*, pp. 169–88.

49. 'Dix-huit notes sur l'ouvrage intitulé *Considérations sur l'art de la guerre*', *Corresp.*, XXXI, p. 365.

50. *Corresp.*, IV, no. 3030, p. 348, to General Zajonchek, Cairo, 29 thermidor an VI—16 August 1798.

51. *Corresp.*, V, no. 4374, p. 574, to General Kléber, Alexandria, 5 fructidor an VII—22 August 1799.

52. 'Campagnes d'Égypte et de Syrie', *Corresp.*, XXX, pp. 83–6.

53. Englund, *Napoleon*, pp. 135–6, 139–40.

54. *Corresp.*, XVII, no. 14117, p. 322, note for General Savary, on mission in Madrid, Bayonne, 19 June 1808.

55. *Corresp.*, XVIII, no. 14499, p. 4, to Joseph Napoléon, Burgos, 20 November 1808.

56. 'Campagnes d'Italie (1796–1797)', *Corresp.*, XXIX, p. 113.

57. *Corresp.*, XIII, no. 10629, p. 61, to the King of Naples, Saint-Cloud, 9 August 1806.

58. *Corresp.*, XVII, no. 14104, p. 315, to Marshal Bessières, Bayonne, 16 June 1808.

59. Montholon, *Récits de la captivité*, II, pp. 462–3.

60. 'Dix-huits notes sur l'ouvrage intitulé *Considérations sur l'art de la guerre*', *Corresp.*, XXXI, p. 340.

61. *Corresp.*, III, no. 2392, p. 465, to the President of the Institut national, Paris, 6 nivôse an VI—26 December 1797.

62. A.-M.-T. de Lort de Sérignan, *Napoléon et les grands généraux de la Révolution et de l'Empire*, Paris, Fontemoing, 1914, pp. 61–4.

63. Juan Cole, *Napoleon's Egypt: Invading the Middle East*, New York, Palgrave Macmillan, 2007.

64. Bruno Colson, 'Napoléon et la guerre irrégulière', *Stratégique*, nos. 93–6, 2009, pp. 227–58; Esdaile, *The Peninsular War*, pp. 352–3 and 357–8.

CONCLUSION TO BOOK VII

1. Don W. Alexander, *Rod of Iron: French Counterinsurgency Policy in Aragon during the Peninsular War*, Wilmington, Del., Scholarly Resources, 1985.

BOOK VIII

1. Clausewitz, *On War*, VIII, 2, pp. 701–2.

2. Clausewitz, *On War*, VIII, 3, p. 706.

CHAPTER I

1. [Roederer], *Autour de Bonaparte*, p. 4.

2. Bonnal, 'La Psychologie militaire de Napoléon', p. 426.

3. AN, 390 AP, MS of 1818 (December), p. 87 (Bertrand, *Cahiers*, II, p. 216).

4. *Corresp.*, XIII, no. 10809, p. 208, to Prince Eugène, Saint-Cloud, 18 September 1806.

5. *Corresp.*, XIII, no. 10810, p. 210, to Joseph, King of Naples, Saint-Cloud, 18 September 1806.

6. 'Ulm—Moreau', *Corresp.*, XXX, p. 409.

7. Colin, *L'Éducation militaire*, pp. 92–6, and 135–7.

8. Wilkinson, *The Rise of General Bonaparte*, pp. 8–9.

9. Bourcet, *Principes de la guerre de montagnes*, p. 88.

10. *Corresp.*, XVII, no. 14307, p. 502, to Marshal Berthier, Saint-Cloud, 8 September 1808.

11. 'Observations sur les opérations militaires des campagnes de 1796 et 1797 en Italie', *Corresp.*, XXIX, p. 335.

12. *Corresp.*, XIII, no. 10845, pp. 239–40, to the King of Holland, Saint-Cloud, 20 September 1806.

13. Béraud, *La Révolution militaire napoléonienne*, I, p. 245.

14. *Corresp.*, XIII, no. 10920, pp. 293–4, to the King of Holland, Mainz, 30 September 1806.

15. 'Dix-huit notes sur l'ouvrage intitulé *Considérations sur l'art de la guerre*', *Corresp.*, XXXI, p. 417.

16. The Austrian body responsible for directing wars.

17. *Mémoires*, Montholon, IV, pp. 336–7; italics in the original.

18. *Corresp.*, XIII, no. 10726, p. 131, to General Dejean, Saint-Cloud, 3 September 1806.

19. *Corresp.*, XXIII, no. 18769, p. 469, to Jérôme Napoléon, commanding the 5th, 7th, and 8th corps of the Grande Armée, Thorn, 5 June 1812.

CHAPTER 2

1. Clausewitz, *On War*, VIII, 3, p. 708.

2. Clausewitz, *On War*, VIII, 4, p. 720.

3. Clausewitz, *On War*, VIII, 4, p. 723.

4. Strachan, *Carl von Clausewitz's* On War, p. 132; Echevarria II, *Clausewitz and Contemporary War*, pp. 181–2.

5. AN, 390 AP 25, MS of May 1817 (18 May), p. 4 (Bertrand, *Cahiers*, I, p. 225).

6. Clausewitz, *On War*, I, 1, p. 101.

7. He was commander-in-chief of the armies of Italy and Naples, was the victor at Pastrengo on 26 March, but defeated at Magnano on 5 April. He resigned his command on 26 April.

8. 'Précis des événements militaires arrivés pendant les six premiers mois de 1799', *Corresp.*, XXX, p. 261.

9. Clausewitz, *On War*, VIII, 9, p. 748.

10. 'Ulm—Moreau', *Corresp.*, XXX, pp. 409–10.

11. We have supported the authenticity of this text in the Introduction.

12. This sentence has become emblematic of the Napoleonic art of war and has been cited many times by Hubert Camon and those influenced by him.

13. Astonishing anticipation of the insurrection of 1808.
14. The British were indeed to use Portugal to land Wellington's troops.
15. The paragraph has been struck out in the manuscript, probably by Augustin Robespierre.
16. SHD/DAT, 17 C 2, 'Note sur la position politique et militaire de nos armées de Piémont et d'Espagne', n.p. n.d., but doubtless in the headquarters at Loano, 25 messidor an II—13 July 1794, pp. 2–5.
17. Casanova, *Napoléon et la pensée de son temps*, p. 142.
18. Strachan, *Carl von Clausewitz's* On War, p. 89.
19. Colin, *L'Éducation militaire*, p. 354.
20. Henri Guillaume, 'Chasteler', *Biographie nationale*, IV, Brussels, Thiry, 1873, p. 34.
21. Pierre Berthezène, *Souvenirs militaires de la République et de l'Empire*, publiés par son fils, 2 vols, Paris, Dumaine, 1855, II, p. 309 (new edn with preface by N. Griffon de Pleineville, Le Livre chez vous, 2005).
22. Berthezène, *Souvenirs militaires*, II, pp. 270, 273, 277, 289, 297. Berthezène's citation is repeated, without any reference, by Marshal Foch (*Des principes de la guerre. Conférences faites en 1900 à l'École supérieure de guerre*, 4th edn, Paris and Nancy, Berger-Levrault, 1917, p. 55) and by Jean Guitton (*La Pensée et la guerre*, p. 75).
23. AN, 390 AP 25, MS of January 1818, p. 95 (Bertrand, *Cahiers*, II, pp. 54–5).

CHAPTER 3

1. Clausewitz, *On War*, VIII, 6, p. 731.
2. SHD/DAT, 17 C 2, 'Note sur la position politique et militaire de nos armées de Piémont et d'Espagne', n.p. n.d., but doubtless in the headquarters of Loano, 25 messidor an II (13 July 1794), p. 1.
3. O'Meara, *Napoleon in Exile*, II, pp. 71–2.
4. *Corresp.*, III, no. 2231, p. 320, to the provisional government of Genoa, Passariano, 4th complementary day an V—20 September 1797.
5. AN, 390 AP 25, MS of January–September 1819 (January), p. 60 (Bertrand, *Cahiers*, II, p. 242).
6. Thierry Lentz, *La France et l'Europe de Napoléon 1804–1814*, Paris, Fayard, 2007, pp. 151–5.
7. Laurent Henninger, 'Military Revolutions and Military History', in Matthew Huges and William J. Philpott (eds), *Palgrave Advances in Modern Military History*, Basingstoke, Palgrave Macmillan, 2006, p. 10.
8. [Thibaudeau], *Mémoires*, pp. 77–80; italics in the original.
9. Gilbert Bodinier, 'Officiers et soldats de l'armée impériale face à Napoléon', *Napoléon de l'histoire à la légende*, proceedings of the conference of 30 November and 1 December 1999 at the Musée de l'Armée, Paris, Éditions in Forma,

Maisonneuve et Larose, 2000, pp. 211–32; Gustave Canton, *Napoléon antimilitariste*, Paris, Alcan, 1902.

10. [Thibaudeau], *Mémoires*, pp. 107–8.
11. *Corresp.*, XII, no. 10254, p. 387, to General Junot, Saint-Cloud, 21 May 1806.
12. *Corresp.*, XVII, no. 13882, p. 119, to Fouché, Bayonne, 14 May 1808.
13. Bertaud, *Quand les enfants parlaient de gloire*, pp. 99–126.

CHAPTER 4

1. *Corresp.*, I, no. 83, p. 104, note on the army of Italy, Paris, 29 nivôse an IV—19 January 1796.
2. *Corresp. gén.*, I, no. 553, p. 375, to Barras, n.p. n.d. [late April 1796].
3. *Corresp.*, I, no. 420, p. 278, to the executive Directory, Lodi, 25 floréal an IV—14 May 1796.
4. *Corresp.*, I, no. 664, pp. 419–20, to the executive Directory, Bologna, 3 messidor an IV—21 June 1796.
5. Antommarchi, *Derniers Momens de Napoléon*, II, p. 94.
6. The figures are almost precise. Napoleon is referring to the Battle of La Hougue (28 May–2 June 1692), where Louis XIV demanded that Admiral de Tourville give battle, when the latter had still not concentrated all his forces.
7. [Jean-Baptiste Jourdan], *Précis des opérations de l'armée du Danube sous les ordres du général Jourdan. Extrait des mémoires manuscrits de ce général*, Paris, Charles, an VIII.
8. *Mémoires*, Montholon, IV, pp. 316–22; italics in the original.
9. Strachan, *Carl von Clausewitz's On War*, pp. 168, 170, 176; Echevarria II, *Clausewitz and Contemporary War*, pp. 92–5.
10. *Corresp.*, XXI, no. 17317, p. 379, to the Prince of Neuchâtel and Wagram, Paris, 2 February 1811; italics in the original.

CHAPTER 5

1. Agathon-Jean-François Fain, *Manuscrit de mil huit cent douze, contenant le précis des événemens de cette année, pour servir à l'histoire de l'empereur Napoléon*, 2 vols, Brussels, Tarlier, 1827, I, pp. 270–2.
2. Clausewitz, *On War*, VIII, 9, pp. 759–60.
3. *Corresp.*, XXIV, no. 19213, p. 221, to Alexander I, Moscow, 20 September 1812.
4. [Caulaincourt], *Mémoires*, II, p. 82.
5. Michael Broers, 'The Concept of "Total War" in the Revolutionary-Napoleonic Period', *War in History*, 15/3, 2008, pp. 247–68.
6. AN, 390 AP 25, MS of April 1817 (2 April), p. 1 (Bertrand, *Cahiers*, I, pp. 208–9).
7. Las Cases, *Mémorial*, Vol. 17, pp. 199–201 (Dunan edn, II, pp. 516–17).

8. Clausewitz, *On War*, VIII, 3, p. 717.

9. Charles-Humbert-Marie Vincent, *Mémorial de l'île d'Elbe*, in *Mémoires de tous. Collection de souvenirs contemporains tendant à établir la vérité dans l'histoire*, 6 vols, Levavasseur, 1834–7, III, p. 166. Most of the memoirs in this collection are apocryphal or highly suspect (Tulard, Garnier, Fierro, and d'Huart, *Nouvelle Bibliographie critique*, p. 203). The general of engineers Vincent, who was still a colonel in May 1814, did indeed exist, however, and several details of his *Mémorial de l'île d'Elbe* (his long stay on Saint Domingue, his activities in Tuscany) correspond accurately to his career (Danielle and Bernard Quintin, *Dictionnaire des colonels de Napoléon*, Paris, SPM, 1996, p. 875). This is insufficient still to prove the authenticity of these words of Napoleon's, for Vincent provides a patriotic, committed account, very favourable to him. But there does not seem to be any doubt for Antoine Casanova, *Napoléon et la pensée de son temps*, p. 127.

CONCLUSION

1. Charles J. Esdaile, 'De-Constructing the French Wars: Napoleon as Anti-Strategist', *Journal of Strategic Studies*, 31, 2008–4, pp. 515–52.

2. Lentz, *L'Effondrement du système napoléonien*, pp. 624–5.

3. John A. Lynn, 'Does Napoleon Really Have Much to Teach Us?', *XXXI Congreso internacional de historia militar*, Madrid, Spanish Commission for Military History, 2005, pp. 599–608. Lynn also contrasts the death of the Sun King at Versailles at the age of 77, surrounded by his family, with that of the deposed emperor on Saint Helena, aged 52, 'poisoned by a member of his entourage'. By uncritically adopting the hypothesis of poisoning, John Lynn undermines his position.

4. Paul W. Schroeder, *The Transformation of European Politics 1763–1848*, Oxford, Oxford University Press, 1998 and 'Napoleon's Foreign Policy: A Criminal Enterprise', *Journal of Military History*, 54, 1990–2, pp. 147–61.

5. Riley, *Napoleon as a General*, pp. 22–3.

6. Lort de Sérignan, *Napoléon et les grands généraux*, pp. 113–14.

7. Henri Bonnal, *Conférences de stratégie et de tactique générales*, École supérieure de guerre, 1892–3, p. 8, quoted in Ganniers, 'Napoléon chef d'armée', p. 525.

8. Lort de Sérignan, *Napoléon et les grands généraux*, p. 109.

9. Edward N. Luttwak, *Strategy: The Logic of War and Peace*, Cambridge, Mass., Belknap Press, 1987.

10. Chandler, 'General Introduction', p. 37.

11. Mark Herman et al., *Military Advantage in History*, Fall Church, Virg., Information Assurance Technology Analysis Center and Washington, DC, Office of the Secretary of Defense for Net Assessment, July 2002.

12. Herman et al., *Military Advantage in History*, pp. 66 and 72–5.

13. Guitton, *La Pensée et la guerre*, p. 60.

14. Bell, *The First Total War*, p. 307.

15. Benoît Durieux, *Relire* De la guerre *de Clausewitz*, Paris, Économica, 2005, p. 46.

16. Bruno Colson, 'Clausewitz for Every War. Review Article', *War in History*, 18/2, 2011, pp. 249–61.

17. Jay Luvaas, 'Student as Teacher: Clausewitz on Frederick the Great and Napoleon', *Journal of Strategic Studies*, 9, 1986–2 and 3, pp. 150–70.

18. Andreas Herberg-Rothe, *Clausewitz's Puzzle: The Political Theory of War*, Oxford, Oxford University Press, 2007, pp. 2 and 15.

19. Strachan, *Carl von Clausewitz's* On War, p. 84.

20. Strachan, *Carl von Clausewitz's* On War, pp. 93–5 and 99.

21. René Girard, *Achever Clausewitz. Entretiens avec Benoît Chantre*, Paris, Carnets Nord, 2007, pp. 42–4, 246, 258.

Sources and Bibliography

MANUSCRIPT SOURCES

Archives nationales, Paris (AN).

Archives privées, 314, Fonds Gourgaud, carton 30 (314 AP 30), Journal de Sainte-Hélène.

Archives privées, 390, Fonds Bertrand, carton 25 (390 AP 25), Cahiers de Sainte-Hélène.

Archives du Service historique de la Défense, département de l'armée de terre (SHD/DAT), Vincennes.

17 C 2—Napoleon's correspondence 3 July 1793–30 April 1795.

PRINTED SOURCES

Titles without an author refer to texts by Napoleon.

Ali (Louis-Étienne Saint-Denis, known as), *Souvenirs sur l'empereur Napoléon*, introduction and notes by C. Bourrachot, Paris, Arléa, 2000 (1st edn, 1926).

Antommarchi, François, *Derniers momens de Napoléon, ou Complément du Mémorial de Sainte-Hélène*, 2 vols, Brussels, Tarlier, 1825.

[Barante, Amable-Guillaume-Prosper de], *Souvenirs du baron de Barante de l'Académie française (1782–1866)*, published by Cl. de Barante, 8 vols, Paris, Calmann-Lévy, 1890.

Bergé, Pierre et al., *Manuscrits, autographes et très beaux livres anciens et modernes*, Drouot, 7 December 2004 (sales catalogue, item no. 22 of which is *Les Mémoires*, a manuscript dictated by Napoleon to Generals Bertrand, Gourgaud, and Montholon, in-folio of 84 pages, approximately 40 of which are in Napoleon's hand in ink or pencil).

Bertrand, Henri-Gatien, *Cahiers de Sainte-Hélène*, manuscript deciphered and annotated by Paul Fleuriot de Langle, 3 vols, Paris, Sulliver and Albin Michel, 1949, 1951, and 1959.

Bonaparte, Napoléon, *Correspondance générale*, 11 vols published to November 2014, Paris, Fayard, 2004–.

Bülow, Adam Heinrich Dietrich von, *Histoire des campagnes de Hohenlinden et de Marengo* [trans. from German by C. L. Sevelinges], containing notes on this work made by Napoleon in 1819 on Saint Helena, ordered and published by Brevet Major Emmett, London, Whittaker, Treacher & Arnot, 1831.

[Caulaincourt, Armand-Louis-Augustin de], *Mémoires du général de Caulaincourt, duc de Vicence, grand écuyer de l'Empereur*, introduction and notes by Jean Hanoteau, 3 vols, Paris, Plon, 1933.

Chaptal, Jean-Antoine, *Mes souvenirs sur Napoléon*, published by his great-grandson A. Chaptal, Paris, Plon, 1893. New edition presented and annotated by Patrice Gueniffey, Paris, Mercure de France, 2009.

Correspondance de Napoléon Ier publiée par ordre de l'empereur Napoléon III, 32 vols, Paris, Plon and Dumaine, 1858–70. [Vols 29–32 contain the *Œuvres de Napoléon Ier à Sainte-Hélène*.]

Correspondance inédite de Napoléon Ier conservée aux Archives de la Guerre, published by Ernest Picard and Louis Tuetey, 5 vols, Paris, Charles-Lavauzelle, 1912–25.

Dernières lettres inédites de Napoléon Ier, published by Léonce de Brotonne, 2 vols, Paris, Honoré Champion, 1903.

Fain, Agathon-Jean-François, *Manuscrit de mil huit cent douze, contenant le précis des événemens de cette année, pour servir à l'histoire de l'empereur Napoléon*, 2 vols, Brussels, Tarlier, 1827.

[Gohier], *Mémoires de Louis-Jérôme Gohier, président du Directoire au 18 Brumaire*, 2 vols, Paris, Bossange, 1824.

Gourgaud, Gaspard, *Journal de Sainte-Hélène 1815–1818*, expanded edition in accordance with the original text, introduction, and notes by Octave Aubry, 2 vols, Paris, Flammarion, 1944.

Gouvion Saint-Cyr, Laurent, *Mémoires pour servir à l'histoire militaire sous le Directoire, le Consulat et l'Empire*, 4 vols, Paris, Anselin, 1831.

Inédits napoléoniens, published by Arthur Chuquet, 2 vols, Paris, Fontemoing and de Boccard, 1913–19.

[Jomini, Antoine-Henri], *Guerre d'Espagne. Extrait des souvenirs inédits du général Jomini (1810–1814)*, published by Ferdinand Lecomte, Paris, L. Baudouin, 1892.

Las Cases, Emmanuel de, *Mémorial de Sainte-Hélène ou Journal où se trouve consigné, jour par jour, ce qu'a dit et fait Napoléon durant dix-huit mois*, new edn, revised and expanded, 20 vols in 10, Paris, Barbezat, 1830.

Las Cases, Emmanuel de, *Le Mémorial de Sainte-Hélène*, 1st complete critical edn, established and annotated by Marcel Dunan, 2 vols, Paris, Flammarion, 1983.

[Lejeune, Louis-François], *Mémoires du général Lejeune*, published by Germain Bapst, 2 vols, Paris, Firmin-Didot, 1895–6; Éditions du Grenadier, 2001.

Lettres inédites de Napoléon Ier (an VIII–1815), published by Léon Lecestre, 2 vols, Paris, Plon, 1897.

Lettres inédites de Napoléon Ier, published by Léonce de Brotonne, Paris, Honoré Champion, 1898.

[Marchand, Louis], *Mémoires de Marchand, premier valet de chambre et exécuteur testamentaire de l'Empereur*, published in accordance with the original manuscript by Jean Bourguignon and Henry Lachouque, 2 vols, Paris, Plon, 1952–5. New edn in one vol., Paris, Tallandier, 2003.

Marmont, Auguste-Frédéric-Louis Viesse de, *De l'esprit des institutions militaires*, preface by Bruno Colson, Paris, ISC-FRS-Économica, 2001 (1st edn, 1845).

[Marmont, A.-F.-L. Viesse de], *Mémoires du duc de Raguse*, 9 vols, Paris, Perrotin, 1857.

Mémoires pour servir à l'histoire de France, sous Napoléon, écrits à Sainte-Hélène, par les généraux qui ont partagé sa captivité, et publiés sur les manuscrits entièrement corrigés de la main de Napoléon, 8 vols, Paris, Didot and Bossange, 1823–5 (2 vols written by Gourgaud and 6 by Montholon); 2 vols of 'Notes et mélanges', written by Montholon and forming vols 9 and 10, Frankfurt am Main, Sauerländer, 1823 and Paris, Didot and Bossange, Berlin, Reimer and Frankfurt, Sauerländer, 1823.

Metternich, Clément-Wenceslas-Lothaire de, *Mémoires, documents et écrits divers laissés par le prince de Metternich, Chancelier de Cour et d'État*, published by his son Richard de Metternich, categorized and collected by M. A. de Klinkowstroem, 8 vols, Paris, Plon, 1880–4.

Montholon, Charles-François-Tristan de, *Récits de la captivité de l'empereur Napoléon à Sainte-Hélène*, 2 vols, Paris, Paulin, 1847.

Napoléon inconnu: papiers inédits (1786–1793), published by Frédéric Masson and Guido Biagi, 2 vols, Paris, Ollendorff, 1895.

Napoléon Ier, *Manuscrits inédits 1786–1791*, published in accordance with the original autographs by Frédéric Masson and Guido Biagi, Paris, P. Ollendorff, 1907.

'Note sur la position politique et militaire de nos armées de Piémont et d'Espagne', sent by Augustin Robespierre to the Committee of Public Safety, 1 Thermidor Year II (19 July 1794), edited by Edmond Bonnal de Ganges, *Les Représentants du peuple en mission près les armées 1791–1797, d'après le Dépôt de la Guerre, les séances de la Convention, les Archives nationales*, 4 vols, Paris, Arthur Savaète, 1898–9, II, pp. 165–8.

Notes inédites de l'empereur Napoléon Ier sur les Mémoires militaires du général Lloyd, published by Ariste Ducaunnès-Duval, Bordeaux, Imprimerie G. Gounouilhou, 1901 (extracted from vol. XXXV of *Archives historiques de la Gironde*).

'Notes sur la fortification dictées par Napoléon à Sainte-Hélène' [to Gourgaud], *Revue du génie militaire*, Vol. 14, July 1897, pp. 5–20.

O'Meara, Barry E., *Napoleon in Exile, or, A Voice from St. Helena: The Opinions and Reflections of Napoleon on the Most Important Events of his Life and Government in his Own Words*, 2 vols, Philadelphia, Jehu Burton, 1822.

Œuvres de Napoléon Bonaparte, 5 vols, Paris, Panckoucke, 1821.

[Rémusat, Claire-Élisabeth-Jeanne Gravier de Vergennes, comtesse de], *Mémoires de madame de Rémusat 1802–1808*, published by her grandson, 7th edn, 3 vols, Paris, Calmann-Lévy, 1880.

[Roederer, Pierre-Louis], *Œuvres du comte P.-L. Roederer . . .*, published by his son, 8 vols, Paris, Firmin-Didot, 1853–9.

[Roederer, Pierre-Louis], *Autour de Bonaparte. Journal du comte P.-L. Roederer, ministre et conseiller d'État. Notes intimes et politiques d'un familier des Tuileries*, introduction and notes by Maurice Vitrac, Paris, Daragon, 1909.

Supplément à la correspondence de Napoléon Ier: lettres curieuses omises par le comité de publication, rectifications, published by Albert Du Casse, Paris, E. Dentu, 1887.

[Thibaudeau, Antoine-Clair], *Mémoires sur le Consulat. 1799 à 1804*, par un ancien conseiller d'État, Paris, Ponthieu and Co., 1827.

Vincent, Charles-Humbert-Marie, 'Mémorial de l'île d'Elbe, ou détails sur l'arrivée et le séjour de Napoléon dans l'île d'Elbe', in *Mémoires de tous. Collection de souvenirs contemporains tendant à établir la vérité dans l'histoire*, 6 vols, Paris, Levavasseur, 1834–7, III, pp. 153–206.

CRITICAL BIBLIOGRAPHY

This contains a few essential studies, some of which serve to extend the subject. The usual research tools are not included.

Arboit, Gérald, *Napoléon et le renseignement*, Centre français de recherche sur le renseignement, note historique no. 27, August 2009 (<http://www.cf2r.org/images/stories/notes%20historique/nh-27.pdf>).

Béraud, Stéphane, *La Révolution militaire napoléonienne*, 1. *Les Manœuvres* and 2. *Les Batailles*, Paris, Bernard Giovanangeli, 2007 and 2013. [A very clear synthesis, albeit unfortunately not referenced, of Napoleon's operational art—its implementation, its structures—by an expert on enterprise management, two more volumes forthcoming.]

Beyerchen, Alan D., 'Clausewitz, Nonlinearity and the Unpredictability of War', *International Security*, 17/3, Winter 1992–3, pp. 59–90. [Fundamental for understanding both Clausewitz and Napoleon.]

Bonnal, Henri, 'La Psychologie militaire de Napoléon', *Revue hebdomadaire des cours et conférences*, 22 February 1908, pp. 421–40. [By a professor at the École supérieure de guerre. More lucid in its historical analysis than its judgement on service in the field, continued in 1914.]

Camon, Hubert, *Quand et comment Napoléon a conçu son système de manœuvre*, Paris, Berger-Levrault, 1931. [The last work—one of many highly repetitive books—by this general, who devoted his life to a rather overly systematic elucidation of Napoleon's operational art.]

Casanova, Antoine, *Napoléon et la pensée de son temps: une histoire intellectuelle singulière*, Paris, La Boutique de l'Histoire, 2000. [Predominantly concerned with the Revolution and its political aspects, but introduces some interesting perspectives.]

Chardigny, Louis, *L'Homme Napoléon*, Paris, Perrin, 2010. [Imprecise, but nevertheless useful, references.]

Chevallier, Bernard, Dancoisne-Martineau, Michel, and Lentz, Thierry, eds, *Sainte-Hélène, île de mémoire*, Paris, Fayard, 2005. [Superb work—the fullest to date on every aspect of the Saint Helena years.]

Clausewitz, Carl von, *De la Révolution à la Restauration. Écrits et lettres*, selection of texts trans. from the German and presented by Marie-Louise Steinhauser, Paris, Gallimard, 1976.

Clausewitz, Carl von, *On War*, ed. and trans. Michael Howard and Peter Paret, London, Everyman, 1993.

Clausewitz, Carl von, *Théorie du combat*, trans. from the German with a preface by Thomas Lindemann, Paris, ISC-Économica, 1998.

Colin, Jean, *L'Éducation militaire de Napoléon*, Paris, Chapelot, 1900; Teissèdre, 2001 (minus the documentary material). [Developments in the art of war that made Napoleon possible, his early reading, and his early military service.]

Colson, Bruno, *Le Général Rogniat, ingénieur et critique de Napoléon*, Paris, ISC-Économica, 2006. [Discusses the general who prompted the most comments on war from Napoleon.]

Colson, Bruno, 'La place et la nature de la manœuvre dans l'art de la guerre napoléonien', in Christian Malis (ed.), *Guerre et manœuvre. Héritages et renouveau*, Paris, Fondation Saint-Cyr-Économica, 2009, pp. 118–40.

Colson, Bruno, 'Napoléon et le guerre irrégulière', in Hervé Coutau-Bégarie (ed.), *Stratégies irrégulières*, Paris, ISC-IHEDN-Économica, 2010, pp. 345–71.

Connelly, Owen, *Blundering to Glory: Napoleon's Military Campaigns*, 3rd edn, Langham, Md., Rowman & Littlefield, 2006. [A new look, in which Napoleon emerges as much more of a brilliant improviser who made mistakes, but who managed to win until 1809, with due stress on the reduction of his activity in the field from 1812 onwards. The author is a former officer of the US Rangers and professor at the University of South Carolina.]

Duggan, William, *Napoleon's Glance: The Secret of Strategy*, New York, Nation Books, 2002. [The interest and relevance of Napoleon and his decision-making, based on strategic intuition, by a professor at Columbia Business School.]

Duggan, William, *Coup d'œil: Strategic Intuition in Army Planning*, Carlisle, Pa., Strategic Studies Institute, 2005 (<http://www.StrategicStudiesInstitute.army.mil/>; visited 2 December 2009.)

Dwyer, Philip, *Napoleon: The Path to Power, 1769–1799*, London, Bloomsbury, 2007; *Citizen Emperor: Napoleon in Power*, New Haven and London, Yale University Press, 2013. [The author is not fond of his subject, but succeeds in remaining objective and raises good questions.]

Englund, Steven, *Napoleon: A Political Life*, New York, Scribner, 2004. [Excellent on Napoleon the politician. Less rich on military matters, but some good analysis of the psychological bases of the art of command.]

Epstein, Robert M., *Napoleon's Last Victory and the Emergence of Modern War*, Lawrence, Kan., University Press of Kansas, 1994. [New, pertinent view of the change in the conditions of war from 1809 onwards.]

Esdaile, Charles, *Napoleon's Wars: An International History, 1803–1815*, London, Allen Lane, 2007. [Solid, intelligent, and well-argued study, but very British and sometimes makes reference to unreliable French sources.]

Forrest, Alan, *Napoleon*, London, Quercus, 2011. [Rightfully situates the man in his generation and sociological context; one of the best biographies of Napoleon in English.]

Ganniers, Arthur de, 'Napoléon chef d'armée. Sa formation intellectuelle, son apogée, son déclin', *Revue des questions historiques*, 73, 1903, pp. 510–78. [Pseudonym of Lort de Sérignan; admires its subject's qualities and severely criticizes his defects.]

Gonnard, Philippe, *Les Origines de la légende napoléonienne. L'Œuvre historique de Napoléon à Sainte-Hélène*, Paris, Calmann-Lévy, 1906. [Still useful for an appreciation of the complexity of the Saint Helena writings.]

Groote, Wolfgang von and Müller, Klaus Jürgen, eds, *Napoleon I und das Militärwesen seiner Zeit*, Freiburg im Breisgau, Rombach & Co., 1968. [A collection of interesting contributions, of which there is no equivalent in more recent publications.]

Healey, F. G., 'La Bibliothèque de Napoléon à Sainte-Hélène. Documents inédits trouvés parmi les *Hudson Lowe Papers*', *Revue de l'Institut Napoléon*, nos. 73–4, 1959–60, pp. 169–78; no. 75, 1960, pp. 203–13; no. 80, 1961, pp. 79–88.

Jones, Archer, *The Art of War in the Western World*, Oxford, Oxford University Press, 1987. [Well-documented, stimulating synthesis, focused on the permanence of a number of basic manoeuvres—which occasionally tends to simplify historical situations.]

Le Gall, Didier, *Napoléon et le* Mémorial de Sainte-Hélène. *Analyse d'un discours*, Paris, Kimé, 2003. [Based on a highly creditable doctoral thesis.]

Lort de Sérignan, A.-M.-T. de, *Napoléon et les grands généraux de la Révolution et de l'Empire*, Paris, Fontemoing, 1914. [The Saint-Cyr professor contrasts the school of the army of the Rhine with the army of Italy, where Napoleon, genius though he was, did not form any pupils.]

Mascilli Migliorini, Luigi, *Napoléon*, trans. from the Italian, Paris, Perrin, 2004. [Seeks not so much to recount events as to bring out their meaning and implications.]

Motte, Martin, *Les Marches de l'Empereur*, Paris, Éditions LBM, 2007. [Brief but excellent synthesis, particularly on the antecedents of the Napoleonic system of war.]

Riley, Jonathon, *Napoleon as a General*, London, Hambledon Continuum, 2007. [Interesting point of view from a British general familiar with the wars of the twenty-first century.]

Rothenberg, Gunther E., *The Art of Warfare in the Age of Napoleon*, London, Batsford, 1977. [Unmatched, it takes equal account of all the belligerents.]

Strachan, Hew, *Carl von Clausewitz's* On War: *A Biography*, London, Atlantic Books, 2007. [Possibly the best introduction to reading Clausewitz.]

Telp, Claus, *The Evolution of Operational Art, 1740–1813: From Frederick the Great to Napoleon*, London, Frank Cass, 2005. [Clearly shows how operational art emerged in the Napoleonic era and stresses the relations between thought and action.]

Tomiche, Nada, *Napoléon écrivain*, Paris, Armand Colin, 1952. [Good intellectual history of Napoleon: his reading, his education, his working methods in power and on Saint Helena.]

Vachée, Jean-Baptiste, *Napoléon en campagne*, Paris, Berger-Levrault, 1913; Bernard Giovanangeli Éditeur, 2003. [Remains useful for the organization and structures of Napoleon's command.]

Wilkinson, Spenser, *The Rise of General Bonaparte*, Oxford, Clarendon Press, 1930. [Goes up to the Cherasco armistice; more precise than Colin for Napoleon's readings.]

Index